Communication Technology Update and Fundamentals

16th Edition

Communication Technology Update and Fundamentals

16th Edition

Editors

August E. Grant

Jennifer H. Meadows

In association with **Technology Futures, Inc.**

Routledge
Taylor & Francis Group

NEW YORK AND LONDON

Editors:
August E. Grant
Jennifer H. Meadows

Technology Futures, Inc.
Production & Graphics Editor:
Helen Mary V. Marek

Publisher: Ross Wagenhofer
Editorial Assistant: Nicole Salazar
Production Editor: Sian Cahill
Marketing Manager: Lynsey Nurthen

Sixteenth edition published 2018
by Routledge
711 Third Avenue, New York, NY 10017
and by Routledge
2 Park Square, Milton Park, Abingdon, Oxon, OX14 4RN

Routledge is an imprint of the Taylor & Francis Group, an informa business

Publisher's note: This book has been prepared from camera-ready copy provided by the editors. Typeset in Palatino Linotype by H.M.V. Marek, Technology Futures, Inc.

[First edition published by Technology Futures, Inc. 1992]
[Fifteenth edition published by Focal Press 2016]

Library of Congress Cataloguing in Publication Data.
CIP data has been applied for.

HB: 9781138571334
Paper: 9781138571365
eBook: 9780203702871

Table of Contents

Glossary and Updates can be found on the

Communication Technology Update and Fundamentals website

http://www.tfi.com/ctu/

Preface

Great changes in technology are coming at a faster and faster pace, introducing new opportunities, challenges, careers, and fields of study at a rate that hasn't been experienced in human history. Keeping up with these changes can simultaneously provide amusement and befuddlement, as well as economic prosperity and ruin.

That's where you come in. Whether you are trying to plan a lucrative investment or a career in media, or you simply have to pass a particular class in order to graduate, the field of communication technologies has become important enough to you that you are investing in the time to read this book. Be warned: the goal of the authors in this book is to serve all of those needs. To do so, the book starts by explaining the Communication Technology Ecosystem, then applies this ecosystem as a tool to help you understand each of the technologies presented.

This is the 16th edition of this book, and most of the book is changed from the 15th edition. In addition to updating every chapter with the latest developments, we have a first-time chapter exploring eSports (Chapter 19) and a chapter we haven't seen in more than a decade discussing Virtual Reality (Chapter 15). A few other chapters, including Video Games (Chapter 14), Home Video (Chapter 16), ebooks (Chapter 19), and Computers (Chapter 11) have been rewritten from scratch to provide a more contemporary discussion.

One thing shared by all of the contributors to this book is a passion for communication technology. In order to keep this book as current as possible we asked the authors to work under extremely tight deadlines. Authors begin working in late 2017, and most chapters were submitted in February or March 2018 with the final details added in April 2018. Individually, the chapters provide snapshots of the state of the field for individual technologies, but together they present a broad overview of the role that communication technologies play in our everyday lives. The efforts of these authors have produced a remarkable compilation, and we thank them for all their hard work in preparing this volume.

The constant in production of this book is our editor extraordinaire, TFI's Helen Mary V. Marek, who deftly handled all production details, moving all 27 chapters from draft to camera-ready in weeks. Helen Mary also provided on-demand graphics production, adding visual elements to help make the content more understandable. Our editorial and marketing team at Routledge, including Ross Wagenhoffer and Nicole Salazar, ensured that production and promotion of the book were as smooth as ever.

We are most grateful to our spouses (and partners in life), Diane Grant and Floyd Meadows for giving us this month every two years so that we can disappear into a haze of bits, pixels, toner, and topics to render the book you are reading right now. They know that a strange compulsion arises every two years, with publication of the book being followed immediately by the satisfaction we get from being part of the process of helping you understand and apply new communications technologies.

You can keep up with developments on technologies discussed in this book by visiting our companion website, where we use the same technologies discussed in the book to make sure you have the latest information. The companion website for the *Communication Technology Update and Fundamentals*: www.tfi.com/ctu. The complete Glossary for the book is on the site, where it is much easier to find individual entries than in the paper version of the book. We have also moved the vast quantity of statistical data on each of the communication technologies that were formerly printed in Chapter 2 to the site. As always, we will periodically update the website to supplement the text with new information and links to a wide variety of information available over the Internet.

Your interest and support is the reason we do this book every two years, and we listen to your suggestions so that we can improve the book after every edition. You are invited to send us updates for the website, ideas for new topics, and other contributions that will inform all members of the community. You are invited to communicate directly with us via email, snail mail, social media, or voice.

Thank you for being part of the CTUF community!

Augie Grant and Jennifer Meadows

April 1, 2018

Augie Grant
School of Journalism and Mass Communications
University of South Carolina
Columbia, SC 29208
Phone: 803.777.4464
augie@sc.edu
Twitter: @augiegrant

Jennifer H. Meadows
Dept. of Media Arts, Design, and Technology
California State University, Chico
Chico, CA 95929-0504
Phone: 530.898.4775
jmeadows@csuchico.edu
Twitter: @mediaartsjen

Section 1

Fundamentals

The Communication Technology Ecosystem

August E. Grant, Ph.D.[*]

Communication technologies are the nervous system of contemporary society, transmitting and distributing sensory and control information and interconnecting a myriad of interdependent units. These technologies are critical to commerce, essential to entertainment, and intertwined in our interpersonal relationships. Because these technologies are so vitally important, any change in communication technologies has the potential to impact virtually every area of society.

One of the hallmarks of the industrial revolution was the introduction of new communication technologies as mechanisms of control that played an important role in almost every area of the production and distribution of manufactured goods (Beniger, 1986). These communication technologies have evolved throughout the past two centuries at an increasingly rapid rate. This evolution shows no signs of slowing, so an understanding of this evolution is vital for any individual wishing to attain or retain a position in business, government, or education.

The economic and political challenges faced by the United States and other countries since the beginning of the new millennium clearly illustrate the central role these communication systems play in our society. Just as the prosperity of the 1990s was credited to advances in technology, the economic challenges that followed were linked as well to a major downturn in the technology sector. Today, communication technology is seen by many as a tool for making more efficient use of a wide range of resources including time and energy.

Communication technologies play as critical a part in our private lives as they do in commerce and control in society. Geographic distances are no longer barriers to relationships thanks to the bridging power of communication technologies. We can also be entertained and informed in ways that were unimaginable a century ago thanks to these technologies—and they continue to evolve and change before our eyes.

This text provides a snapshot of the state of technologies in our society. The individual chapter authors have compiled facts and figures from hundreds of sources to provide the latest information on more than two dozen communication technologies. Each discussion explains the roots and evolution, recent developments, and current status of the technology as of mid-2018. In discussing each technology, we address them from a systematic perspective, looking at a range of factors beyond hardware.

---

[*] J. Rion McKissick Professor of Journalism, School of Journalism and Mass Communications, University of South Carolina (Columbia, South Carolina).

The goal is to help you analyze emerging technologies and be better able to predict which ones will succeed and which ones will fail. That task is more difficult to achieve than it sounds. Let's look at an example of how unpredictable technology can be.

The Alphabet Tale

As this book goes to press in mid-2018, Alphabet, the parent company of Google, is the most valuable media company in the world in terms of market capitalization (the total value of all shares of stock held in the company). To understand how Alphabet attained that lofty position, we have to go back to the late 1990s, when commercial applications of the Internet were taking off. There was no question in the minds of engineers and futurists that the Internet was going to revolutionize the delivery of information, entertainment, and commerce. The big question was how it was going to happen.

Those who saw the Internet as a medium for information distribution knew that advertiser support would be critical to its long-term financial success. They knew that they could always find a small group willing to pay for content, but the majority of people preferred free content. To become a mass medium similar to television, newspapers, and magazines, an Internet advertising industry was needed.

At that time, most Internet advertising was banner ads—horizontal display ads that stretched across most of the screen to attract attention, but took up very little space on the screen. The problem was that most people at that time accessed the Internet using slow, dial-up connections, so advertisers were limited in what they could include in these banners to about a dozen words of text and simple graphics. The dream among advertisers was to be able to use rich media, including full-motion video, audio, animation, and every other trick that makes television advertising so successful.

When broadband Internet access started to spread, advertisers were quick to add rich media to their banners, as well as create other types of ads using graphics, video, and sound. These ads were a little more effective, but many Internet users did not like the intrusive nature of rich media messages.

At about the same time, two Stanford students, Sergey Brin and Larry Page, had developed a new type of search engine, Google, that ranked results on the basis of how often content was referred to or linked from other sites, allowing their computer algorithms to create more robust and relevant search results (in most cases) than having a staff of people indexing Web content. What they needed was a way to pay for the costs of the servers and other technology.

According to Vise & Malseed (2006), their budget did not allow the company, then known as Google, to create and distribute rich media ads. They could do text ads, but they decided to do them differently from other Internet advertising, using computer algorithms to place these small text ads on the search results that were most likely to give the advertisers results. With a credit card, anyone could use this "AdWords" service, specifying the search terms they thought should display their ads, writing the brief ads (less than 100 characters total—just over a dozen words), and even specifying how much they were willing to pay every time someone clicked on their ad. Even more revolutionary, the Google team decided that no one should have to pay for an ad unless a user clicked on it.

For advertisers, it was as close to a no-lose proposition as they could find. Advertisers did not have to pay unless a person was interested enough to click on the ad. They could set a budget that Google computers could follow, and Google provided a control panel for advertisers that gave a set of measures that was a dream for anyone trying to make a campaign more effective. These measures indicated not only the overall effectiveness of the ad, but also the effectiveness of each message, each keyword, and every part of every campaign.

The result was remarkable. Google's share of the search market was not that much greater than the companies that had held the #1 position earlier, but Google was making money—lots of money—from these little text ads. Wall Street investors noticed, and, once Google went public, investors bid up the stock price, spurred by increases in revenues and a very large profit margin. Today, Google's parent company, renamed Alphabet, is involved in a number of other ventures designed to aggregate and deliver content ranging from text to full-motion video, but its little text ads on its Google search engine are still the primary revenue generator.

In retrospect, it was easy to see why Google was such a success. Their little text ads were effective because of context—they always appeared where they would be the most effective. They were not intrusive, so people did not mind the ads on Google pages, and later on other pages that Google served ads to through its "content network." Plus, advertisers had a degree of control, feedback, and accountability that no advertising medium had ever offered before (Grant & Wilkinson, 2007).

So what lessons should we learn from this story? Advertisers have their own set of lessons, but there are a separate set of lessons for those wishing to understand new media. First, no matter how insightful, no one is ever able to predict whether a technology will succeed or fail. Second, success can be due as much to luck as to careful, deliberate planning and investment. Third, simplicity matters—there are few advertising messages as simple as the little text ads you see when doing a Google search.

The Alphabet tale provides an example of the utility of studying individual companies and industries, so the focus throughout this book is on individual technologies. These individual snapshots, however, comprise a larger mosaic representing the communication networks that bind individuals together and enable them to function as a society. No single technology can be understood without understanding the competing and complementary technologies and the larger social environment within which these technologies exist. As discussed in the following section, all of these factors (and others) have been considered in preparing each chapter through application of the "technology ecosystem." Following this discussion, an overview of the remainder of the book is presented.

The Communication Technology Ecosystem

The most obvious aspect of communication technology is the hardware—the physical equipment related to the technology. The hardware is the most tangible part of a technology system, and new technologies typically spring from developments in hardware.

However, understanding communication technology requires more than just studying the hardware. One of the characteristics of today's digital technologies is that most are based upon computer technology, requiring instructions and algorithms more commonly known as "software."

In addition to understanding the hardware and software of the technology, it is just as important to understand the content communicated through the technology system. Some consider the content as another type of software. Regardless of the terminology used, it is critical to understand that digital technologies require a set of instructions (the software) as well as the equipment and content.

Figure 1.1
The Communication Technology Ecosystem

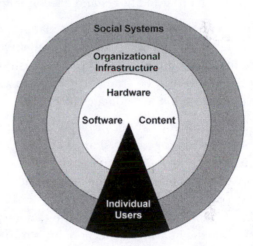

Source: A. E. Grant

The hardware, software, and content must also be studied within a larger context. Rogers' (1986) definition of "communication technology" includes some of these contextual factors, defining it as "the hardware equipment, organizational structures, and social values by which individuals collect, process, and exchange information with other individuals" (p. 2). An even broader range of factors is suggested by Ball-Rokeach (1985) in her media system dependency theory, which suggests that communication media can be understood by analyzing dependency relations within and across levels of analysis, including the individual, organizational, and system levels. Within the system level, Ball-Rokeach identifies three systems for analysis: the media system, the political system, and the economic system.

These two approaches have been synthesized into the "Technology Ecosystem" illustrated in Figure 1.1. The core of the technology ecosystem consists of the hardware, software, and content (as previously defined). Surrounding this core is the organizational infrastructure: the group of organizations involved in the production and distribution of the technology. The next level moving outwards is the system level, including the political, economic, and media systems, as well as other groups of individuals or organizations serving a common set of functions in society. Finally, the individual users of the technology cut across all of the other areas, providing a focus for understanding each one. The basic premise of the technology ecosystem is that all areas of the ecosystem interact and must be examined in order to understand a technology.

(The technology ecosystem is an elaboration of the "umbrella perspective" (Grant, 2010) that was explicated in earlier editions of this book to illustrate the elements that need to be studied in order to understand communication technologies.)

Adding another layer of complexity to each of the areas of the technology ecosystem is also helpful. In order to identify the impact that each individual characteristic of a technology has, the factors within each area of the ecosystem may be identified as "enabling," "limiting," "motivating," and "inhibiting" depending upon the role they play in the technology's diffusion.

Enabling factors are those that make an application possible. For example, the fact that the coaxial cable used to deliver traditional cable television can carry dozens of channels is an enabling factor at the hardware level. Similarly, the decision of policy makers to allocate a portion of the radio frequency spectrum for cellular telephony is an enabling factor at the system level (political system). One starting point to use in examining any technology is to make a list of the underlying factors from each area of the technology ecosystem that make the technology possible in the first place.

Limiting factors are the opposite of enabling factors; they are those factors that create barriers to the adoption or impacts of a technology. A great example

is related to the cellular telephone illustration in the previous paragraph. The fact that the policy makers discussed above initially permitted only two companies to offer cellular telephone service in each market was a system level limitation on that technology. The later introduction of digital technology made it possible for another four companies to compete for mobile phone service. To a consumer, six telephone companies may seem to be more than is needed, but to a start-up company wanting to enter the market, this system-level factor represents a definite limitation. Again, it is useful to apply the technology ecosystem to create a list of factors that limit the adoption, use, or impacts of any specific communication technology.

Motivating factors are a little more complicated. They are those factors that provide a reason for the adoption of a technology. Technologies are not adopted just because they exist. Rather, individuals, organizations, and social systems must have a reason to take advantage of a technology. The desire of local telephone companies for increased profits, combined with the fact that growth in providing local telephone service is limited, is an organizational factor motivating the telcos to enter the markets for new communication technologies. Individual users desiring information more quickly can be motivated to adopt electronic information technologies. If a technology does not have sufficient motivating factors for its use, it cannot be a success.

Inhibiting factors are the opposite of motivating ones, providing a disincentive for adoption or use of a communication technology. An example of an inhibiting factor at the organizational level might be a company's history of bad customer service. Regardless of how useful a new technology might be, if customers don't trust a company, they are not likely to purchase its products or services. One of the most important inhibiting factors for most new technologies is the cost to individual users. Each potential user must decide whether the cost is worth the service, considering their budget and the number of competing technologies. Competition from other technologies is one of the biggest barriers any new (or existing) technology faces. Any factor that works against the success of a technology can be considered an inhibiting factor. As you might guess, there are usually more inhibiting factors for most technologies than motivating ones. And if the motivating factors are more numerous and stronger than the inhibiting factors, it is an easy bet that a technology will be a success.

All four factors—enabling, limiting, motivating, and inhibiting—can be identified at the individual user, organizational, content, and system levels. However, hardware and software can only be enabling or limiting; by themselves, hardware and software do not provide any motivating factors. The motivating factors must always come from the messages transmitted or one of the other areas of the ecosystem.

The final dimension of the technology ecosystem relates to the environment within which communication technologies are introduced and operate. These factors can be termed "external" factors, while ones relating to the technology itself are "internal" factors. In order to understand a communication technology or be able to predict how a technology will diffuse, both internal and external factors must be studied.

Applying the Communication Technology Ecosystem

The best way to understand the communication technology ecosystem is to apply it to a specific technology. One of the fastest diffusing technologies discussed later in this book is the "personal assistant," such as the Amazon Alexa or Google Home—these devices provide a great application of the communication technology ecosystem.

Let's start with the hardware. Most personal assistants are small or medium-sized units, designed to sit on a shelf or table. Studying the hardware reveals that the unit contains multiple speakers, a microphone, some computer circuitry, and a radio transmitter and receiver. Studying the hardware, we can get clues about the functionality of the device, but the key to the functionality is the software.

The software related to the personal assistant enables conversion of speech heard by the microphone into text or other commands that connect to another set of software designed to fulfill the commands given to the system. From the perspective of the user, it doesn't matter whether the device converts speech to commands or whether the device transmits speech to a central computer where the translation takes place—

the device is designed so that it doesn't matter to the user. The important thing that becomes apparent is that the hardware used by the system extends well beyond the device through the Internet to servers that are programmed to deliver answers and content requested through the personal assistant.

So, who owns these servers? To answer that question, we have to look at the organizational infrastructure. It is apparent that there are two distinct sets of organizations involved—one set that makes and distributes the devices themselves to the public and the other that provides the back-end processing power to find answers and deliver content. For the Amazon Alexa, Amazon has designed and arranged for the manufacture of the device. (Note that few companies specialize in making hardware; rather, most communication hardware is made by companies that specialize in manufacturing on a contract basis.) Amazon also owns and controls the servers that interpret and seek answers to questions and commands. But to get to those servers, the commands have to first pass through cable or phone networks owned by other companies, with answers or content provided by servers on the Internet owned by still other companies. At this point, it is helpful to examine the economic relationships among the companies involved. The users' Internet Service Provider (ISP) passes all commands and content from the home device to the cloud-based servers, which are, in turn, connected to servers owned by other companies that deliver content.

So, if a person requests a weather forecast, the servers connect to a weather service for content. A person might also request music, finding themselves connected to Amazon's own music service or to another service such as Pandora or Sirius/XM. A person ordering a pizza will have their message directed to the appropriate pizza delivery service, with the only content returned being a confirmation of the order, perhaps with status updates as the order is fulfilled.

The pizza delivery example is especially important because it demonstrates the economics of the system. The servers used are expensive to purchase and operate, so the company that designs and sells personal assistants has a motivation to contract with individual pizza delivery services to pay a small commission every time someone orders a pizza. Extending this example to multiple other services will help you understand why some services are provided for free but others must be paid, with the pieces of the system working together to spread revenue to all of the companies involved.

The point is that it is not possible to understand the personal assistant without understanding all of the organizations implicated in the operation of the device. And if two organizations decide not to cooperate with each other, content or service may simply not be available.

The potential conflicts among these organizations can move our attention to the next level of the ecosystem, the social system level. The political system, for example, has the potential to enable services by allowing or encouraging collaboration among organizations. Or it can do the opposite, limiting or inhibiting cooperation with regulations. (Net neutrality, discussed in Chapter 5, is a good example of the role played by the political system in enabling or limiting capabilities of technology.) The system of retail stores enables distribution of the personal assistant devices to local retail stores, making it easier for a user to become an "adopter" of the device.

Studying the personal assistant also helps understand the enabling and limiting functions. For example, the fact that Amazon has programmed the Alexa app to accept commands in dozens of languages from Spanish to Klingon is an enabling factor, but the fact that there are dozens of other languages that have not been programming is definitely a limiting factor.

Similarly, the ease of ordering a pizza through your personal assistant is a motivating factor, but having your device not understand your commands is an inhibiting factor.

Finally, examination of the environment gives us more information, including competitive devices, public sentiment, and general economic environment.

All of those details help us to understand how personal assistants work and how companies can profit in many different ways from their use. But we can't fully understand the role that these devices play in the lives of their users without studying the individual user. We can examine what services are used,

why they are used, how often they are used, the impacts of their use, and much more.

Applying the Communication Technology Ecosystem thus allows us to look at a technology, its uses, and its effects by giving a multidimensional perspective that provides a more comprehensive insight than we would get from just examining the hardware or software.

Each communication technology discussed in this book has been analyzed using the technology ecosystem to ensure that all relevant factors have been included in the discussions. As you will see, in most cases, organizational and system-level factors (especially political factors) are more important in the development and adoption of communication technologies than the hardware itself. For example, political forces have, to date, prevented the establishment of a single world standard for high-definition television (HDTV) production and transmission. As individual standards are selected in countries and regions, the standard selected is as likely to be the product of political and economic factors as of technical attributes of the system.

Organizational factors can have similar powerful effects. For example, as discussed in Chapter 4, the entry of a single company, IBM, into the personal computer business in the early 1980s resulted in fundamental changes in the entire industry, dictating standards and anointing an operating system (MS-DOS) as a market leader. Finally, the individuals who adopt (or choose not to adopt) a technology, along with their motivations and the manner in which they use the technology, have profound impacts on the development and success of a technology following its initial introduction.

Perhaps the best indication of the relative importance of organizational and system-level factors is the number of changes individual authors made to the chapters in this book between the time of the initial chapter submission in January 2018 and production of the final, camera-ready text in April 2018. Very little new information was added regarding hardware, but numerous changes were made due to developments at the organizational and system levels.

To facilitate your understanding of all of the elements related to the technologies explored, each chapter in this book has been written from the perspective of the technology ecosystem. The individual writers have endeavored to update developments in each area to the extent possible in the brief summaries provided. Obviously, not every technology experienced developments in each area of the ecosystem, so each report is limited to areas in which relatively recent developments have taken place.

Why Study New Technologies?

One constant in the study of media is that new technologies seem to get more attention than traditional, established technologies. There are many reasons for the attention. New technologies are more dynamic and evolve more quickly, with greater potential to cause change in other parts of the media system. Perhaps the reason for our attention is the natural attraction that humans have to motion, a characteristic inherited from our most distant ancestors.

There are a number of other reasons for studying new technologies. Maybe you want to make a lot of money—and there is a lot of money to be made (and lost!) on new technologies. If you are planning a career in the media, you may simply be interested in knowing how the media are changing and evolving, and how those changes will affect your career.

Or you might want to learn lessons from the failure of new communication technologies so you can avoid failure in your own career, investments, etc. Simply put, the majority of new technologies introduced do not succeed in the market. Some fail because the technology itself was not attractive to consumers (such as the 1980s' attempt to provide AM stereo radio). Some fail because they were far ahead of the market, such as Qube, the first interactive cable television system, introduced in the 1970s. Others failed because of bad timing or aggressive marketing from competitors that succeeded despite inferior technology.

The final reason for studying new communication technologies is to identify patterns of adoption, effects, economics, and competition so that we can be prepared to understand, use, and/or compete with the next generation of media. Virtually every new technology discussed in this book is going to be one of those "traditional, established technologies" in a few short years, but there will always be another generation of new media to challenge the status quo.

Overview of Book

The key to getting the most out of this book is therefore to pay as much attention as possible to the reasons that some technologies succeed and others fail. To that end, this book provides you with a number of tools you can apply to virtually any new technology that comes along. These tools are explored in the first five chapters, which we refer to as the *Communication Technology Fundamentals*. You might be tempted to skip over these to get to the latest developments about the individual technologies that are making an impact today, but you will be much better equipped to learn lessons from these technologies if you are armed with these tools.

The first of these is the "technology ecosystem" discussed previously that broadens attention from the technology itself to the users, organizations, and system surrounding that technology. To that end, each of the technologies explored in this book provides details about all of the elements of the ecosystem.

Of course, studying the history of each technology can help you find patterns and apply them to different technologies, times, and places. In addition to including a brief history of each technology, the next chapter, A History of Communication Technologies, provides a broad overview of most of the technologies discussed later in the book, allowing comparisons along a number of dimensions: the year introduced, growth rate, number of current users, etc. This chapter highlights commonalties in the evolution of individual technologies, as well as presents the "big picture" before we delve into the details. By focusing on the number of users over time, this chapter also provides a useful basis of comparison across technologies.

Another useful tool in identifying patterns across technologies is the application of theories related to new communication technologies. By definition, theories are general statements that identify the underlying mechanisms for adoption and effects of these new technologies. Chapter 3 provides an overview of a wide range of these theories and provides a set of analytic perspectives that you can apply to both the technologies in this book and any new technologies that follow.

The structure of communication industries is then addressed in Chapter 4. This chapter then explores the complexity of organizational relationships, along with the need to differentiate between the companies that make the technologies and those that sell the technologies. The most important force at the system level of the ecosystem, regulation, is introduced in Chapter 5.

These introductory chapters provide a structure and a set of analytic tools that define the study of communication technologies. Following this introduction, the book then addresses the individual technologies.

The technologies discussed in this book are organized into three sections: Electronic Mass Media, Computers & Consumer Electronics, and Networking Technologies. These three are not necessarily exclusive; for example, Digital Signage could be classified as either an electronic mass medium or a computer technology. The ultimate decision regarding where to put each technology was made by determining which set of current technologies most closely resemble the technology. Thus, Digital Signage was classified with electronic mass media. This process also locates the discussion of a cable television technology—cable modems—in the Broadband and Home Networks chapter in the Networking Technologies section.

Each chapter is followed by a brief bibliography that represents a broad overview of literally hundreds of books and articles that provide details about these technologies. It is hoped that the reader will not only use these references but will examine the list of source material to determine the best places to find newer information since the publication of this *Update*.

To help you find your place in this emerging technology ecosystem, each technology chapter includes a paragraph or two discussing how you can get a job in that area of technology. And to help you imagine the future, some authors have also added their prediction of what that technology will be like in 2033—or fifteen years after this book is published. The goal is not to be perfectly accurate, but rather to show you some of the possibilities that could emerge in that time frame.

Most of the technologies discussed in this book are continually evolving. As this book was completed, many technological developments were announced but not released, corporate mergers were under discussion, and regulations had been proposed but not passed. Our goal is for the chapters in this book to establish a basic understanding of the structure, functions, and background for each technology, and for the supplementary Internet site to provide brief synopses of the latest developments for each technology. (The address for the website is www.tfi.com/ctu.)

The final chapter returns to the "big picture" presented in this book, attempting to place these discussions in a larger context, exploring the process of starting a company to exploit or profit from these technologies. Any text such as this one can never be fully comprehensive, but ideally this text will provide you with a broad overview of the current developments in communication technology.

Bibliography

Ball-Rokeach, S. J. (1985). The origins of media system dependency: A sociological perspective. *Communication Research, 12* (4), 485-510.

Beniger, J. (1986). *The control revolution.* Cambridge, MA: Harvard University Press.

Grant, A. E. (2010). Introduction to communication technologies. In A. E. Grant & J. H. Meadows (Eds.) *Communication Technology Update and Fundamentals (12th ed).* Boston: Focal Press.

Grant, A. E. & Wilkinson, J. S. (2007, February). Lessons for communication technologies from Web advertising. Paper presented to the Mid-Winter Conference of the Association of Educators in Journalism and Mass Communication, Reno.

Rogers, E. M. (1986). *Communication technology: The new media in society.* New York: Free Press.

Vise, D. & Malseed, M. (2006). *The Google story: Inside the hottest business, media, and technology success of our time.* New York: Delta.

2

A History of Communication Technology

Yicheng Zhu, Ph.D.*

The other chapters in this book provide details regarding the history of one or more communication technologies. However, one needs to understand that history works, in some ways, like a telescope. The closer an observer looks at the details, i.e. the particular human behaviors that changed communication technologies, the less they can grasp the *big picture*.

This chapter attempts to provide the *big picture* by discussing recent advancements along with a review of happenings "before we were born." Without the understanding of the collective memory of the trailblazers of communication technology, we will be "children forever" when we make interpretations and implications from history records. (Cicero, 1876).

We will visit the print era, the electronic era, and the digital era in this chapter. To provide a useful perspective, we compare numerical statistics of adoption and use of these technologies across time. To that end, this chapter follows patterns adopted in previous summaries of trends in U.S. communications media (Brown & Bryant, 1989; Brown, 1996, 1998, 2000, 2002, 2004, 2006, 2008, 2010, 2012, 2014; Zhu & Brown, 2016). Nonmonetary units are reported when possible, although dollar expenditures appear as supplementary measures. A notable exception is the de facto standard of measuring motion picture acceptance in the market: box office receipts.

Government sources are preferred for consistency in this chapter. However, they have recently become more volatile in terms of format, measurement and focus due to the shortened life circle of technologies (for example, some sources don't distinguish laptops from tablets when calculating PC shipments). Readers should use caution in interpreting data for individual years and instead emphasize the trends over several years. One limitation of this government data is the lag time before statistics are reported, with the most recent data being a year or more older. The companion website for this book (www.tfi.com/ctu) reports more detailed statistics than could be printed in this chapter.

---

* Doctoral candidate in the School of Journalism and Mass Communications at the University of South Carolina (Columbia, SC).
(Zhu and the editors acknowledge the contributions of the late Dan Brown, Ph.D., who created the first versions of this chapter and the related figures and tables).

Communication technologies are evolving at a much faster pace today than they used to be, and the way in which we differentiate technologies is more about concepts rather than products. For example, audiocassettes and compact discs seem doomed in the face of rapid adoption of newer forms of digital audio recordings. But what fundamentally changed our daily experience is the surge of individual power brought by technological convenience: digitized audios empowered our mobility and efficiency both at work or at play. Quadraphonic sound, CB radios, 8-track audiotapes, and 8mm film cameras ceased to exist as standalone products in the marketplace, and we exclude them, not because they disappeared, but because their concepts were converted or integrated into newer and larger concepts. This chapter traces trends that reveal clues about what has happened and what may happen in the use of respective media forms.

To illustrate the growth rates and specific statistics regarding each technology, a large set of tables and figures have been placed on the companion website for this book at www.tfi.com/ctu. Your understanding of each technology will be aided by referring to the website as you read each section.

The Print Era

Printing began in China thousands of years before Johann Gutenberg developed the movable type printing press in 1455 in Germany. Gutenberg's press triggered a revolution that began an industry that remained stable for another 600 years (Rawlinson, 2011).

Printing in the United States grew from a one-issue newspaper in 1690 to become the largest print industry in the world (U.S. Department of Commerce/International Trade Association, 2000). This enterprise includes newspapers, periodicals, books, directories, greeting cards, and other print media.

Newspapers

Publick Occurrences, Both Foreign and Domestick was the first newspaper produced in North America, appearing in 1690 (Lee, 1917). Table 2.1 and Figure 2.1 from the companion website (www.tfi.com/ctu) for this book show that U.S. newspaper firms and newspaper circulation had extremely slow growth until the 1800s. Early growth suffered from relatively low literacy rates and the lack of discretionary cash among the bulk of the population. The progress of the industrial revolution brought money for workers and improved mechanized printing processes. Lower newspaper prices and the practice of deriving revenue from advertisers encouraged significant growth beginning in the 1830s. Newspapers made the transition from the realm of the educated and wealthy elite to a mass medium serving a wider range of people from this period through the Civil War era (Huntzicker, 1999).

The Mexican and Civil Wars stimulated public demand for news by the middle 1800s, and modern journalism practices, such as assigning reporters to cover specific stories and topics, began to emerge. Circulation wars among big city newspapers in the 1880s featured sensational writing about outrageous stories. Both the number of newspaper firms and newspaper circulation began to soar. Although the number of firms would level off in the 20th century, circulation continued to rise.

The number of morning newspapers more than doubled after 1950, despite a 16% drop in the number of daily newspapers over that period. Overall newspaper circulation remained higher at the start of the new millennium than in 1950, although it inched downward throughout the 1990s. Although circulation actually increased in many developing nations, both U.S. newspaper circulation and the number of U.S. newspaper firms are lower today than the respective figures posted in the early 1990s. Many newspapers that operated for decades are now defunct, and many others offer only online electronic versions.

The newspaper industry shrunk as we entered the 21st century when new technologies such as the Internet became popular outlets for advertising. Newspaper publishing revenue declined from $29 billion in 2010 to $23 billion in 2017 (US Census Bureau, 2016). In the meantime, percentage of revenue generated from online newspapers rose from 5.6% to 15.2%. (US Census Bureau, 2016). Advertising as a source of revenue for newspaper publishers dropped by 27% from $20.4 billion to $14.9 billion (U.S. Census Bureau, 2017).

Figure 2.A
Communication Technology Timeline

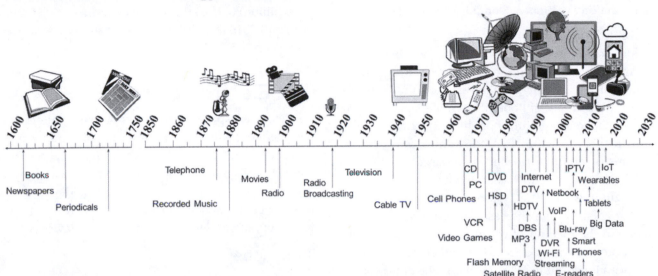

Source: Technology Futures, Inc.

Periodicals

"The first colonial magazines appeared in Philadelphia in 1741, about 50 years after the first newspapers" (Campbell, 2002, p. 310). Few Americans could read in that era, and periodicals were costly to produce and circulate. Magazines were often subsidized and distributed by special interest groups, such as churches (Huntzicker, 1999). The *Saturday Evening Post*, the longest running magazine in U.S. history, began in 1821 and became the first magazine to both target women as an audience and to be distributed to a national audience. By 1850, nearly 600 magazines were operating.

By early in the 20th century, national magazines became popular with advertisers who wanted to reach wide audiences. No other medium offered such opportunity. However, by the middle of the century, the many successful national magazines began dying in the face of advertiser preferences for the new medium of television and the increasing costs of periodical distribution. Magazines turned to smaller niche audiences that were more effectively targeted. Table 2.2, Figure 2.2, and Figure 2.3 on the companion website (www.tfi.com/ctu) show the number of American periodical titles by year, revealing that the number of new periodical titles nearly doubled from 1958 to 1960.

Single copy magazine sales were mired in a long period of decline in 2009 when circulation fell by 17.2%. However, subscription circulation fell by only 5.9%. In 2010, the Audit Bureau of Circulation reported that, among the 522 magazine titles monitored by the Bureau, the number of magazine titles in the United States fell by 8.7% (Agnese, 2011).

In 2010, 20,707 consumer magazines were published in North America, reaching a paid circulation of $8.8 billion. Subscriptions accounted for $6.2 billion (71%) of that circulation. During that year, 193 new North American magazines began publishing, but 176 magazines closed. Many print magazines were also available in digital form, and many had eliminated print circulation in favor of digital publishing. In 2009, 81 North American magazines moved online, but the number of additional magazines that went online in 2010 dropped to 28 (Agnese, 2011).

Books

Stephen Daye printed the first book in colonial America, *The Bay Psalm Book*, in 1640 (Campbell, 2002). Books remained relatively expensive and rare until after the printing process benefited from the industrial revolution. Linotype machines developed in the 1880s allowed for mechanical typesetting. After World War II, the popularity of paperback books helped the industry expand. The U.S. book publishing industry includes 87,000 publishers, most of which are small businesses. Many of these literally operate as "mom-and-pop desktop operations" (Peters & Donald, 2007, p. 11).

Table 2.3 and Figures 2.3 and 2.4 from the companion website (www.tfi.com/ctu) show new book titles published by year from the late 1800s through 2016. While times of war negatively affected the book industry, the number of book titles in the U.S. has been generally increasing with short-lived fluctuations like those in 1983-1986 and 1997-1999. The U.S. Bureau of the Census reports furnished data based on material from R. R. Bowker, which changed its reporting methods beginning with the 1998 report. Ink and Grabois (2000) explained the increase as resulting from the change in the method of counting titles "that results in a more accurate portrayal of the current state of American book publishing" (p. 508). The older counting process included only books included by the Library of Congress Cataloging in Publication program. This program included publishing by the largest American publishing companies, but omitted such books as "inexpensive editions, annuals, and much of the output of small presses and self-publishers" (Ink & Grabois, 2000, p. 509). Ink and Grabois observed that the U.S. ISBN (International Standard Book Number) Agency assigned more than 10,000 new ISBN publisher prefixes annually.

Books have long been available for reading via computers, but dedicated e-book readers have transformed the reading experience by bringing many readers into the digital era. By the end of 2009, 3.7 million Americans were reading e-books. In 2010, the readership grew to more than 10.3 million, an increase of 178%, and surveys reported by the Book Industry Study Group (BISG) reported that 20% of respondents had stopped buying printed books in favor of e-books within a year. By July 2010, Amazon reported that sales of e-books surpassed that of print hardcover sales for the first time, with "143 e-books sold for every 100 print hardcover books" (Dillon, 2011, p. 5). From mid-December 2011 through January 2012, the proportion of Americans owning both e-book readers and tablet computers nearly doubled from 10% to 19%, with 29% owning at least one of the devices (Rainie, 2012). In January 2014, e-book penetration rate in the U.S. reached 32% (Pew Research Center,

2014), and 51% of U.S. households owned a tablet in April 2015 (Nielsen, 2015a). However, e-book sales revenue in the United States reached its peak in 2014 ($1.6 billion) and continued to drop in 2015 ($1.4 billion) and 2016 ($1.1 billion) (Association of American Publishers, 2017).

The Electronic Era

The telegraph transitioned from the print era to a new period by introducing a means of sending messages far more rapidly than was previously possible. Soon, Americans and people around the world enjoyed a world enhanced by such electronic media as wired telephones, motion pictures, audio recording, radio, television, cable television, and satellite television.

Telephone

With the telephone, Alexander Graham Bell became the first to transmit speech electronically in 1876. By June 30, 1877, 230 telephones were in use, and the number rose to 1,300 by the end of August, mostly to avoid the need for a skilled interpreter of telegraph messages. The first switching office connected three company offices in Boston beginning on May 17, 1877, reflecting a focus on business rather than residential use during the telephone's early decades. Hotels became early adopters of telephones as they sought to reduce the costs of employing human messengers, and New York's 100 largest hotels had 21,000 telephones by 1909. After 1894, non-business telephone use became common, in part because business use lowered the cost of telephone service. By 1902, 2,315,000 telephones were in service in the United States (Aronson, 1977). Table 2.4 and Figure 2.4 on the companion website (www.tfi.com/ctu) document the growth to near ubiquity of telephones in U.S. households and the expanding presence of wireless telephones.

Wireless Telephones

Guglielmo Marconi sent the first wireless data messages in 1895. The growing popularity of telephony led many to experiment with Marconi's radio technology as another means for interpersonal communication. By the 1920s, Detroit police cars had mobile radiophones for voice communication (ITU, 1999). The Bell system offered radio telephone service in 1946 in St. Louis, the first of 25 cities to receive the service. Bell engineers divided reception areas into cells in 1947, but cellular telephones that switched effectively among cells as callers moved did not arrive until the 1970s. The first call on a portable, handheld cell phone occurred in 1973. However, in 1981, only 24 people in New York City could use their mobile phones at the same time, and only 700 customers could have active contracts. To increase the number of people who could receive service, the Federal Communications Commission (FCC) began offering cellular telephone system licenses by lottery in June 1982 (Murray, 2001). Other countries, such as Japan in 1979 and Saudi Arabia in 1982, operated cellular systems earlier than the United States (ITU, 1999).

The U.S. Congress promoted a more advanced group of mobile communication services in 1993 by creating a classification that became known as Commercial Mobile Radio Service. This classification allowed for consistent regulatory oversight of these technologies and encouraged commercial competition among providers (FCC, 2005). By the end of 1996, about 44 million Americans subscribed to wireless telephone services (U.S. Bureau of the Census, 2008).

The new century brought an explosion of wireless telephones, and phones morphed into multipurpose devices (i.e., smartphones) with capabilities previously limited to computers. By the end of 2016, wireless phone penetration in the United States reached 395.9 million subscribers (CTIA, 2017), it further trended up to 417.5 million by 2017 (FierceWireless, 2017). *CTIA-The Wireless Association* (CTIA, 2017) reported that more than half (50.8%) of all American households were wireless-only by the end of 2016 up from about 10% in 2006. By 2017, worldwide shipments of smartphones exceeded 1.5 billion units (IDC, 2017), five times the quantity shipped in 2010 (IDC as cited by Amobi, 2013). As the Chinese smartphone market gradually matured, IDC (2015a) also forecasted that India would replace China as the leading driver of shipment increases.

Nevertheless, a factor that may trouble the advancement of smartphone sales and development is trade policy conflicts. Some governments are setting up barriers to foreign smartphone imports for trade or national security reasons. Without these obstacles, technological competition could have been fairer in the global market place and may lead to higher penetration rates and revenues for companies, with better services and products for ordinary customers.

Motion Pictures

In the 1890s, George Eastman improved on work by and patents purchased from Hannibal Goodwin in 1889 to produce workable motion picture film. The Lumière brothers projected moving pictures in a Paris café in 1895, hosting 2,500 people nightly at their movies. William Dickson, an assistant to Thomas Edison, developed the kinetograph, an early motion picture camera, and the kinetoscope, a motion picture viewing system. A New York movie house opened in 1894, offering moviegoers several coin-fed kinetoscopes. Edison's Vitascope, which expanded the length of films over those shown via kinetoscopes and allowed larger audiences to simultaneously see the moving images, appeared in public for the first time in 1896. In France in that same year, Georges Méliès started the first motion picture theater. Short movies became part of public entertainment in a variety of American venues by 1900 (Campbell, 2002), and average weekly movie attendance reached 40 million people by 1922.

Average weekly motion picture theater attendance, as shown in Table 2.5 and Figure 2.6 on the companion website (www.tfi.com/ctu), increased annually from the earliest available census reports on the subject in 1922 until 1930. After falling dramatically during the Great Depression, attendance regained growth in 1934 and continued until 1937. Slight declines in the prewar years were followed by a period of strength and stability throughout the World War II years. After the end of the war, average weekly attendance reached its greatest heights: 90 million attendees weekly from 1946 through 1949. After the introduction of television, weekly attendance would never again reach these levels.

Although a brief period of leveling off occurred in the late 1950s and early 1960s, average weekly attendance continued to plummet until a small recovery began in 1972. This recovery signaled a period of relative stability that lasted into the 1990s. Through the last decade of the century, average weekly attendance enjoyed small but steady gains.

Box office revenues, which declined generally for 20 years after the beginning of television, began a recovery in the late 1960s, then began to skyrocket in the 1970s. The explosion continued until after the turn of the new century. However, much of the increase in revenues came from increases in ticket prices and inflation, rather than from increased popularity of films with audiences, and total motion picture revenue from box office receipts declined during recent years, as studios realized revenues from television and videocassettes (U.S. Department of Commerce/International Trade Association, 2000).

As shown in Table 2.5 on the companion website (www.tfi.com/ctu), American movie fans spent an average of 12 hours per person per year from 1993 through 1997 going to theaters. That average stabilized through the first decade of the 21st century (U.S. Bureau of the Census, 2010), despite the growing popularity of watching movies at home with new digital tools. In 2011, movie rental companies were thriving, with *Netflix* boasting 25 million subscribers and *Redbox* having 32,000 rental kiosks in the United States (Amobi, 2011b). However, recent physical sales and rental of home entertainment content suffered from the rise of web streaming services and consumer behavior change (Digital Entertainment Group, 2017). *Redbox* kiosk rentals started to decline in 2013 ($1.97 billion) to $1.76 billion in 2015 (Outerwall, 2016).

The record-breaking success of *Avatar* in 2009 as a 3D motion picture triggered a spate of followers who tried to revive the technology that was a brief hit in the 1950s. *Avatar* earned more than $761 million at American box offices and nearly $2.8 billion worldwide.

In the United States, nearly 8,000 of 39,500 theater screens were set up for 3D at the end of 2010, half of them having been installed in that year. The ticket prices for 3D films ran 20-30% higher than that of 2D films, and 3D films comprised 20% of the new films released. Nevertheless, American audiences preferred subsequent 2D films to 3D competitors, although 3D response remained strong outside the United States, where 61% of the world's 22,000 3D screens were installed. In 2014, there were 64,905 3D screens worldwide, except for the Asian Pacific region (55% annual growth), the annual growth rates of 3D screen numbers have stabilized around 6%-10% (HIS quoted in

MPAA, 2015). Another factor in the lack of success of 3D in America might have been the trend toward viewing movies at home, often with digital playback. In 2010, home video purchases and rentals reached $18.8 billion in North America, compared with only $10.6 billion spent at theaters (Amobi, 2011b). U.S. home entertainment spending rose to $20.8 billion in 2017, with revenues in the physical market shrinking ($12 billion in 2016) and digital subscriptions to web streaming (e.g. Netflix) soaring (Digital Entertainment Group, 2018). In the meantime, the 2014 U.S. domestic box office slipped 1.4% to $10.44 billion (Nash Information Services quoted in Willens, 2015) and remained at a similar level for the next three years (Digital Entertainment Group, 2017).

Globally, the Asian Pacific region and Latin America have been the main contributors to global box office revenue since 2004 (MPAA, 2017). And the bloom of the Chinese movie market has been a major reason for the increase of global revenue. Chinese box office revenue continued to soar to $6.78 billion in 2015, this figure was only $1.51 billion in 2011, and a 48.7% annual growth rate in 2015 was also a new historical record for the Chinese movie market (State Administration of Press, Publication, Radio, Film and TV, 2016).

Audio Recording

Thomas Edison expanded on experiments from the 1850s by Leon Scott de Martinville to produce a talking machine or phonograph in 1877 that played back sound recordings from etchings in tin foil. Edison later replaced the foil with wax. In the 1880s, Emile Berliner created the first flat records from metal and shellac designed to play on his gramophone, providing mass production of recordings. The early standard recordings played at 78 revolutions per minute (rpm). After shellac became a scarce commodity because of World War II, records were manufactured from polyvinyl plastic. In 1948, CBS Records produced the long-playing record that turned at 33-1/3 rpm, extending the playing time from three to four minutes to 10 minutes. *RCA* countered in 1949 with 45 rpm records that were incompatible with machines that played other formats. After a five-year war of formats, record players were manufactured that would play recordings at all of the speeds (Campbell, 2002).

The Germans used plastic magnetic tape for sound recording during World War II. After the Americans confiscated some of the tapes, the technology was adopted and improved, becoming a boon for Western audio editing and multiple track recordings that played on bulky reel-to-reel machines. By the 1960s, the reels were encased in plastic cases, variously known as 8-track tapes and compact audio cassettes, which would prove to be deadly competition in the 1970s for single song records playing at 45 rpm and long-playing albums playing at 33-1/3 rpm. Thomas Stockholm began recording sound digitally in the 1970s, and the introduction of compact disc (CD) recordings in 1983 decimated the sales performance of earlier analog media types (Campbell, 2002). Tables 2.6 and 2.6A and Figures 2.7 and 2.7A on the companion website (www.tfi.com/ctu) show that total unit sales of recorded music generally increased from the early 1970s through 2004 and kept declining after that mainly because of the rapid decline in CD sales. Figure 2.7a shows trends in downloaded music.

The 21st century saw an explosion in new digital delivery systems for music. Digital audio players, which had their first limited popularity in 1998 (Beaumont, 2008), hit a new gear of growth with the 2001 introduction of the Apple iPod, which increased the storage capacity and became responsible for about 19% of music sales within its first decade. Apple's online iTunes store followed in 2003, soon becoming the world's largest music seller (Amobi, 2009). However, as shown in Table 2.6A, notwithstanding with the drop in CD sales, a new upward trend was prompted by the emergence of paid online music subscriptions, sound exchange, synchronization, etc. Revenue-wise, paid subscriptions generated $2.26 billion, while CD revenue declined to $1.17 billion. Again, we are witnessing another "technological dynasty" after the "CD dynasty", the "cassette dynasty" and the "LP/EP dynasty" of recorded music (RIAA, 2017).

Radio

Guglielmo Marconi's wireless messages in 1895 on his father's estate led to his establishing a British company to profit from ship-to-ship and ship-to-shore messaging. He formed a U.S. subsidiary in 1899 that would become the American Marconi Company. Reginald A. Fessenden and Lee De Forest independently transmitted voice by means of wireless radio in 1906, and a radio broadcast from the stage of a performance by Enrico Caruso occurred in 1910. Various U.S. companies and Marconi's British company owned important patents that were necessary to the development of the infant industry, so the U.S. firms, including AT&T formed the Radio Corporation of America (RCA) to buy the patent rights from Marconi.

The debate still rages over the question of who became the first broadcaster among KDKA in Pittsburgh (Pennsylvania), WHA in Madison (Wisconsin), WWJ in Detroit (Michigan), and KQW in San Jose (California). In 1919, Dr. Frank Conrad of Westinghouse broadcast music from his phonograph in his garage in East Pittsburgh. Westinghouse's KDKA in Pittsburgh announced the presidential election returns over the airwaves on November 2, 1920. By January 1, 1922, the Secretary of Commerce had issued 30 broadcast licenses, and the number of licensees swelled to 556 by early 1923. By 1924, RCA owned a station in New York, and Westinghouse expanded to Chicago, Philadelphia, and Boston. In 1922, AT&T withdrew from RCA and started WEAF in New York, the first radio station supported by commercials. In 1923, AT&T linked WEAF with WNAC in Boston by the company's telephone lines for a simultaneous program. This began the first network, which grew to 26 stations by 1925. RCA linked its stations with telegraph lines, which failed to match the voice quality of the transmissions of AT&T. However, AT&T wanted out of the new business and sold WEAF in 1926 to the National Broadcasting Company, a subsidiary of RCA (White, 1971).

The 1930 penetration of radio sets in American households reached 40%, then approximately doubled over the next 10 years, passing 90% by 1947 (Brown, 2006). Table 2.7 and Figure 2.8, on the companion website (www.tfi.com/ctu), show the rapid rate of increase in the number of radio households from 1922 through the early 1980s, when the rate of increase declined. The increases resumed until 1993, when they began to level off.

Although thousands of radio stations were transmitting via the Internet by 2000, Channel1031.com became the first station to cease using FM and move exclusively to the Internet in September 2000 (Raphael, 2000). Many other stations were operating only on the

Internet when questions about fees for commercial performers and royalties for music played on the Web arose. In 2002, the Librarian of Congress set royalty rates for Internet transmissions of sound recordings (U.S. Copyright Office, 2003). A federal court upheld the right of the Copyright Office to establish fees on streaming music over the Internet (*Bonneville v. Peters*, 2001).

In March 2001, the first two American digital audio satellites were launched, offering the promise of hundreds of satellite radio channels (Associated Press, 2001). Consumers were expected to pay about $9.95 per month for access to commercial-free programming that would be targeted to automobile receivers. The system included amplification from about 1,300 ground antennas. By the end of 2003, about 1.6 million satellite radio subscribers tuned to the two top providers, XM and Sirius (Schaeffler, 2004). These two players merged soon before the 2008 stock market crisis, during which the new company, Sirius XM Radio, lost nearly all of its stock value. In 2011, the service was used by 20.5 million subscribers, with its market value beginning to recover (Sirius XM Radio, 2011). The company continues to attract new subscribers and reported its highest subscriber growth since 2007 in 2015 with a 30% growth rate to 29.6 million subscribers; this number continued to grow to more than 32.7 million in 2017 (Sirius XM Radio, 2016; 2018).

Television

Paul Nipkow invented a scanning disk device in the 1880s that provided the basis from which other inventions would develop into television. In 1927, Philo Farnsworth became the first to electronically transmit a picture over the air. Fittingly, he transmitted the image of a dollar sign. In 1930, he received a patent for the first electronic television, one of many patents for which RCA would be forced, after court challenges, to negotiate. By 1932, Vladimir Zworykin discovered a means of converting light rays into electronic signals that could be transmitted and reconstructed at a receiving device. RCA offered the first public demonstration of television at the 1939 World's Fair.

The FCC designated 13 channels in 1941 for use in transmitting black-and-white television, and the com-

mission issued almost 100 television station broadcasting licenses before placing a freeze on new licenses in 1948. The freeze offered time to settle technical issues, and it ran longer because of U.S. involvement in the Korean War (Campbell, 2002). As shown in Table 2.8 on the companion website (www.tfi.com/ctu), nearly 4 million households had television sets by 1950, a 9% penetration rate that would escalate to 87% a decade later. Penetration has remained steady at about 98% since 1980 until a recent small slide to about 96% in 2014. Figure 2.8 illustrates the meteoric rise in the number of households with television by year from 1946 through 2015. In 2010, 288.5 million Americans had televisions, up by 0.8% from 2009, and average monthly time spent viewing reached 158 hours and 47 minutes, an increase of 0.2% from the previous year (Amobi, 2011a). In 2015, Nielsen estimated that 296 million persons age 2 and older lived in 116.3 million homes that have TV, which showed that "Both the universe of U.S. television homes and the potential TV audience in those homes continue to grow" (Nielsen, 2015b). In 2014, Americans were spending an average of 141 hours per month watching TV and paid more attention to online video streaming and Over-The-Top TV service such as Netflix (IDATE quoted in Statista, 2016).

By the 1980s, Japanese high-definition television (HDTV) increased the potential resolution to more than 1,100 lines of data in a television picture. This increase enabled a much higher-quality image to be transmitted with less electromagnetic spectrum space per signal. In 1996, the FCC approved a digital television transmission standard and authorized broadcast television stations a second channel for a 10-year period to allow the transition to HDTV. As discussed in Chapter 6, that transition made all older analog television sets obsolete because they cannot process HDTV signals (Campbell, 2002).

The FCC (2002) initially set May 2002 as the deadline by which all U.S. commercial television broadcasters were required to be broadcasting digital television signals. Progress toward digital television broadcasting fell short of FCC requirements that all affiliates of the top four networks in the top 10 markets transmit digital signals by May 1, 1999.

Within the 10 largest television markets, all except one network affiliate had begun HDTV broadcasts by

August 1, 2001. By that date, 83% of American television stations had received construction permits for HDTV facilities or a license to broadcast HDTV signals (FCC, 2002). HDTV penetration into the home marketplace would remain slow for the first few years of the 21st century, in part because of the high price of the television sets.

Although 3D television sets were available in 2010, little sales success occurred. The sets were quite expensive, not much 3D television content was available, and the required 3D viewing glasses were inconvenient to wear (Amobi, 2011b).

During the fall 2011-12 television season, The Nielsen Company reported that the number of households with televisions in the United States dropped for the first time since the company began such monitoring in the 1970s. The decline to 114.7 million from 115.9 million television households represented a 2.2% decline, leaving the television penetration at 96.7%. Explanations for the reversal of the long-running trend included the economic recession, but the decline could represent a transition to digital access in which viewers were getting TV from devices other than television sets (Wallenstein, 2011). But the number of television households has since increased again, with Nielsen reporting 116.3 million U.S. television households in 2015 (Nielsen, 2015b).

Cable Television

Cable television began as a means to overcome poor reception for broadcast television signals. John Watson claimed to have developed a master antenna system in 1948, but his records were lost in a fire. Robert J. Tarlton of Lansford (Pennsylvania) and Ed Parsons of Astoria (Oregon) set up working systems in 1949 that used a single antenna to receive programming over the air and distribute it via coaxial cable to multiple users (Baldwin & McVoy, 1983). At first, the FCC chose not to regulate cable, but after the new medium appeared to offer a threat to broadcasters, cable became the focus of heavy government regulation. Under the Reagan administration, attitudes swung toward deregulation, and cable began to flourish. Table 2.9 and Figure 2.9 on the companion website (www.tfi.com/ctu) show the growth of cable systems and subscribers, with penetration remaining below

25% until 1981, but passing the 50% mark before the 1980s ended.

In the first decade of the 21st century, cable customers began receiving access to such options as Internet access, digital video, video on demand, DVRs, HDTV, and telephone services. By fall 2010, 3.2 million (9%) televisions were connected to the Internet (Amobi, 2011b). The success of digital cable led to the FCC decision to eliminate analog broadcast television as of February 17, 2009. However, in September 2007, the FCC unanimously required cable television operators to continue to provide carriage of local television stations that demand it in both analog and digital formats for three years after the conversion date. This action was designed to provide uninterrupted local station service to all cable television subscribers, protecting the 40 million (35%) U.S. households that remained analog-only (Amobi & Kolb, 2007).

Telephone service became widespread via cable during the early years of the 21st century. For years, some cable television operators offered circuit-switched telephone service, attracting 3.6 million subscribers by the end of 2004. Also by that time, the industry offered telephone services via voice over Internet protocol (VoIP) to 38% of cable households, attracting 600,000 subscribers. That number grew to 1.2 million by July 2005 (Amobi, 2005).

The growth of digital cable in the first decade of the new century also saw the growth of video-on-demand (VOD), offering cable television customers the ability to order programming for immediate viewing.

Cable penetration declined in the United States after 2000, as illustrated in Figure 2.9 on the companion website (www.tfi.com/ctu). However, estimated use of a combination of cable and satellite television increased steadily over the same period (U.S. Bureau of the Census, 2008).

Worldwide, pay television flourished in the new century, especially in the digital market. From 2009 to 2014, pay TV subscriptions increased from 648 million households to 923.5 million households. (Statista, 2016). This increase was and is expected to be led by rapid growth in the Asia Pacific region and moderate growth in Latin America and Africa, while pay TV subscription slowly declined in North America and

Eastern Europe (Digital TV Research, 2017). At the same time, the number of pay television subscribers in the U.S. was falling, from more than 95 million in 2012 to 94.2 million in 2015 (Dreier, 2016).

Direct Broadcast Satellite and Other Cable TV Competitors

Satellite technology began in the 1940s, but HBO became the first service to use it for distributing entertainment content in 1976 when the company sent programming to its cable affiliates (Amobi & Kolb, 2007). Other networks soon followed this lead, and individual broadcast stations (WTBS, WGN, WWOR, and WPIX) used satellites in the 1970s to expand their audiences beyond their local markets by distributing their signals to cable operators around the U.S.

Competitors for the cable industry include a variety of technologies. Annual FCC reports distinguish between home satellite dish (HSD) and direct broadcast satellite (DBS) systems. Both are included as MVPDs (multi-channel video program distributors), which include cable television, wireless cable systems called multichannel multipoint distribution services (MMDS), and private cable systems called satellite master antenna television (SMATV). Table 2.10 and Figure 2.10 on the companion website for this book (www.tfi.com/ctu), show trends in home satellite dish, DBS, MMDS, and SMATV (or PCO, Private Cable Operator) subscribers. However, the FCC (2013a) noted that little public data was available for the dwindling services of HSD, MMDS, and PCO, citing SNL Kagan conclusions that those services accounted for less than 1% of MVPDs and were expected to continue declining over the coming decade.

In developed markets like the U.S., Internet-based Over-the-Top TV services such as Netflix, Hulu and SlingTV have grown substantially since 2015. SlingTV's subscriber totals grew from 169,000 in March 2015 to 523,000 in the end of 2015 (Ramachandran, 2016). The number of Netflix subscribers grew to 52.77 million in the U.S. in the third quarter of 2017 (Netflix, 2017) and its international subscribers are accumulating even faster, adding up to 109.25 million subscribers worldwide (Netflix, 2017). In the company's 2016 long term market view, Netflix reported: "People love TV content, but they don't love the linear TV experience, where channels present programs only at particular times on non-portable screens with complicated remote controls" (Netflix, 2016). It is possible that the concept of TV is again being redefined by people's need for cord-cutting and screen convergence. From 2014 to 2017, share of cord-cutters/nevers among all U.S. TV households grew from 18.8% to 24.6% (Convergence Consulting Group, 2017).

Home Video

Although VCRs became available to the public in the late 1970s, competing technical standards slowed the adoption of the new devices. After the longer taping capacity of the VHS format won greater public acceptance over the higher-quality images of the Betamax, the popularity of home recording and playback rapidly accelerated, as shown in Table 2.11 and Figure 2.11 on the companion website (www.tfi.com/ctu).

By 2004, rental spending for videotapes and DVDs reached $24.5 billion, far surpassing the $9.4 billion spent for tickets to motion pictures in that year. During 2005, DVD sales increased by 400% over the $4 billion figure for 2000 to $15.7 billion. However, the annual rate of growth reversed direction and slowed that year to 45% and again the following year to 2%. VHS sales amounted to less than $300 million in 2006 (Amobi & Donald, 2007).

Factors in the decline of VHS and DVD use included growth in cable and satellite video-on-demand services, growth of broadband video availability, digital downloading of content, and the transition to DVD Blu-ray format (Amobi, 2009). The competing new formats for playing high-definition content was similar to the one waged in the early years of VCR development between the Betamax and VHS formats. Similarly, in early DVD player development, companies touting competing standards settled a dispute by agreeing to share royalties with the creator of the winning format. Until early 2008, the competition between proponents of the HD-DVD and Blu-ray formats for playing high-definition DVD content remained unresolved, and some studios were planning to distribute motion pictures in both formats. Blu-ray seemed to emerge the victor in 2008 when large companies (e.g., Time Warner, Walmart, Netflix) declared allegiance to that format. By July 2010, Blu-ray penetration reached 17% of American households

(Gruenwedel, 2010), and 170 million Blu-ray discs shipped that year (Amobi, 2011b).

Digital video recorders (DVRs, also called personal video recorders, PVRs) debuted during 2000, and about 500,000 units were sold by the end of 2001 (FCC, 2002). The devices save video content on computer hard drives, allowing fast-forwarding, reversing, and pausing of live television; replay of limited minutes of previously displayed live television; automatic recording of all first-run episodes; automatic recording logs; and superior quality to that of analog VCRs. Multiple tuner models allow viewers to watch one program, while recording others simultaneously.

DVR providers generate additional revenues by charging households monthly fees, and satellite DVR households tend to be less likely to drop their satellite subscriptions. Perhaps the most fundamental importance of DVRs is the ability of consumers to make their own programming decisions about when and what they watch. This flexibility threatens the revenue base of network television in several ways, including empowering viewers to skip commercials. Advertiser responses included sponsorships and product placements within programming (Amobi, 2005).

Reflecting the popularity of the DVR, time shifting was practiced in 2010 by 107.1 million American households (up 13.2% from 2009). Time shifted viewing increased by 12.2% in 2010 from 2009 to an average of 10 hours and 46 minutes monthly (Amobi, 2011a). As shown in Table 2.11 on the companion website (www.tfi.com/ctu), its penetration rate also bounced back to 45.1% in 2014, however, it declined again in 2016 (Plastics Industry Association, 2017).

The Digital Era

The digital era represents a transition in modes of delivery of mediated content. Although the tools of using digital media may have changed, in many cases, the content remains remarkably stable. With other media, such as social media, the digital content fostered new modes of communicating. This section contains histories of the development of computers and the Internet. Segments of earlier discussions could be

considered part of the digital era, such as audio recording, HDTV, films on DVD, etc., but discussions of those segments remain under earlier eras.

Computers

The history of computing traces its origins back thousands of years to such practices as using bones as counters (Hofstra University, 2000). Intel introduced the first microprocessor in 1971. The MITS Altair, with an 8080 processor and 256 bytes of RAM (random access memory), sold for $498 in 1975, introducing the desktop computer to individuals. In 1977, Radio Shack offered the TRS80 home computer, and the Apple II set a new standard for personal computing, selling for $1,298. Other companies began introducing personal computers, and, by 1978, 212,000 personal computers were shipped for sale.

Early business adoption of computers served primarily to assist practices such as accounting. When computers became small enough to sit on office desktops in the 1980s, word processing became popular and fueled interest in home computers. With the growth of networking and the Internet in the 1990s, both businesses and consumers began buying computers in large numbers. Computer shipments around the world grew annually by more than 20% between 1991 and 1995 (Kessler, 2007).

By 1997, the majority of American households with annual incomes greater than $50,000 owned a personal computer. At the time, those computers sold for about $2,000, exceeding the reach of lower income groups. By the late 1990s, prices dropped below $1,000 per system (Kessler, 2007), and American households passed the 60% penetration mark within a couple of years (U.S. Bureau of the Census, 2008).

Table 2.12 and Figure 2.12 on the companion website (www.tfi.com/ctu) trace the rapid and steady rise in American computer shipments and home penetration. After 2006, U.S. PC shipments started to decline and with penetration reaching 65.3% in 2015 (IDC, 2016). By 1998, 42.1% of American households owned personal computers (U.S. Bureau of the Census, 2006). After the start of the 21st century, personal computer prices declined, and penetration increased from 63% in 2000 to 77% in 2008 (Forrester Research as cited in Kessler, 2011). Worldwide personal computer sales increased by 34% from 287 million in 2008 to 385 million in 2011 (IDC as cited Kessler, 2011). However, this number suffered from a 3.2% decrease in 2012 (IDC, 2016). This downward trend continued, in 2016 worldwide shipments were down to 260 million, lower than in 2008 (IDC, 2015b).

IDC (2017) also predicted that the downward trend would continue. However, while pointing out that shipments of desktops and slate tables would decline, it predicted that shipments of detachable tablets would increase from 21.5 million units in 2016 to 37.1 units in 2021. It is important to note that media convergence is present even with PCs, as an analyst from IDC pointed out: "a silver lining is that the industry has continued to refine the more mobile aspects of personal computers—contributing to higher growth in Convertible & Ultraslim Notebooks" (IDC, 2015c).

Internet

The Internet began in the 1960s with ARPANET, or the Advanced Research Projects Agency (ARPA) network project, under the auspices of the U.S. Defense Department. The project intended to serve the military and researchers with multiple paths of linking computers together for sharing data in a system that would remain operational even when traditional communications might become unavailable. Early users—mostly university and research lab personnel—took advantage of electronic mail and posting information on computer bulletin boards. Usage increased dramatically in 1982 after the National Science Foundation (NSF) supported high-speed linkage of multiple locations around the United States. After the collapse of the Soviet Union in the late 1980s, military users abandoned ARPANET, but private users continued to use it and multimedia transmissions of audio and video became possible once this content could be digitized. More than 150,000 regional computer networks and 95 million computer servers hosted data for Internet users (Campbell, 2002).

Penetration and Usage

During the first decade of the 21st century, the Internet became the primary reason that consumers purchased new computers (Kessler, 2007). Cable modems and digital subscriber line (DSL) telephone line connections increased among home users as the means for connecting to the Internet, as more than half of American households accessed the Internet with high-speed broadband connections.

Tables 2.13 and 2.14 and Figures 2.13 and 2.14 on the companion website (www.tfi.com/ctu) show trends in Internet penetration in the United States. By 2008, 74 million (63%) American households had high-speed broadband access (Kessler, 2011). In June 2013, 91,342,000 fixed broadband subscriptions and 299,447,000 million American subscribers had wireless broadband subscriptions (OECD, 2013). In 2012, although the number of DSL+ users decreased, the numbers on other fronts are showing growth (FCC, 2013b). In 2013, however, subscription to all types of Internet connections declined, and resulted in a 6% decline in total penetration (FCC, 2014). In recent years, however, the penetration gradually climbed up to 76.2% in 2016, this rate is the highest recorded Internet penetration in the U.S. so far (ITU, 2017).

Synthesis

Although this chapter has emphasized the importance of quantitative records of technology adoption, any understanding and interpretation of these numbers should consider the social contexts and structures in each historical period.

Early visions of the Internet (see Chapter 23) did not include the emphasis on entertainment and information to the general public that has emerged. The combination of this new medium with older media belongs to a phenomenon called *convergence*, referring to the merging of functions of old and new media (Grant, 2009). By 2002, the FCC (2002) reported that the most important type of convergence related to video content is the joining of Internet services. The report also noted that companies from many business areas were providing a variety of video, data, and other communications services.

The word *convergence* itself excludes the idea that technologies are monolithic constructs that symbolize the separation between two or more types of products. Rather, we see that older media, including both print and electronic types, have gone through transitions from their original forms into digital media.

In this way, many of the core assumptions and social roles of older media have converged with new desires and imaginations of society, and then take form in new media (Castells, 2011). Such technological convergence is compatible with both the power structure of the current society and that of a desired society (Habermas, 1991) which (was) inspired (by) the process of liberal movements, democratic transitions, capitalist motivations and globalization (Ravenhill, 2014).

For example, print media, with the help of a complex writing system, largely protected the centrality and authority of government power in ancient China. When government announcements were replaced by folklore announcements (i.e. the press), and when the printing press abandoned its paper form and went online, organizational power and authority decreased to be replaced with individual power.

However, this change does not suggest that the core functions of print media have changed: governmental power is now represented by individual power in democratic societies and it has become more invisible with technological advancements. "Print media" still carries an assumption of authority whether it is distributed using ink or electricity; the difference is that such authority is more interactive.

Just as media forms began converging nearly a century ago when radios and record players merged in the same appliance, media in recent years have been converging at a more rapid pace. As the popularity of print media generally declined throughout the 1990s, the popularity of the Internet grew rapidly, particularly with the increase in high-speed broadband connections, for which adoption rates achieved comparability with previous new communications media. Consumer flexibility through the use of digital media became the dominant media consumption theme during the first decade of the new century.

Bibliography

Agnese, J. (2011, October 20). Industry surveys: Publishing & advertising. In E. M. Bossong-Martines (Ed.), *Standard & Poor's industrysSurveys, Vol. 2.*

Amobi, T. N. (2005, December 8). Industry surveys: Broadcasting, cable, and satellite industry survey. In E. M. Bossong-Martines (Ed.), *Standard & Poor's industry surveys, 173* (49), Section 2.

Amobi, T. N. (2009). Industry surveys: Movies and home entertainment. In E. M. Bossong-Martines (Ed.), *Standard & Poor's industry surveys, 177* (38), Section 2.

Amobi, T. N. (2011a). Industry surveys: Broadcasting, cable, and satellite. In E. M. Bossong-Martines (Ed.), *Standard & Poor's industry surveys, Vol. 1.*

Amobi, T. N. (2011b). Industry surveys: Movies & entertainment. In E. M. Bossong-Martines (Ed.), *Standard & Poor's industry surveys, Vol. 2.*

Amobi, T. N. (2013). Industry surveys: Broadcasting, cable, & satellite. In E. M. Bossong-Martines (Ed.), *Standard & Poor's industry surveys, Vol. 1.*

Amobi, T. N. & Donald, W. H. (2007, September 20). Industry surveys: Movies and home entertainment. In E. M. Bossong-Martines (Ed.), *Standard & Poor's industry surveys, 175* (38), Section 2.

Amobi, T. N. & Kolb, E. (2007, December 13). Industry surveys: Broadcasting, cable & satellite. In E. M. Bossong-Martines (Ed.), *Standard & Poor's industry surveys, 175* (50), Section 1.

Aronson, S. (1977). Bell's electrical toy: What's the use? The sociology of early telephone usage. In I. Pool (Ed.). *The social impact of the telephone.* Cambridge, MA: The MIT Press, 15-39.

Association of American Publishers. (2017). Publisher Revenue for Trade Books Up 10.2% in November 2016. Retrieved January 15, 2018, from http://newsroom.publishers.org/publisher-revenue-for-trade-books-up-102-in-november-2016/.

Associated Press. (2001, March 20). Audio satellite launched into orbit. *New York Times.* Retrieved from http://www.nytimes.com/aponline/national/AP-Satellite-Radio.html?ex=986113045& ei=1&en=7af33c7805ed8853.

Baldwin, T. & McVoy, D. (1983). *Cable communication.* Englewood Cliffs, NJ: Prentice-Hall.

Beaumont, C. (2008, May 10). Dancing to the digital tune As the MP3 turns 10. *The Daily Telegraph*, p. 19. Retrieved from LexisNexis Database.

Bonneville International Corp., et al. v. Marybeth Peters, as Register of Copyrights, et al. Civ. No. 01-0408, 153 F. Supp.2d 763 (E.D. Pa., August 1, 2001).

Brown, D. & Bryant, J. (1989). An annotated statistical abstract of communications media in the United States. In J. Salvaggio & J. Bryant (Eds.), *Media use in the information age: Emerging patterns of adoption and consumer use.* Hillsdale, NJ: Lawrence Erlbaum Associates, 259-302.

Brown, D. (1996). A statistical update of selected American communications media. In Grant, A. E. (Ed.), *Communication Technology Update* (5th ed.). Boston, MA: Focal Press, 327-355.

Brown, D. (1998). Trends in selected U. S. communications media. In Grant, A. E. & Meadows, J. H. (Eds.), Communication Technology Update (6th ed.). Boston, MA: Focal Press, 279-305.

Brown, D. (2000). Trends in selected U. S. communications media. In Grant, A. E. & Meadows, J. H. (Eds.), *Communication Technology Update* (7th ed.). Boston, MA: Focal Press, 299-324.

Brown, D. (2002). Communication technology timeline. In A. E. Grant & J. H. Meadows (Eds.), *Communication technology update* (8th ed.) Boston: Focal Press, 7-45.

Brown, D. (2004). Communication technology timeline. In A. E. Grant & J. H. Meadows (Eds.). *Communication technology update* (9th ed.). Boston: Focal Press, 7-46.

Brown, D. (2006). Communication technology timeline. In A. E. Grant & J. H. Meadows (Eds.), *Communication technology update* (10th ed.). Boston: Focal Press. 7-46.

Brown, D. (2008). Historical perspectives on communication technology. In A. E. Grant & J. H. Meadows (Eds.), *Communication technology update and fundamentals* (11th ed.). Boston: Focal Press. 11-42.

Brown, D. (2010). Historical perspectives on communication technology. In A. E. Grant & J. H. Meadows (Eds.), *Communication technology update and fundamentals* (12th ed.). Boston: Focal Press. 9-46.

Brown, D. (2012). Historical perspectives on communication technology. In A. E. Grant & J. H. Meadows (Eds.), *Communication technology update and fundamentals* (13th ed.). Boston: Focal Press. 9-24.

Brown, D. (2014). Historical perspectives on communication technology. In A. E. Grant & J. H. Meadows (Eds.), *Communication technology update and fundamentals* (14th ed.). Boston: Focal Press. 9-20.

Campbell, R. (2002). *Media & culture.* Boston, MA: Bedford/St. Martins.

Castells, M. (2011). The rise of the network society: The information age: Economy, society, and culture (Vol. 1). John Wiley & Sons.

Cicero, M. T. (1876). *Orator ad M. Brutum.* BG Teubner.

Convergence Consulting Group. (2017). The Battle for the North American (US/Canada) Couch Potato: Online & Traditional TV and Movie Distribution. Retrieved January 14, 2018, from http://www.convergenceonline.com/downloads/New-Content2017.pdf?lbisphpreq=1.

CTIA. (2017). Annual Wireless Industry Survey. Retrieved January 15, 2018, from https://www.ctia.org/industry-data/ctia-annual-wireless-industry-survey.

Digital Entertainment Group. (2017). DEG Year End 2016 Home Entertainment Report. from http://degonline.org/portfolio_page/deg-year-end-2016-home-entertainment-report/.

Digital Entertainment Group. (2018). DEG Year End 2017 Home Entertainment Report. from http://degonline.org/portfolio_page/deg-year-end-2017-home-entertainment-report/.

Digital TV Research. (2017). Digital TV Research's July 2017 newsletter. Retrieved January 14, 2018, from https://www.digitaltvresearch.com/press-releases?id=204.

Dillon, D. (2011). E-books pose major challenge for publishers, libraries. In D. Bogart (Ed.), *Library and book trade almanac* (pp. 3-16). Medford, NJ: Information Today, Inc.

Dreier, T. (2016, March 11). Pay TV industry losses increase to 385,000 subscribers in 2015. Retrieved from: http://www.streamingmedia.com/Articles/Editorial/Featured-Articles/Pay-TV-Industry-Losses-Increase-to-385000-Subscribers-in-2015-109711.aspx.

Federal Communications Commission. (2002, January 14). *In the matter of annual assessment of the status of competition in the market for the delivery of video programming* (eighth annual report). CS Docket No. 01-129. Washington, DC 20554.

Federal Communications Commission. (2005). *In the matter of Implementation of Section 6002(b) of the Omnibus Budget Reconciliation Act of 1993: Annual report and analysis of competitive market conditions with respect to commercial mobile services* (10th report). WT Docket No. 05-71. Washington, DC 20554.

Federal Communications Commission. (2013a, July 22). *In the matter of annual assessment of the status of competition in the market for the delivery of video programming* (fifteenth annual report). MB Docket No. 12-203. Washington, DC 20554.

Federal Communications Commission. (2013b, December). Internet access services: Status as of December 31, 2012. Retrieved from http://hraunfoss.fcc.gov/edocs_public/attachmatch/DOC-324884A1.pdf.

Federal Communications Commission. (2014, October). Internet access services: Status as of December 31, 2013. Retrieved from https://transition.fcc.gov/Daily_Releases/Daily_Business/2014/db1016/DOC-329973A1.pdf.

FierceWireless. (2017). How Verizon, AT&T, T-Mobile, Sprint and more stacked up in Q3 2017: The top 7 carriers. Retrieved January 12, 2018, from https://www.fiercewireless.com/wireless/how-verizon-at-t-t-mobile-sprint-and-more-stacked-up-q3-2017-top-7-carriers.

Grant, A. E. (2009). Introduction: Dimensions of media convergence. In Grant, A E. and Wilkinson, J. S. (Eds.) *Understanding media convergence: The state of the field*. New York: Oxford University Press.

Gruenwedel, E. (2010). Report: Blu's household penetration reaches 17%. *Home Media Magazine, 32,* 40.

Habermas, J. (1991). The structural transformation of the public sphere: An inquiry into a category of bourgeois society. MIT press.

Hofstra University. (2000). *Chronology of computing history*. Retrieved from http://www.hofstra.edu/pdf/CompHist_9812tla1.pdf.

Huntzicker, W. (1999). *The popular press, 1833-1865*. Westport, CT: Greenwood Press.

Ink, G. & Grabois, A. (2000). *Book title output and average prices: 1998 final and 1999 preliminary figures*, 45th edition. D. Bogart (Ed.). New Providence, NJ: R. R. Bowker, 508-513.

International Data Corporation. (2015a). *Worldwide Smartphone Market Will See the First Single-Digit Growth Year on Record, According to IDC* [Press release]. Retrieved from http://www.idc.com/getdoc.jsp?containerId=prUS40664915.

International Data Corporation. (2015b). *Worldwide PC Shipments Will Continue to Decline into 2016 as the Short-Term Outlook Softens, According to IDC* [Press release]. Retrieved from http://www.idc.com/getdoc.jsp?containerId=prUS40704015.

International Data Corporation. (2015c). *PC Shipments Expected to Shrink Through 2016 as Currency Devaluations and Inventory Constraints Worsens Outlook, According to IDC* [Press release]. Retrieved from http://www.idc.com/getdoc.jsp?containerId=prUS25866615.

International Data Corporation. (2016). *United States Quarterly PC Tracker.* Retrieved from http://www.idc.com/tracker/showproductinfo.jsp?prod_id=141.

International Data Corporation. (2017). Worldwide Quarterly Mobile Phone Tracker. Retrieved from https://www.idc.com/tracker/showproductinfo.jsp?prod_id=37.

International Telecommunications Union. (1999). *World telecommunications report 1999*. Geneva, Switzerland: Author.

ITU. (2017). Percentage of Individuals using the Internet. Retrieved January 15, 2018, from http://www.itu.int/en/ITU-D/Statistics/Documents/statistics/2017/Individuals_Internet_2000-2016.xls.

Kessler, S. H. (2007, August 26). Industry surveys: Computers: Hardware. In E M. Bossong-Martines (Ed.), *Standard & Poor's industry surveys, 175* (17), Section 2.

Kessler, S. H. (2011). Industry surveys: Computers: Consumer services and the Internet. In E. M. Bossong-Martines (Ed.), *Standard & Poor's industry surveys, Vol. 1.*

Lee, J. (1917). *History of American journalism*. Boston: Houghton Mifflin.

Moorman, J. G. (2013, July). Industry surveys: Publishing and advertising. In E. M. Bossong-Martines (Ed.), Standard & Poor's industry surveys, Vol 2.

MPAA. (2015). *2014 Theatrical Market Statistics* [Press release]. Retrieved from http://www.mpaa.org/wp-content/uploads/2015/03/MPAA-Theatrical-Market-Statistics-2014.pdf.

MPAA. (2017). *2016 Theatrical Market Statistics* [Press release]. Retrieved from https://www.mpaa.org/wp-content/uploads/2017/03/2016-Theatrical-Market-Statistics-Report-2.pdf.

Murray, J. (2001). *Wireless nation: The frenzied launch of the cellular revolution in America*. Cambridge, MA: Perseus Publishing.

Nielsen. (2015a). *Q1 2015 Local Watch Report: Where You Live and Its Impact On Your Choices.* [Press release] Retrieved from http://www.nielsen.com/us/en/insights/reports/2015/q1-2015-local-watch-report-where-you-live-and-its-impact-on-your-choices.html.

Nielsen. (2015b). *Nielsen Estimates 116.3 Million TV Homes In The U.S., Up 0.4%.* [Press release] Retrieved from http://www.nielsen.com/us/en/insights/news/2014/nielsen-estimates-116-3-million-tv-homes-in-the-us.html.

Netflix. (2016). *Netflix's View: Internet TV is replacing linear TV*. Retrieved from http://ir.netflix.com/long-term-view.cfmOECD. (2013). Broadband and telecom. Retrieved from http://www.oecd.org/internet/broadband/oecdbroadbandportal.htm.

Netflix. (2017). Consolidated Segment Information. Retrieved January 14, 2018, from https://ir.netflix.com/financial-statements.

OECD. (2013). Broadband and telecom. Retrieved from http://www.oecd.org/internet/broadband/oecdbroadbandportal.htm.

Outerwall. (2016). Outerwall Annual Report to U.S. Security and Exchange Commission. Retrieved January 13, 2018, from http://d1lge852tjjqow.cloudfront.net/CIK-0000941604/a5b4e8c7-6f8f-428d-b176-d9723dcbc772.pdf?noexit=true .

Peters, J. & Donald, W. H. (2007). Industry surveys: Publishing. In E. M. Bossong-Martines (Ed.), *Standard & Poor's industry surveys*. *175* (36). Section 1.

Pew Research Center. (2014, January 16). *E-Reading Rises as Device Ownership Jumps*. Retrieved from http://www.pewinternet.org/files/2014/01/PIP_E-reading_011614.pdf.

Plastics Industry Association. (2017). Watching: Consumer Technology Plastics' Innovative Chapter in The Consumer Technology Story. from http://www.plasticsindustry.org/sites/plastics.dev/files/PlasticsMarketWatchConsumer TechnologyWebVersion.pdf.

Rainie, L. (2012, January 23). Tablet and e-book reader ownership nearly doubles over the holiday gift-giving period. *Pew Research Center's Internet & American Life Project*. Retrieved from http://pewinternet.org/Reports/2012/E-readers-and-tablets.aspx.

Ramachandran, S. (2016, February 18). Dish Network's Sling TV Has More Than 600,000 Subscribers. *Wall Street Journal*. Retrieved http://www.wsj.com/article_email/dish-networks-sling-tv-has-more-than-600-000-subscribers-1455825689-

Raphael, J. (2000, September 4). Radio station leaves earth and enters cyberspace. Trading the FM dial for a digital stream. *New York Times*. Retrieved from http://www.nytimes.com/library/tech/00/ 09/biztech/articles/04radio.html.

Ravenhill, J. (2014). Global political economy. Oxford University Press.

Rawlinson, N. (2011, April 28). Books vs ebooks. *Computer Act!ve*. Retrieved from General OneFile database.

The Recording Industry Association of America. (2017). *U.S. Sales Database*. from https://www.riaa.com/u-s-sales-database/

Schaeffler, J. (2004, February 2). The real satellite radio boom begins. *Satellite News, 27* (5). Retrieved from Lexis-Nexis.

Sirius XM Radio Poised for Growth, Finally. (2011, May 10). *Newsmax*.

Sirius XM Radio. (2016, February 2). SiriusXM Reports Fourth Quarter and Full-Year 2015 Results [Press release]. Retrieved from http://s2.q4cdn.com/835250846/files/doc_financials/annual2015/SiriusXM-Reports-Fourth-Quarter-and-Full-Year-2015-Results.pdf.

Sirius XM Radio. (2018, January 10). SiriusXM Beats 2017 Subscriber Guidance; Issues 2018 Subscriber and Financial Guidance [Press release]. Retrieved from http://investor.siriusxm.com/investor-overview/press-releases/press-release-details/2018/SiriusXM-Beats-2017-Subscriber-Guidance-Issues-2018-Subscriber-and-Financial-Guidance/default.aspx.

State Administration of Press, Publication, Radio, Film and Television of The People's Republic of China. (2016, January 3). *China Box Office Surpassed 44 Billion Yuan in 2015*. Retrieved from http://www.sarft.gov.cn/art/2016/1/3/art_160_29536.html.

Statista. (2016). *Average daily TV viewing time per person in selected countries worldwide in 2014 (in minutes)*. Retrieved from http://www.statista.com/statistics/276748/average-daily-tv-viewing-time-per-person-in-selected-countries/.

U.S. Bureau of the Census. (2006). *Statistical abstract of the United States: 2006* (125th Ed.). Washington, DC: U.S. Government Printing Office.

U.S. Bureau of the Census. (2008). *Statistical abstract of the United States: 2008* (127th Ed.). Washington, DC: U.S. Government Printing Office. Retrieved from http://www.census.gov/compendia/statab/.

U.S. Bureau of the Census. (2010). *Statistical abstract of the United States: 2008* (129th Ed.). Washington, DC: U.S. Government Printing Office. Retrieved from http://www.census.gov/compendia/statab/.

US Census Bureau. (2016). 2015 Annual Services Report. Retrieved January 15, 2018, from http://www2.census.gov/services/sas/data/Historical/sas-15.xls.

US Census Bureau. (2017). 2016 Annual Services Report. Retrieved January 15, 2018, from http://www2.census.gov/services/sas/data/table4.xls.

U.S. Copyright Office. (2003). *106th Annual report of the Register of Copyrights for the fiscal year ending September 30, 2003*. Washington, DC: Library of Congress.

U.S. Department of Commerce/International Trade Association. (2000). *U.S. industry and trade outlook 2000*. New York: McGraw-Hill.

Wallenstein, A. (2011, September 16). Tube squeeze: economy, tech spur drop in TV homes. *Daily Variety*, 3.

White, L. (1971). *The American radio*. New York: Arno Press.

Willens, M. (2015, January 7). Home Entertainment 2014: US DVD Sales and Rentals Crater, Digital Subscriptions Soar. *International Business Times*. Retrieved from http://www.ibtimes.com/home-entertainment-2014-us-dvd-sales-rentals-crater-digital-subscriptions-soar-1776440.

Zhu, Y. & Brown, D. (2016). A history of communication technology. In A. E. Grant & J. H. Meadows (Eds.), *Communication technology update and fundamentals*. Taylor & Francis. 9-24.

3

Understanding Communication Technologies

Jennifer H. Meadows, Ph.D.*

Think back just 20 years ago. In many ways, we were still living in an analog world. There were no smart phones or free Wi-Fi hotspots, no Netflix or Hulu, and you couldn't just grab a Lyft. The latest communication technologies at that time were Palm Pilots and DVDs, and those were cutting edge. Communication technologies are in a constant state of change. New ones are developed while others fade. This book was created to help you understand these technologies, but there is a set of tools that will not only help you understand them, but also understand the next generation of technologies.

All of the communication technologies explored in this book have a number of characteristics in common, including how their adoption spreads from a small group of highly interested users to the general public (or not), what the effects of these technologies are upon the people who use them (and on society in general), and how these technologies affect each other.

For more than a century, researchers have studied adoption, effects, and other aspects of new technologies, identifying patterns that are common across dissimilar technologies, and proposing theories of technology adoption and effects. These theories have proven to be valuable to entrepreneurs seeking to develop new technologies, regulators who want to control those technologies, and everyone else who just wants to understand them.

The utility of these theories is that they allow you to apply lessons from one technology to another or from old technologies to new technologies. The easiest way to understand the role played by the technologies explored in this book is to have a set of theories you can apply to virtually any technology you discuss. The purpose of this chapter is to give you these tools by introducing you to the theories.

The technology ecosystem discussed in Chapter 1 is a useful framework for studying communication technologies, but it is not a theory. This perspective is

---

* Professor and Chair, Department of Media Arts, Design, and Technology, California State University, Chico (Chico, California).

a good starting point to begin to understand communication technologies because it targets your attention at a number of different levels that might not be immediately obvious: hardware, software, content, organizational infrastructure, social systems, and, finally, the user.

Understanding each of these levels is aided by knowing a number of theoretical perspectives that can help us understand the different sections of the ecosystem for these technologies. Theoretical approaches are useful in understanding the origins of the information-based economy in which we now live, why some technologies take off while others fail, the impacts and effects of technologies, and the economics of the communication technology marketplace.

The Information Society and the Control Revolution

Our economy used to be based on tangible products such as coal, lumber, and steel. This is no longer the case as information is now the basis of our economy. Information industries include education; research and development; creating informational goods such as computer software, banking, insurance; and even entertainment and news (Beniger, 1986).

Information is different from other commodities like coffee and pork bellies, which are known as "private goods." Instead, information is a "public good" because it is intangible, lacks a physical presence, and can be sold as many times as demand allows without regard to consumption.

For example, if 10 sweaters are sold, then 10 sweaters must be manufactured using raw materials. If 10 subscriptions to an online news service are sold, there is no need to create a different story for each user; 10—or 10,000—subscriptions can be sold without additional raw materials.

This difference actually gets to the heart of a common misunderstanding about ownership of information that falls into a field known as "intellectual property rights." A common example is the purchase of a digital music download. A person may believe that because they purchased the music, that they can copy and distribute that music to others. Just because

the information (the music) was purchased doesn't mean they own the song and performance (intellectual property).

Several theorists have studied the development of the information society, including its origin. Beniger (1986) argues that there was a control revolution: "A complex of rapid changes in the technological and economic arrangements by which information is collected, stored, processed, and communicated and through which formal or programmed decisions might affect social control" (p. 52). In other words, as society progressed, technologies were created to help control information. For example, information was centralized by mass media.

In addition, as more and more information is created and distributed, new technologies must be developed to control that information. For example, with the explosion of information available over the Internet, search engines were developed to help users find relevant information.

Another important point is that information is power, and there is power in giving information away. Power can also be gained by withholding information. At different times in modern history, governments have blocked access to information or controlled information dissemination to maintain power.

Adoption

Why do some technologies succeed while others fail? This question is addressed by a number of theoretical approaches including the diffusion of innovations, social information processing theory, critical mass theory, the theory of planned behavior, the technology acceptance model, and more.

Diffusion of Innovations

The diffusion of innovations, also referred to as diffusion theory, was developed by Everett Rogers (1962; 2003). This theory tries to explain how an innovation is communicated over time through different channels to members of a social system. There are four main aspects of this approach.

First, there is the innovation. In the case of communication technologies, the innovation is some technology that is perceived as new. Rogers also defines characteristics of innovations: relative advantage, compatibility, complexity, trialability, and observability.

So, if someone is deciding to purchase a new mobile phone, characteristics would include the relative advantage over other mobile phones; whether or not the mobile phone is compatible with the existing needs of the user; how complex it is to use; whether or not the potential user can try it out; and whether or not the potential user can see others using the new mobile phone with successful results.

Information about an innovation is communicated through different channels. Mass media is good for awareness knowledge. For example, each new iPhone has Web content, television commercials, and print advertising announcing its existence and its features.

Interpersonal channels are also an important means of communication about innovations. These interactions generally involve subjective evaluations of the innovation. For example, a person might ask some friends how they like their new iPhones.

Rogers (2003) outlines the decision-making process a potential user goes through before adopting an innovation. This is a five-step process.

The first step is knowledge. You find out there is a new mobile phone available and learn about its new features. The next step is persuasion—the formation of a positive attitude about the innovation. The third step is when you decide to accept or reject the innovation. Implementation is the fourth step. Finally, confirmation occurs when you decide that you made the correct decision.

Another stage discussed by Rogers (2003) and others is "reinvention," the process by which a person who adopts a technology begins to use it for purposes other than those intended by the original inventor. For example, mobile phones were initially designed for calling other people regardless of location, but users have found ways to use them for a wide variety of applications ranging from alarm clocks to personal calendars and flashlights.

Have you ever noticed that some people are the first to have the new technology gadget, while others refuse to adopt a proven successful technology?

Adopters can be categorized into different groups according to how soon or late they adopt an innovation.

The first to adopt are the innovators. Innovators are special because they are willing to take a risk adopting something new that may fail. Next come the early adopters, the early majority, and then the late majority, followed by the last category, the laggards. In terms of percentages, innovators make up the first 2.5% percent of adopters, early adopters are the next 13.5%, early majority follows with 34%, late majority are the next 34%, and laggards are the last 16%.

Adopters can also be described in terms of ideal types. Innovators are venturesome. These are people who like to take risks and can deal with failure. Early adopters are respectable. They are valued opinion leaders in the community and role models for others. Early majority adopters are deliberate. They adopt just before the average person and are an important link between the innovators, early adopters, and everyone else. The late majority are skeptical. They are hesitant to adopt innovations and often adopt because they pressured. Laggards are the last to adopt and often are isolated with no opinion leadership. They are suspicious and resistant to change. Other factors that affect adoption include education, social status, social mobility, finances, and willingness to use credit (Rogers, 2003).

Adoption of an innovation does not usually occur all at once; it happens over time. This is called the rate of adoption. The rate of adoption generally follows an S-shaped "diffusion curve" where the X-axis is time and the Y-axis is percent of adopters. You can note the different adopter categories along the diffusion curve.

Figure 3.1 shows a diffusion curve. See how the innovators are at the very beginning of the curve, and the laggards are at the end. The steepness of the curve depends on how quickly an innovation is adopted. For example, DVD has a steeper curve than VCR because DVD players were adopted at a faster rate than VCRs.

Also, different types of decision processes lead to faster adoption. Voluntary adoption is slower than collective decisions, which, in turn, are slower than authority decisions. For example, a company may let its workers decide whether to use a new software package, the employees may agree collectively to use that software, or finally, the management may decide that everyone at the company is going to use the software. In most cases, voluntary adoption would take the longest, and a management dictate would result in the swiftest adoption.

Figure 3.1
Innovation Adoption Rate

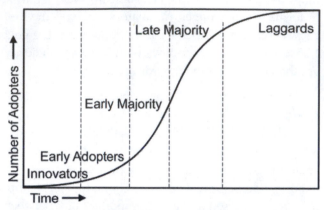

Source: Technology Futures, Inc.

Moore (2001) further explored diffusion of innovations and high-tech marketing in *Crossing the Chasm*. He noted there are gaps between the innovators and the early adopters, the early adopters and the early majority, and the early majority and late majority.

For a technology's adoption to move from innovators to the early adopters the technology must show a major new benefit. Innovators are visionaries that take the risk of adopting something new such as virtual home assistants.

Early adopters then must see the new benefit of virtual home assistants before adopting. The chasm between early adopters and early majority is the greatest of these gaps. Early adopters are still visionary and want to be change agents. They don't mind dealing with the troubles and glitches that come along with a new technology. Early adopters were likely to use a beta version of a new service or product.

The early majority, on the other hand, are pragmatists and want to see some improvement in productivity—something tangible. Moving from serving the visionaries to serving the pragmatists is difficult; hence Moore's description of "crossing the chasm."

Finally, there is a smaller gap between the early majority and the late majority. Unlike the early majority, the late majority reacts to the technical demands

on the users. The early majority is more comfortable working with technology. So, the early majority would be comfortable using a virtual home assistant like the Amazon Echo but the late majority is put off by the perceived technical demands. The technology must alleviate this concern before late majority adoption.

Another perspective on adoption can be found in most marketing textbooks (e.g., Kottler & Keller, 2011): the product lifecycle. As illustrated in Figure 3.2, the product lifecycle extends the diffusion curve to include the maturity and decline of the technology. This perspective provides a more complete picture of a technology because it focuses our attention beyond the initial use of the technology to the time that the technology is in regular use, and ultimately, disappears from market. Remember laserdisc or Myspace?

Considering the short lifespan of many communication technologies, it may be just as useful to study the entire lifespan of a technology rather than just the process of adoption.

Figure 3.2
Product Lifecycle

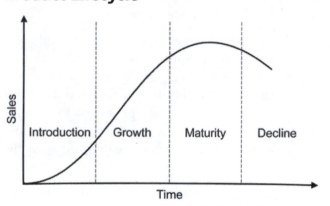

Source: Technology Futures, Inc.

Other Theories of Adoption

Other theorists have attempted to explain the process of adoption as well. Among the most notable perspectives in this regard are the Theory of Planned Behavior (TPB), the Technology Acceptance Model (TAM) and the Unified Theory of Acceptance and Use of Technology. These models emerged from the need to identify factors that can help predict future adoption of a new technology when there is no history of adoption or use of the technology.

The Theory of Planned Behavior (TPB) (Ajzen, 1991; Ajzen & Fishbein, 1980; Fishbein & Ajzen, 1975) presumes that a suitable predictor of future behavior is "behavioral intention," a cognitive rather than behavioral variable that represents an individual's plans for adopting (or not adopting!) an innovation. Behavioral intentions are, in turn, predicted by attitudes toward the innovation and the innovators.

The Technology Acceptance Model (Davis, 1986; Davis, Bagozzi & Warshaw, 1989) elaborates on TPB by adding factors that may predict attitudes toward an innovation and behavioral intentions, including perceived usefulness, perceived ease of use, and external variables.

The Unified Theory of Acceptance and Use of Technology attempts to combine elements of several theories of technology adoptions including the Theory of Planned Behavior, the Technology Acceptance model, and Diffusions of Innovations (Venkatesh, Morris, Davis; & Davis (2003). The authors argue that there are four core determinates of user acceptance and behavior. These are performance expectancy, effort expectancy, social influence, and facilitating conditions.

A substantial body of research has demonstrated the efficacy of these factors in predicting behavioral intentions (e.g., Jiang, Hsu & Klein, 2000; Chau & Hu, 2002, but much more research is needed regarding the link between behavioral intentions and actual adoption at a later point in time.

Another theory that expands upon Rogers' diffusion theory is presented in Chapter 4. Grant's pre-diffusion theory identifies organizational functions that must be served before any consumer adoption of the technology can take place.

Critical Mass Theory

Have you ever wondered who had the first email address or the first telephone? Who did they communicate with? Interactive technologies such as telephony, social networking and email become more and more useful as more people adopt these technologies. There have to be some innovators and early adopters who are willing to take the risk to try a new interactive technology.

These users are the "critical mass," a small segment of the population that chooses to make big contributions to the public good (Markus, 1987). In general terms, any social process involving actions by individuals that benefit others is known as "collective action." In this case, the technologies become more useful if everyone in the system is using the technology, a goal known as "universal access."

Ultimately, universal access means that you can reach anyone through some communication technology. For example, in the United States, the landline phone system reaches almost everywhere, and everyone benefits from this technology, although a small segment of the population initially chose to adopt the telephone to get the ball rolling. There is a stage in the diffusion process that an interactive medium has to reach in order for adoption to take off. This is the critical mass. Interestingly, 49% of homes in the US are mobile phone only but the important factor to remember is that even early mobile phone users could call anyone on a land line so the technology didn't have the same critical hurdle to overcome (CTIA, 2018).

Another conceptualization of critical mass theory is the "tipping point" (Gladwell, 2002). Here is an example. The videophone never took off, in part, because it never reached critical mass. The videophone was not really any better than a regular phone unless the person you were calling also had a videophone. If there were not enough people you knew who had videophones, then you might not adopt it because it was not worth it.

On the other hand, if most of your regular contacts had videophones, then that critical mass of users might drive you to adopt the videophone. Critical mass is an important aspect to consider for the adoption of any interactive technology. Think of the iPhone's Facetime App. The only people who can use Facetime are iPhone users. Skype, on the other hand, works on all mobile operating systems so it doesn't have the same limitations as Facetime.

Another good example is facsimile or fax technology. The first method of sending images over wires was invented in the '40s—the 1840s—by Alexander Bain, who proposed using a system of electrical pendulums to send images over wires (Robinson, 1986). Within a few decades, the technology was adopted by the newspaper industry to send photos over wires, but the technology was limited to a small number of news organizations.

The development of technical standards in the 1960s brought the fax machine to corporate America, which generally ignored the technology because few businesses knew of another business that had a fax machine.

Adoption of the fax took place two machines at a time, with those two usually being purchased to communicate with each other, but rarely used to communicate with additional receivers. By the 1980s, enough businesses had fax machines that could communicate with each other that many businesses started buying fax machines one at a time.

As soon as the critical mass point was reached, fax machine adoption increased to the point that it became referred to as the first technology adopted out of fear of embarrassment that someone would ask, "What's your fax number?" (Wathne & Leos, 1993). In less than two years, the fax machine became a business necessity.

Social Information Processing

Another way to look at how and why people choose to use or not use a technology is social information processing. This theory begins by critiquing rational choice models, which presume that people make adoption decisions and other evaluations of technologies based upon objective characteristics of the technology. In order to understand social information processing, you first have to look at a few rational choice models.

One model, social presence theory, categorizes communication media based on a continuum of how the medium "facilitates awareness of the other person and interpersonal relationships during the interaction" (Fulk, et al., 1990, p. 118).

Communication is most efficient when the social presence level of the medium best matches the interpersonal relationship required for the task at hand. For example, most people would propose marriage face-to-face instead of using a text message.

Another rational choice model is information richness theory. In this theory, media are also arranged on a continuum of richness in four areas: speed of feedback, types of channels employed, personalness of source, and richness of language carried (Fulk, et al., 1990). Face-to-face communications is the highest in social presence and information richness.

In information richness theory, the communication medium chosen is related to message ambiguity. If the message is ambiguous, then a richer medium is chosen. In this case, teaching someone how to dance would be better with an online video that illustrates the steps rather than just an audio podcast that describes the steps.

Social information processing theory goes beyond the rational choice models because it states that perceptions of media are "in part, subjective and socially constructed." Although people may use objective standards in choosing communication media, use is also determined by subjective factors such as the attitudes of coworkers about the media and vicarious learning, or watching others' experiences

Social influence is strongest in ambiguous situations. For example, the less people know about a medium, then the more likely they are to rely on social information in deciding to use it (Fulk, et al., 1987).

Think about whether you prefer a Macintosh or a Windows-based computer. Although you can probably list objective differences between the two, many of the important factors in your choice are based upon subjective factors such as which one is owned by friends and coworkers, the perceived usefulness of the computer, and advice you receive from people who can help you set up and maintain your computer. Think of the iPhone vs Android debate or Instagram vs Snapchat.

In the end, these social factors probably play a much more important role in your decision than "objective" factors such as processor speed, memory capacity and other technical specifications.

Impacts & Effects

Do video games make players violent? Do users seek out social networking sites for social interactions? These are some of the questions that theories of impacts or effects try to answer.

To begin, Rogers (1986) provides a useful typology of impacts. Impacts can be grouped into three dichotomies: desirable and undesirable, direct and indirect, and anticipated and unanticipated.

Desirable impacts are the functional impacts of a technology. For example, a desirable impact of social networking is the ability to connect with friends and family. An undesirable impact is one that is dysfunctional, such as bullying or stalking.

Direct impacts are changes that happen in immediate response to a technology. A direct impact of wireless telephony is the ability to make calls while driving. An indirect impact is a byproduct of the direct impact. To illustrate, laws against driving and using a handheld wireless phone are an impact of the direct impact described above.

Anticipated impacts are the intended impacts of a technology. An anticipated impact of text messaging is to communicate without audio. An unanticipated impact is an unintended impact, such as people sending text messages in a movie theater and annoying other patrons. Often, the desirable, direct, and anticipated impacts are the same and are considered first. Then, the undesirable, indirect, and unanticipated impacts are noted later.

Here is an example using email. A desirable, anticipated, and direct impact of email is to be able to quickly send a message to multiple people at the same time. An undesirable, indirect, and unanticipated impact of email is spam—unwanted email clogging the inboxes of millions of users.

Uses and Gratifications

Uses and gratifications research is a descriptive approach that gives insight into what people do with technology. This approach sees the users as actively seeking to use different media to fulfill different needs (Rubin, 2002). The perspective focuses on "(1) the social and psychological origins of (2) needs, which generate (3) expectations of (4) the mass media or other sources, which lead to (5) differential patterns of media exposure (or engagement in other activities), resulting in (6) needs gratifications and (7) other consequences, perhaps mostly unintended ones" (Katz, et al., 1974, p. 20).

Uses and gratifications research surveys audiences about why they choose to use different types of media. For example, uses and gratifications of television studies have found that people watch television for information, relaxation, to pass time, by habit, excitement, and for social utility (Rubin, 2002).

This approach is also useful for comparing the uses and gratifications between media, as illustrated by studies of the World Wide Web (www and television gratifications that found that, although there are some similarities such as entertainment and to pass time, they are also very different on other variables such as companionship, where the Web was much lower than for television (Ferguson & Perse, 2000). Uses and gratifications studies have examined a multitude of communication technologies including mobile phones (Wei, 2006), radio (Towers, 1987), satellite television (Etefa, 2005), and social media (Raacke & Bonds-Raacke, 2008; Phua, Jin & Kim, 2017).

Media System Dependency Theory

Often confused with uses and gratifications, media system dependency theory is "an ecological theory that attempts to explore and explain the role of media in society by examining dependency relations within and across levels of analysis" (Grant, et al., 1991, p. 774). The key to this theory is the focus it provides on the dependency relationships that result from the interplay between resources and goals.

The theory suggests that, in order to understand the role of a medium, you have to look at relationships at multiple levels of analysis, including the individual level—the audience, the organizational level, the media system level, and society in general.

These dependency relationships can be symmetrical or asymmetrical. For example, the dependency relationship between audiences and network television is asymmetrical because an individual audience member may depend more on network television to reach his or her goal than the television networks depend on that one audience member to reach their goals.

A typology of individual media dependency relations was developed by Ball-Rokeach & DeFleur (1976) to help understand the range of goals that individuals have when they use the media. There are six dimensions: social understanding, self-understanding, action orientation, interaction orientation, solitary play, and social play.

Social understanding is learning about the world around you, while self-understanding is learning about yourself. Action orientation is learning about specific behaviors, while interaction orientation is about learning about specific behaviors involving other people. Solitary play is entertaining yourself alone, while social play is using media as a focus for social interaction.

Research on individual media system dependency relationships has demonstrated that people have different dependency relationships with different media. For example, Meadows (1997) found that women had stronger social understanding dependencies for television than magazines, but stronger self-understanding dependencies for magazines than television.

In the early days of television shopping (when it was considered "new technology"), Grant, et al. (1991) applied media system dependency theory to the phenomenon. Their analysis explored two dimensions: how TV shopping changed organizational dependency relations within the television industry and how and why individual users watched television shopping programs.

By applying a theory that addressed multiple levels of analysis, a greater understanding of the new technology was obtained than if a theory that focused on only one level had been applied.

Social Learning Theory/Social Cognitive Theory

Social learning theory focuses on how people learn by modeling others (Bandura, 2001). This observational learning occurs when watching another person model the behavior. It also happens with symbolic modeling, modeling that happens by watching the behavior modeled on a television or computer screen. So, a person can learn how to fry an egg by watching another person fry an egg in person or on a video.

Learning happens within a social context. People learn by watching others, but they may or may not perform the behavior. Learning happens, though, whether the behavior is imitated or not.

Reinforcement and punishment play a role in whether or not the modeled behavior is performed. If the behavior is reinforced, then the learner is more likely to perform the behavior. For example, if a student is successful using online resources for a presentation, other students watching the presentation will be more likely to use online resources.

On the other hand, if the action is punished, then the modeling is less likely to result in the behavior. To illustrate, if a character drives drunk and gets arrested on a television program, then that modeled behavior is less likely to be performed by viewers of that program.

Reinforcement and punishment is not that simple though. This is where cognition comes in—learners think about the consequences of performing that behavior. This is why a person may play *Grand Theft Auto* and steal cars in the videogame, but will not then go out and steal a car in real life. Self-regulation is an important factor. Self-efficacy is another important dimension: learners must believe that they can perform the behavior.

Social learning/cognitive theory, then, is a useful framework for examining not only the effects of communication media, but also the adoption of communication technologies (Bandura, 2001).

The content that is consumed through communication technologies contains symbolic models of behavior that are both functional and dysfunctional. If viewers model the behavior in the content, then some form of observational learning is occurring.

A lot of advertising works this way. A celebrity uses a new shampoo and then is admired by others. This message models a positive reinforcement of using the shampoo. Cognitively, the viewer then thinks about the consequences of using the shampoo.

Modeling can happen with live models and symbolic models. For example, a person can watch another playing *Just Dance, 2016*, a videogame where the player has to mimic the dance moves of an avatar in the game. The other player considers the consequences of this modeling.

In addition, if the other person had not played with this gaming system, watching the other person play with the system and enjoy the experience will make it more likely that he or she will adopt the system. Therefore, social learning/cognitive theory can be used to facilitate the adoption of new technologies and to understand why some technologies are adopted and why some are adopted faster than others (Bandura, 2001).

Economic Theories

Thus far, the theories and perspectives discussed have dealt mainly with individual users and communication technologies. How do users decide to adopt a technology? What impacts will a technology have on a user?

Theory, though, can also be applied to organizational infrastructure and the overall technology market. Here, two approaches will be addressed: the theory of the long tail that presents a new way of looking at digital content and how it is distributed and sold, and the principle of relative constancy that examines what happens to the marketplace when new media products are introduced.

The Theory of the Long Tail

Former *Wired Magazine* editor Chris Anderson developed the theory of the long tail. While some claim this is not a "theory," it is nonetheless a useful framework for understanding new media markets.

This theory begins with the realization that there are not any huge hit movies, television shows, and albums like there used to be. What counts as a hit TV show today, for example, would be a failed show just 15 years ago.

One of the reasons for this is choice: 40 years ago, viewers had a choice of only a few television channels. Today, you can have hundreds of channels of video

programming on cable or satellite and limitless amounts of video programming on the Internet.

New communication technologies are giving users access to niche content. There is more music, video, video games, news, etc. than ever before because the distribution is no longer limited to the traditional mass media of over-the-air broadcasting, newspapers, etc. or the shelf space at a local retailer.

The theory states that, "our culture and economy are increasingly shifting away from a focus on a relatively small number of 'hits' at the headend of the demand curve and toward a huge number of niches in the tail" (Anderson, n.d.).

Figure 3.3 shows a traditional demand curve; most of the hits are at the head of the curve, but there is still demand as you go into the tail. There is a demand for niche content and there are opportunities for businesses that deliver content in the long tail.

Figure 3.3
The Long Tail

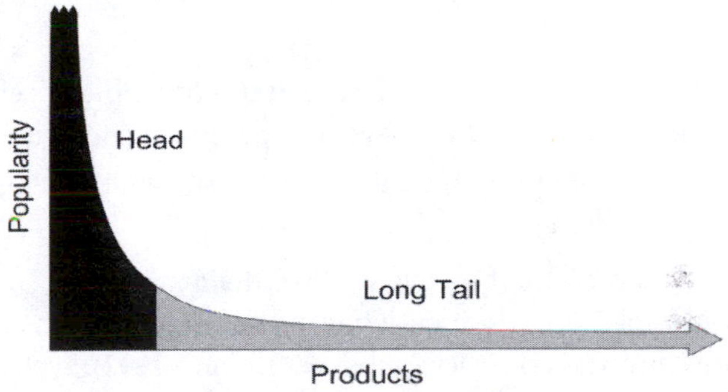

Source: Anderson (n.d.)

Both physical media and traditional retail have limitations. For example, there is only so much shelf space in the store. Therefore, the store, in order to maximize profit, is only going to stock the products most likely to sell. Digital content and distribution changes this.

For example, Amazon and Netflix can have huge inventories of hard-to-find titles, as opposed to a Red Box kiosk, which has to have duplicate inventories at each location. All digital services, such as the iTunes store, completely eliminate physical media. You purchase and download the content digitally, and there is no need for a warehouse to store DVDs and CDs.

Because of these efficiencies, these businesses can better serve niche markets. Taken one at a time, these niche markets may not generate significant revenue but when they are aggregated, these markets are significant.

Anderson (2006) suggests rules for long tail businesses. Make everything available, lower the price, and help people find it. Traditional media are responding to these services. For example, Nintendo is making classic games available for download. Network television is putting up entire series of television programming on the Internet.

The audience is changing, and expectations for content selection and availability are changing. The audience today, Anderson argues, wants what they want, when they want it, and how they want it.

The Principle of Relative Constancy

So now that people have all of this choice of content, delivery mode, etc., what happens to older media? Do people just keep adding new entertainment media, or do they adjust by dropping one form in favor of another?

This question is at the core of the principle of relative constancy, which says that people spend a constant fraction of their disposable income on mass media over time.

People do, however, alter their spending on mass media categories when new services/products are introduced (McCombs & Nolan, 1992). What this means is that, if a new media technology is introduced, in order for adoption to happen, the new technology has to be compelling enough for the adopter to give up something else.

For example, a person who signs up for Netflix may spend less money on movie tickets. A Spotify user will spend less money purchasing music downloads or CDs. So, when considering a new media technology, the relative advantage it has over existing service must be considered, along with other characteristics of the technology discussed earlier in this chapter. Remember, the money users spend on any new technology has to come from somewhere.

Critical & Cultural Theories

Most of the theories discussed above are derived from sociological perspectives on media and media use, relying upon quantitative analysis and the study of individuals to identify the dimensions of the technologies and the audience that help us understand these technologies. An alternate perspective can be found in critical and cultural studies, which provide different perspectives that help us understand the role of media in society.

Critical and cultural studies make greater use of qualitative analysis, including more analysis of macro-level factors such as media ownership and influences on the structure and content of media.

Marxist scholars, for example, focus on the underlying economic system and the division between those who control and benefit from the means of production and those who actually produce and consume goods and services.

This tradition considers that ownership and control of media and technology can affect the type of content provided. Scholars including Herman & Chomsky (2008) and Bagdikian (2000) have provided a wide range of evidence and analysis informing our understanding of these structural factors.

Feminist theory provides another set of critical perspectives that can help us understand the role and function of communication technology in society. Broadly speaking, feminist theory encompasses a broad range of factors including bodies, race, class, gender, language, pleasure, power, and sexual division of labor (Kolmar & Bartkowski, 2005).

New technologies represent an especially interesting challenge to the existing power relationships in media (Allen, 1999), and application of feminist theory has the potential to explicate the roles and power relationships in these media as well as to prescribe new models of organizational structure that take advantage of the shift in power relationships offered by interactive media (Grant, Meadows, & Storm, 2009).

Cultural studies also address the manner in which content is created and understood. Semiotics, for example, differentiates among "signs," "signifiers," and "signified" (Eco, 1979), explicating the manner in which meaning is attached to content. Scholars such as Hall (2010) have elaborated the processes by which content is encoded and decoded, helping us to understand the complexities of interpreting content in a culturally diverse environment.

Critical and cultural studies put the focus on a wide range of systemic and subjective factors that challenge conventional interpretations and understandings of media and technologies.

Conclusion

This chapter has provided a brief overview of several theoretical approaches to understanding communication technology. As you work through the book, consider theories of adoption, effects, economics, and critical/cultural studies and how they can inform you about each technology and allow you to apply lessons from one technology to others. For more in-depth discussions of these theoretical approaches, check out the sources cited in the bibliography.

Bibliography

Ajzen, I. (1991). The theory of planned behavior. *Organizational Behavior and Human Decision Processes, 50,* 179–211.

Ajzen, I. & Fishbein, M. (1980). *Understanding attitudes and predicting social behavior.* Englewood Cliffs, NJ: Prentice-Hall.

Allen, A. (1999). *The power of feminist theory.* Boulder, CO: Westview Press.

Anderson, C. (n.d.). *About me.* Retrieved from http://www.thelongtail.com/about.html.

Anderson, C. (2006). *The long tail: Why the future of business is selling less of more.* New York, NY: Hyperion.

Ball-Rokeach, S. & DeFleur, M. (1976). A dependency model of mass-media effects. *Communication Research, 3,* 1 3-21.

Bagdikian, B. H. (2000). *The media monopoly* (Vol. 6). Boston: Beacon Press.

Bandura, A. (2001). Social cognitive theory of mass communication. *Media Psychology, 3,* 265-299.

Beniger, J. (1986). The information society: Technological and economic origins. In S. Ball-Rokeach & M. Cantor (Eds.). *Media, audience, and social structure.* Newbury Park, NJ: Sage, pp. 51-70.

Chau, P. Y. K. & Hu, P. J. (2002). Examining a model of information technology acceptance by individual professionals: An exploratory study. *Journal of Management Information Systems, 18*(4), 191–229.

CTIA (2018). Your Wireless Life. Retrieved from http://www.ctia.org/your-wireless-life/how-wireless-works/wireless-quick-facts.

Davis, F. D. (1986). *A technology acceptance model for empirically testing new end-user information systems: Theory and results.* Unpublished doctoral dissertation, Massachusetts Institute of Technology, Cambridge.

Davis, F. D., Bagozzi, R. P. & Warshaw, P. R. (1989). User acceptance of computer technology: A comparison of two theoretical models. *Management Science, 35*(8), 982–1003.

Eco, U. (1979). *A theory of semiotics* (Vol. 217). Indiana University Press.

Etefa, A. (2005). *Arabic satellite channels in the U.S.: Uses & gratifications.* Paper presented at the annual meeting of the International Communication Association, New York. Retrieved from http://www.allacademic.com/ meta/p14246_index.html.

Ferguson, D. & Perse, E. (2000, Spring). The World Wide Web as a functional alternative to television. *Journal of Broadcasting and Electronic Media. 44* (2), 155-174.

Fishbein, M. & Ajzen, I. (1975). *Belief, attitude, intention and behavior: An introduction to theory and research.* Reading, MA: Addison-Wesley.

Fulk, J., Schmitz, J. & Steinfield, C. W. (1990). A social influence model of technology use. In J. Fulk & C. Steinfield (Eds.), *Organizations and communication technology.* Thousand Oaks, CA: Sage, pp. 117-140.

Fulk, J., Steinfield, C., Schmitz, J. & Power, J. (1987). A social information processing model of media use in organizations. *Communication Research, 14* (5), 529-552.

Gladwell, M. (2002). *The tipping point: How little things can make a big difference.* New York: Back Bay Books.

Grant, A. E., Guthrie, K. & Ball-Rokeach, S. (1991). Television shopping: A media system dependency perspective. *Communication Research, 18* (6), 773-798.

Grant, A. E., Meadows, J. H., & Storm, E. J. (2009). A feminist perspective on convergence. In Grant, A. E. & Wilkinson, J. S. (Eds.) *Understanding media convergence: The state of the field.* New York: Oxford University Press.

Hall, S. (2010). Encoding, decoding1. *Social theory: Power and identity in the global era, 2,* 569.

Herman, E. S., & Chomsky, N. (2008). *Manufacturing consent: The political economy of the mass media.* Random House.

Jiang, J. J., Hsu, M. & Klein, G. (2000). E-commerce user behavior model: An empirical study. *Human Systems Management,* 19, 265–276.

Katz, E., Blumler, J. & Gurevitch, M. (1974). Utilization of mass communication by the individual. In J. Blumler & E. Katz (Eds.). *The uses of mass communication: Current perspectives on gratifications research.* Beverly Hills: Sage.

Kolmar, W., & Bartkowski, F. (Ed.). (2005). *Feminist theory: A reader.* Boston: McGraw-Hill Higher Education.

Kottler, P. and Keller, K. L. (2011) *Marketing Management (14th ed.)* Englewood Cliffs, NJ: Prentice-Hall.

Markus, M. (1987, October). Toward a "critical mass" theory of interactive media. *Communication Research, 14* (5), 497-511.

McCombs, M. & Nolan, J. (1992, Summer). The relative constancy approach to consumer spending for media. *Journal of Media Economics,* 43-52.

Meadows, J. H. (1997, May). *Body image, women, and media: A media system dependency theory perspective.* Paper presented to the 1997 Mass Communication Division of the International Communication Association Annual Meeting, Montreal, Quebec, Canada.

Moore, G. (2001). *Crossing the Chasm.* New York: Harper Business.

Phua, J., Jin, S. and Kim, J (2017). Uses and gratifications of social networking sites for bridging and bonding social capital: a comparison of Facebook, Twitter, Instagram, and Snapchat. *Computers in Human Behavior,* 72, 115-122.

Raacke, J. and Bonds-Raacke, J (2008). MySpace and Facebook: Applying the Uses and Gratificaitons Theory to exploring friend-networking sites. *CyperPsychology & Behavior,* 11, 2, 169-174.

Robinson, L. (1986). *The facts on fax.* Dallas: Steve Davis Publishing.

Rogers, E. (1962). *Diffusion of Innovations.* New York: Free Press.

Rogers, E. (1986). *Communication technology: The new media in society.* New York: Free Press.

Rogers, E. (2003). *Diffusion of Innovations, 5th Edition.* New York: Free Press.

Rubin, A. (2002). The uses-and-gratifications perspective of media effects. In J. Bryant & D. Zillmann (Eds.). *Media effects: Advances in theory and research.* Mahwah, NJ: Lawrence Earlbaum Associates, pp. 525-548.

Towers, W. (1987, May 18-21). *Replicating perceived helpfulness of radio news and some uses and gratifications.* Paper presented at the Annual Meeting of the Eastern Communication Association, Syracuse, New York.

Venkatesh, V, Morris, M, Davis, G, and Davis, F. (2003). User acceptance of information technology: Toward a unified View. MSI Quarterly, 27, 3, 425-478.

Wathne, E. & Leos, C. R. (1993). Facsimile machines. In A. E. Grant & K. T. Wilkinson (Eds.). *Communication Technology Update: 1993-1994.* Austin: Technology Futures, Inc.

Wei, R. (2006). Staying connected while on the move. *New Media and Society, 8* (1), 53-72.

4

The Structure of the Communication Industries

August E. Grant, Ph.D.*

The first factor that many people consider when studying communication technologies is changes in the equipment and use of the technology. But, as discussed in Chapter 1, it is equally important to study and understand all areas of the technology ecosystem. In editing the *Communication Technology Update* for 25 years, one factor stands out as having the greatest amount of short-term change: the organizational infrastructure of the technology.

The continual flux in the organizational structure of communication industries makes this area the most dynamic area of technology to study. "New" technologies that make a major impact come along only a few times a decade. New products that make a major impact come along once or twice a year. Organizational shifts are constantly happening, making it almost impossible to know all of the players at any given time.

Even though the players are changing, the organizational structure of communication industries is relatively stable. The best way to understand the industry, given the rapid pace of acquisitions, mergers, startups, and failures, is to understand its organizational functions. This chapter addresses the organizational structure and explores the functions of those industries, which will help you to understand the individual technologies discussed throughout this book.

In the process of using organizational functions to analyze specific technologies, you must consider that these functions cross national as well as technological boundaries. Most hardware is designed in one country, manufactured in another, and sold around the globe. Although there are cultural and regulatory differences addressed in the individual technology chapters later in the book, the organizational functions discussed in this chapter are common internationally.

What's in a Name? The AT&T Story

A good illustration of the importance of understanding organizational functions comes from analyzing the history of AT&T, one of the biggest names in communication of all time. When you hear the name

---

* J. Rion McKissick Professor of Journalism, School of Journalism and Mass Communications, University of South Carolina (Columbia, South Carolina).

"AT&T," what do you think of? Your answer probably depends on how old you are and where you live. If you live in Florida, you may know AT&T as your local phone company. In New York, it is the name of one of the leading mobile telephone companies. If you are older than 60, you might think of the company's old nickname "Ma Bell."

The Birth of AT&T

In the study of communication technology over the last century, no name is as prominent as AT&T. But the company known today as AT&T is an awkward descendent of the company that once held a monopoly on long-distance telephone service and a near monopoly on local telephone service through the first four decades of the 20th century. The AT&T story is a story of visionaries, mergers, divestiture, and rebirth.

Alexander Graham Bell invented his version of the telephone in 1876, although historians note that he barely beat his competitors to the patent office. His invention soon became an important force in business communication, but diffusion of the telephone was inhibited by the fact that, within 20 years, thousands of entrepreneurs established competing companies to provide telephone service in major cities. Initially, these telephone systems were not interconnected, making the choice of telephone company a difficult one, with some businesses needing two or more local phone providers to connect with their clients.

The visionary who solved the problem was Theodore Vail, who realized that the most important function was the interconnection of these telephone companies. Vail led American Telephone & Telegraph to provide the needed interconnection, negotiating with the U.S. government to provide "universal service" under heavy regulation in return for the right to operate as a monopoly. Vail brought as many local telephone companies as he could into AT&T, which evolved under the eye of the federal government as a behemoth with three divisions:

- *AT&T Long Lines*—the company that had a virtual monopoly on long distance telephony in the United States.

- *The Bell System*—Local telephone companies providing service to 90% of U.S. subscribers.

- *Western Electric*—A manufacturing company that made equipment needed by the other two divisions, from telephones to switches. (Bell Labs was a part of Western Electric.)

As a monopoly that was generally regulated on a rate-of-return basis (making a fixed profit percentage), AT&T had little incentive—other than that provided by regulators—to hold down costs. The more the company spent, the more it had to charge to make its profit, which grew in proportion with expenses. As a result, the U.S. telephone industry became the envy of the world, known for "five nines" of reliability; that is, the telephone network was designed to be available to users 99.999% of the time. The company also spent millions every year on basic research, with its Bell Labs responsible for the invention of many of the most important technologies of the 20th century, including the transistor and the laser.

Divestiture

The monopoly suffered a series of challenges in the 1960s and 1970s that began to break AT&T's monopoly control. First, AT&T lost a suit brought by the Hush-a-Phone company, which made a plastic mouthpiece that fit over the AT&T telephone mouthpiece to make it easier to hear a call made in a noisy area (*Hush-a-phone v. AT&T*, 1955; *Hush-a-phone v. U.S.*, 1956). (The idea of a company having to win a lawsuit in order to sell such an innocent item seems frivolous today, but this suit was the first major crack in AT&T's monopoly armor.) Soon, MCI won a suit to provide long-distance service between St. Louis and Chicago, allowing businesses to bypass AT&T's long lines (*Microwave Communications, Inc.*, 1969).

Since the 1920s, the Department of Justice (DOJ) had challenged aspects of AT&T's monopoly control, earning a series of consent decrees to limit AT&T's market power and constrain corporate behavior. By the 1970s, it was clear to the antitrust attorneys that AT&T's ownership of Western Electric inhibited innovation, and the DOJ attempted to force AT&T to divest itself of its manufacturing arm. In a surprising move, AT&T proposed a different divestiture, spinning off all of its local telephone companies into seven new "Baby Bells," keeping the now-competitive long distance service and manufacturing arms. The DOJ agreed, and a new AT&T was born (Dizard, 1989).

Cycles of Expansion and Contraction

After divestiture, the leaner, "new" AT&T attempted to compete in many markets with mixed success; AT&T long distance service remained a national leader, but few bought overpriced AT&T personal computers. In the meantime, the seven Baby Bells focused on serving their local markets, with most named after the region they served. Nynex served New York and the extreme northeast states, Bell Atlantic served the mid-Atlantic states, BellSouth served the southeastern states, Ameritech served the midwest, Southwestern Bell served south central states, U S West served a set of western states, and Pacific Telesis served California and the far western states.

Over the next two decades, consolidation occurred among these Baby Bells. Nynex and Bell Atlantic merged to create Verizon. U S West was purchased by Qwest Communication and renamed after its new parent, which was, in turn, acquired by CenturyLink in 2010. As discussed below, Southwestern Bell was the most aggressive Baby Bell, ultimately reuniting more than half of the Baby Bells.

In the meantime, AT&T entered the 1990s with a repeating cycle of growth and decline. It acquired NCR Computers in 1991 and McCaw Communications (then the largest U.S. cellular phone company) in 1993. Then, in 1995, it divested itself of its manufacturing arm (which became Lucent Technologies) and the computer company (which took the NCR name). It grew again in 1998 by acquiring TCI, the largest U.S. cable TV company, renaming it AT&T Broadband, and then acquired another cable company, MediaOne. In 2001, it sold AT&T Broadband to Comcast, and it spun off its wireless interests into an independent company (AT&T Wireless), which was later acquired by Cingular (a wireless phone company co-owned by Baby Bells SBC and BellSouth) (AT&T, 2008).

The only parts of AT&T remaining were the long distance telephone network and the business services, resulting in a company that was a fraction of the size of the AT&T behemoth that had a near monopoly on telephony in the United States two decades earlier.

Under the leadership of Edward Whitacre, Southwestern Bell became one of the most formidable players in the telecommunications industry. With a visionary style not seen in the telephone industry since the days of Theodore Vail, Whitacre led Southwestern Bell to acquire Baby Bells Pacific Telesis and Ameritech (and a handful of other, smaller telephone companies), renaming itself SBC. Ultimately, SBC merged with BellSouth and purchased what was left of AT&T, then renamed the company AT&T, an interesting case comparable to a child adopting its parent.

Today's AT&T is a dramatically different company with a dramatically different culture than its parent, but the company serves most of the same markets in a much more competitive environment. The lesson is that it is not enough to know the technologies or the company names; you also have to know the history of both in order to understand the role that a company plays in the marketplace.

Functions within the Industries

The AT&T story is an extreme example of the complexity of communication industries. These industries are easier to understand by breaking their functions into categories that are common across most of the segments of these industries. Let's start with the heart of the technology ecosystem introduced in Chapter 1, the hardware, software, and content.

For this discussion, let's use the same definitions used in Chapter 1, with *hardware* referring to the physical equipment used, *software* referring to instructions used by the hardware to manipulate content, and *content* referring to the messages transmitted using these technologies. Some companies produce both hardware and software, ensuring compatibility between the equipment and programming, but few companies produce both equipment and content.

The next distinction has to be made between production and distribution of both equipment and content. As these names imply, companies involved in production engage in the manufacture of equipment or content, and companies involved in distribution are the intermediaries between production and consumers. It is a common practice for some companies to be involved in both production and distribution, but, as discussed below, a large number of companies choose to focus on one or the other.

These two dimensions interact, resulting in separate functions of equipment production, equipment distribution, content production, and content distribution. As discussed below, distribution can be further broken down into national and local distribution. The following section introduces these dimensions, which then help us identify the role played by specific companies in communication industries.

One other note: These functions are hierarchical, with production coming before distribution in all cases. Let's say you are interested in creating a new type of telephone, perhaps a "high-definition telephone." You know that there is a market, and you want to be the person who sells it to consumers. But you cannot do so until someone first makes the device. Production always comes before distribution, but you cannot have successful production unless you also have distribution—hence the hierarchy in the model. Figure 4.1 illustrates the general pattern, using the U.S. television industry as an example.

Hardware

When you think of hardware, you typically envision the equipment you handle to use a communication technology. But it is also important to note that there is a second type of hardware for most communication industries—the equipment used to make the content. Although most consumers do not deal with this equipment, it plays a critical role in the system.

Content Production Hardware

Production hardware is usually more expensive and specialized than other types. Examples in the television industry include TV cameras, microphones, and editing equipment. A successful piece of production equipment might sell only a few hundred or a few thousand units, compared with tens of thousands to millions of units for consumer equipment. The profit margin on production equipment is usually much higher than on consumer equipment, making it a lucrative market for manufacturing companies.

Consumer Hardware

Consumer hardware is the easiest to identify. It includes anything from a television to a mobile phone or DirecTV satellite dish. A common term used to identify consumer hardware in consumer electronics industries is CPE, which stands for customer premises equipment. An interesting side note is that many companies do not actually make their own products, but instead hire manufacturing facilities to make products they design, shipping them directly to distributors. For example, Microsoft does not manufacture the Xbox One; Flextronics and Foxconn do. As you consider communication technology hardware, consider the lesson from Chapter 1—people are not usually motivated to buy equipment because of the equipment itself, but because of the content it enables, from the conversations (voice and text!) on a wireless phone to the information and entertainment provided by a high-definition television (HDTV) receiver.

Distribution

After a product is manufactured, it has to get to consumers. In the simplest case, the manufacturer sells directly to the consumer, perhaps through a company-owned store or a website. In most cases, however, a product will go through multiple organizations, most often with a wholesaler buying it from the manufacturer and selling it, with a mark-up, to a retail store, which also marks up the price before selling it to a consumer. The key point is that few manufacturers control their own distribution channels, instead relying on other companies to get their products to consumers.

Figure 4.1

Structure of the Traditional Broadcast TV Industry

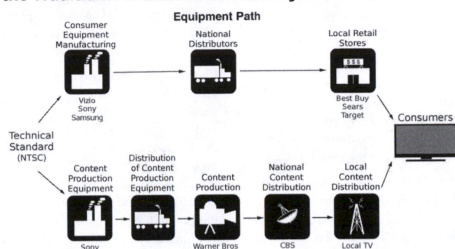

Source: R. Grant & G. Martin

Content Path: Production and Distribution

The process that media content goes through to get to consumers is a little more complicated than the process for hardware. The first step is the production of the content itself. Whether the product is movies, music, news, images, etc., some type of equipment must be manufactured and distributed to the individuals or companies who are going to create the content. (That hardware production and distribution goes through a similar process to the one discussed above.) The content must then be created, duplicated, and distributed to consumers or other end users.

The distribution process for media content/software follows the same pattern for hardware. Usually there will be multiple layers of distribution, a national wholesaler that sells the content to a local retailer, which in turn sells it to a consumer.

Disintermediation

Although many products go through multiple layers of distribution to get to consumers, information technologies have also been applied to reduce the complexity of distribution. The process of eliminating layers of distribution is called disintermediation

(Kottler & Keller, 2005); examples abound of companies that use the Internet to get around traditional distribution systems to sell directly to consumers.

Netflix is a great example. Traditionally, digital videodiscs (DVDs) of a movie were sold by the studio to a national distributor, which then delivered them to thousands of individual movie rental stores, which, in turn rented or sold them to consumers. (Note: The largest video stores bought directly from the studio, handling both national and local distribution.)

Netflix cut one step out of the distribution process, directly bridging the movie studio and the consumer. (As discussed below, iTunes serves the same function for the music industry, simplifying music distribution.) The result of getting rid of one middleman is greater profit for the companies involved, lower costs to the consumer, or both. The "disruption," in this case, was the demise of the rental store.

Illustrations: HDTV and HD Radio

The emergence of digital broadcasting provides two excellent illustrations of the complexity of the organizational structure of media industries. HDTV and its distant cousin HD radio have had a difficult time penetrating the market because of the need for so many organizational functions to be served before consumers can adopt the technology.

Let's start with the simpler one: HD radio. As illustrated in Figure 4.2, this technology allows existing radio stations to broadcast their current programming (albeit with much higher fidelity), so no changes are needed in the software production area of the model. The only change needed in the software path is that radio stations simply need to add a digital transmitter. The complexity is related to the consumer hardware needed to receive HD radio signals. One set of companies makes the radios, another distributes the radios to retail stores and other distribution channels, then stores and distributors have to agree to sell them.

The radio industry has therefore taken an active role in pushing diffusion of HD radios throughout the hardware path. In addition to airing thousands of radio commercials promoting HD radio, the industry is promoting distribution of HD radios in new cars (because so much radio listening is done in automobiles).

As discussed in Chapter 8, adoption of HD radio has begun, but has been slow because listeners see little advantage in the new technology. However, if the number of receivers increases, broadcasters will have a reason to begin using the additional channels available with HD. As with FM radio, programming and receiver sales have to *both* be in place before consumer adoption takes place. Also, as with FM, the technology may take decades to take off.

The same structure is inherent in the adoption of HDTV, as illustrated in Figure 4.3. Before the first consumer adoption could take place, both programming and receivers (consumer hardware) had to be available. Because a high percentage of primetime television programming was recorded on 35mm film at the time HDTV receivers first went on sale in the United States, that programming could easily be transmitted in high-definition, providing a nucleus of available programming. On the other hand, local news and network programs shot on video required entirely new production and editing equipment before they could be distributed to consumers in high-definition.

Figure 4.2
Structure of the HD Radio Industry

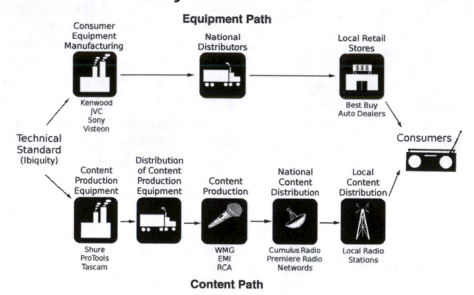

Source: R. Grant & G. Martin

Figure 4.3

Structure of the HDTV Industry

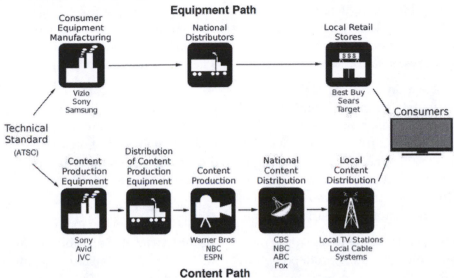

Source: R. Grant & G. Martin

As discussed in Chapter 6, the big force behind the diffusion of HDTV and digital TV was a set of regulations issued by the Federal Communications Commission (FCC) that first required stations in the largest markets to begin broadcasting digital signals, then required that all television receivers include the capability to receive digital signals, and finally required that all full-power analog television broadcasting cease on June 12, 2009. In short, the FCC implemented mandates ensuring production and distribution of digital television, easing the path toward both digital TV and HDTV.

As you will read in Chapter 6, this cycle will repeat itself over the next few years as the television industry adopts an improved, but incompatible, format known as UltraHDTV (or 4K).

The success of this format will require modifications of virtually all areas of the television system diagrammed in Figure 4.3, with new production equipment manufactured and then used to produce content in the new format, and then distributed to consumers—who are not likely to buy until all of these functions are served.

From Target to iTunes

One of the best examples of the importance of distribution comes from an analysis of the popular music industry. In the last century, music was recorded on physical media such as CDs and audio tapes and then shipped to retail stores for sale directly to consumers. At one time, the top three U.S. retailers of music were Target, Walmart, and Best Buy.

Once digital music formats that could be distributed over the Internet were introduced in the late 1990s, dozens of start-up companies created online stores to sell music directly to consumers. The problem was that few of these online stores offered the top-selling music. Record companies were leery of the lack of control they had over digital distribution, leaving most of these companies to offer a marginal assortment of music.

The situation changed in 2003 when Apple introduced the iTunes store to provide content for its iPods, which had sold slowly since appearing on the market in 2001. Apple obtained contracts with major record companies that allowed them to provide most of the music that was in high demand.

Initially, record companies resisted the iTunes distribution model that allowed a consumer to buy a single song for $0.99; they preferred that a person have to buy an entire album of music for $13 to $20 to get the one or two songs they wanted.

Record company delays spurred consumers to create and use file-sharing services that allowed listeners to get the music for free—and the record companies ended up losing billions of dollars. Soon, the $0.99 iTunes model began to look very attractive to the record companies, and they trusted Apple's digital rights management system to protect their music.

Today, as discussed in Chapter 8, iTunes is the number one music retailer in the United States. The music is similar, but the distribution of music today is dramatically different from what it was in 2001. The change took years of experimentation, and the successful business model that emerged required cooperation from dozens of separate companies serving different roles in production and distribution.

Two more points should be made regarding distribution. First, there is typically more profit potential and less risk in being a distributor than a creator (of either hardware or software) because the investment is less and distributors typically earn a percentage of the value of what they sell. Second, distribution channels can become very complicated when multiple layers of distribution are involved; the easiest way to unravel these layers is simply to "follow the money."

Importance of Distribution

As the above discussion indicates, distributors are just as important to new technologies as manufacturers and service providers. When studying these technologies, and the reasons for success or failure, the distribution process (including economics) must be examined as thoroughly as the product itself.

Diffusion Threshold

Analysis of the elements in Figure 4.1 reveals an interesting conundrum—there cannot be any consumer adoption of a new technology until all of the production and distribution functions are served, along both the hardware and software paths.

This observation adds a new dimension to Rogers' (2003) diffusion theory, discussed in the previous chapter. The point at which all functions are served has been identified as the "diffusion threshold," the point at which diffusion of the technology can begin (Grant, 1990).

It is easier for a technology to "take off" and begin diffusing if a single company provides a number of different functions, perhaps combining production and distribution, or providing both national and local distribution. The technical term for owning multiple functions in an industry is "vertical integration," and a vertically integrated company has a disproportionate degree of power and control in the marketplace.

Vertical integration is easier said than done, however, because the "core competencies" needed for production and distribution are so different. A company that is great at manufacturing may not have the resources needed to sell the product to end consumers.

Let's consider the next generation of television, UltraHD, again. A company such as Vizio might handle the first level of distribution, from the manufacturing plant to the retail store, but they do not own and operate their own stores—that is a very different business. They are certainly not involved in owning the television stations or cable channels that distribute UltraHD programming; that function is served by another set of organizations.

In order for UltraHD to become popular, one organization (or set of organizations) has to make the television receivers, another has to get those televisions into stores, a third has to operate the stores, a fourth has to make UltraHD cameras and technical equipment to produce content, and a fifth has to package and distribute the content to viewers, through the Internet using Internet protocol television (IPTV), cable television systems, or over-the-air.

Most companies that would like to grow are more interested in applying their core competencies by buying up competitors and commanding a greater market share, a process known as "horizontal integration." For example, it makes more sense for a company that makes televisions to grow by making other electronics rather than by buying television stations. Similarly, a company that already owns television stations will probably choose to grow by buying more television stations or cable networks rather than by starting to make and sell television receivers.

The complexity of the structure of most communication industries prevents any one company from serving every needed role. Because so many organizations have to be involved in providing a new technology, many new technologies end up failing.

The lesson is that understanding how a new communication technology makes it to market requires comparatively little understanding of the technology itself compared with the understanding needed of the industry in general, especially the distribution processes.

A "Blue" Lesson

One of the best examples of the need to understand (and perhaps exploit) all of the paths illustrated in Figure 4.1 comes from the earliest days of the personal computer. When the PC was invented in the 1970s, most manufacturers used their own operating systems, so that applications and content could not easily be transferred from one type of computer to other types. Many of these manufacturers realized that they needed to find a standard operating system that would allow the same programs and content to be used on computers from different manufacturers, and they agreed on an operating system called CP/M.

Before CP/M could become a standard, however, IBM, the largest U.S. computer manufacturer—mainframe computers, that is—decided to enter the personal computer market. "Big Blue," as IBM was known (for its blue logo and its dominance in mainframe computers, typewriters, and other business equipment) determined that its core competency was making hardware, and they looked for a company to provide them an operating system that would work on their computers. They chose a then-little-known operating system called MS-DOS, from a small start-up company called Microsoft.

IBM's open architecture allowed other companies to make compatible computers, and dozens of companies entered the market to compete with Big Blue. For a time, IBM dominated the personal computer market, but, over time, competitors steadily made inroads on the market. (Ultimately, IBM sold its personal computer manufacturing business in 2006 to Lenovo, a Chinese company.)

The one thing that most of these competitors had in common was that they used Microsoft's operating systems. Microsoft grew… and grew… and grew. (It is also interesting to note that, although Microsoft has dominated the market for software with its operating systems and productivity software such as Office, it has been a consistent failure in most areas of hardware manufacturing. Notable failures include its routers and home networking hardware, keyboards and mice, and WebTV hardware. The only major success Microsoft has had in manufacturing hardware is with its Xbox video game system, discussed in Chapter 15.)

The lesson is that there is opportunity in all areas of production and distribution of communication technologies. All aspects of production and distribution must be studied in order to understand communication technologies. Companies have to know their own core competencies, but a company can often improve its ability to introduce a new technology by controlling more than one function in the adoption path.

What are the Industries?

We need to begin our study of communication technologies by defining the industries involved in providing communication-related services in one form or another. Broadly speaking, these can be divided into:

- *Mass media*—including books, newspapers, periodicals, movies, radio, and television.
- *Telecommunications*—including networking and all kinds of telephony (landlines, long distance, wireless, and voice over Internet protocol).
- *Computers*—including hardware (desktops, laptops, tablets, etc.) and software.
- *Consumer electronics*—including audio and video electronics, video games, and cameras.
- *Internet*—including enabling equipment, network providers, content providers, and services.

These industries were introduced in Chapter 2 and are discussed in more detail in the individual chapters that follow.

At one point, these industries were distinct, with companies focusing on one or two industries. The opportunity provided by digital media and convergence enables companies to operate in numerous industries, and many companies are looking for synergies across industries.

Table 4.1

Examples of Major Communication Company Industries, 2018

	TV/Film/Video Production	TV/Film/Video Distribution	Print	Telephone	Wireless	Internet
AT&T		●●●		●●●	●●●	●●●
Comcast	●●●	●●●		●●		●●●
Disney	●●●	●●●	●			●●
Google		●			●	●●●
21st Century Fox	●●●	●●●	●●●			●●
Sony	●●●	●				●
Time Warner, Inc.	●●●	●				●
Charter Cable		●●●		●●		●●●
Verizon	●	●		●●●	●●●	●●●
Viacom	●●●	●●				●

The number of dots is proportional to the importance of this business to each company.   *Source:* A. Grant (2018)

Table 4.1 lists examples of well-known companies in the communication industries, some of which work across many industries, and some of which are (as of this writing) focused on a single industry. Part of the fun in reading this chart is seeing how much has changed since the book was printed in mid-2018.

There is a risk in discussing specific organizations in a book such as this one; in the time between when the book is written and when it is published, there are certain to be changes in the organizational structure of the industries. For example, as this chapter was being written in early 2018, Netflix and DirecTV were distributing UltraHD television through online services. By the time you read this, some broadcasters are sure to have adopted a set of technical standards known as ATSC 3.0, allowing them to deliver UltraHD signals directly to viewers.

Fortunately, mergers and takeovers that revolutionize an industry do not happen that often—only a couple a year! The major players are more likely to acquire other companies than to be acquired, so it is fairly safe (but not completely safe) to identify the major players and then analyze the industries in which they are doing business.

As in the AT&T story earlier in this chapter, the specific businesses a company is in can change dramatically over the course of a few years.

Future Trends

The focus of this book is on changing technologies. It should be clear that some of the most important changes to track are changes in the organizational structure of media industries. The remainder of this chapter projects organizational trends to watch to help you predict the trajectory of existing and future technologies.

Disappearing Newspapers

For decades, newspapers were the dominant mass medium, commanding revenues, consumer attention, and significant political and economic power. Since the dramatic drop in newspaper revenues and subscriptions began in 2005, newspaper publishers have been reconsidering their core business. Some prognosticators have even predicted the demise of the printed newspaper completely, forcing newspaper companies to plan for digital distribution of their news and advertisements.

Before starting the countdown clock, it is necessary to define what we mean by a "newspaper publisher." If a newspaper publisher is defined as an organization that communicates and obtains revenue by smearing ink on dead trees, then we can easily predict a steady decline in that business. If, however, a newspaper publisher is defined as an organization that gathers news and advertising messages, distributing them via a wide range of available media, then newspaper publishers should be quite healthy through the century.

The current problem is that there is no comparable revenue model for delivery of news and advertising through new media that approaches the revenues available from smearing ink on dead trees. It is a bad news/good news situation.

The bad news is that traditional newspaper readership and revenues are both declining. Readership is suffering because of competition from the Web and other new media, with younger cohorts increasingly ignoring print in favor of other sources.

Advertising revenues are suffering for two reasons. The decline in readership and competition from new media are impacting revenues from display advertising. More significant is the loss in revenues from classified advertising, which at one point comprised up to one-third of newspaper revenues.

The good news is that newspapers remain profitable, at least on a cash flow basis, with gross margins of 10% to 20%. This profit margin is one that many industries would envy. But many newspaper companies borrowed extensively to expand their reach, with interest payments often exceeding these gross profits. Stockholders in newspaper publishers have been used to much higher profit margins, and the stock prices of newspaper companies have been punished for the decline in profits.

Some companies reacted by divesting themselves of their newspapers in favor of TV and new media investments. Some newspaper publishers used the opportunity to buy up other newspapers; consider McClatchy's 2006 purchase of the majority of Knight-Ridder's newspapers (McClatchy, 2008) or the Digital First Media purchase of Freedom Communications, Inc. in 2016 (Gleason, 2016).

Advertiser-Supported Media

For advertiser-supported media organizations, the primary concern is the impact of the Internet and other new media on revenues. As discussed above, some of the loss in revenues is due to loss of advertising dollars (including classified advertising), but that loss is not experienced equally by all advertiser-supported media.

The Internet is especially attractive to advertisers because online advertising systems have the most comprehensive reporting of any advertising medium. For example, advertisers using Google AdWords (discussed in Chapter 1) gets comprehensive reports on the effectiveness of every message—but "effectiveness" is defined by these advertisers as an immediate response such as a click-through.

As Grant & Wilkinson (2007) discuss, not all advertising is "call-to-action" advertising. There is another type of equally important advertising—image advertising, which does not demand immediate results, but rather works over time to build brand identity, increasing the likelihood of a purchase.

Any medium can carry any type of advertising, but image advertising is more common on television (especially national television) and magazines, and call-to-action advertising is more common in newspapers. As a result, newspapers, at least in the short term, are more likely to be impacted by the increase in Internet advertising.

Interestingly, local advertising is more likely to be call-to-action advertising, but local advertisers have been slower than national advertisers to move to the Internet, most likely because of the global reach of the Internet. This paradox could be seen as an opportunity for an entrepreneur wishing to earn a million or two by exploiting a new advertising market (Wilkinson, Grant & Fisher, 2012).

The "Mobile Revolution"

Another trend that can help you analyze media organizations is the shift toward mobile communication technologies. Companies that are positioned to produce and distribute content and technology that further enable the "mobile revolution" are likely to have increased prospects for growth.

Areas to watch include mobile Internet access (involving new hardware and software, provided by a mixture of existing and new organizations), mobile advertising, new applications of GPS technology, and new applications designed to take advantage of Internet access available anytime, anywhere.

Consumers—Time Spent Using Media

Another piece of good news for media organizations is the fact that the amount of time consumers are spending with media is increasing, with much of that increase coming from simultaneous media use (Papper, et al., 2009). Advertiser-supported media thus have more "audience" to sell, and subscription-based media have more prospects for revenue.

Furthermore, new technologies are increasingly targeting specific messages at specific consumers, increasing the efficiency of message delivery for advertisers and potentially reducing the clutter of irrelevant advertising for consumers. Already, advertising services such as Google's Double-Click and Google's Ad-Words provide ads that are targeted to a specific person or the specific content on a Web page, greatly increasing their effectiveness.

Imagine a future where every commercial on TV that you see is targeted—and is interesting—to you! Technically, it is possible, but the lessons of previous technologies suggest that the road to customized advertising will be a meandering one.

Principle of Relative Constancy

The potential revenue from consumers is limited because they devote a fixed proportion of their disposable income to media, the phenomenon discussed in Chapter 3 as the "Principle of Relative Constancy." The implication is emerging companies and technologies have to wrest market share and revenue from established companies. To do that, they can't be just as good as the incumbents. Rather, they have to be faster, smaller, less expensive, or in some way better so that consumers will have the motivation to shift spending from existing media.

Conclusions

The structure of the media system may be the most dynamic area in the study of new communication technologies, with new industries and organizations constantly emerging and merging. In the following chapters, organizational developments are therefore given significant attention. Be warned, however; between the time these chapters are written and published, there is likely to be some change in the organizational structure of each technology discussed. To keep up with these developments, visit the *Communication Technology Update and Fundamentals* home page at www.tfi.com/ctu.

Bibliography

AT&T. (2008). *Milestones in AT&T history*. Retrieved from http://www.corp.att.com/history/ milestones.html.

Dizard, W. (1989). *The coming information age: An overview of technology, economics, and politics, 2nd ed.* New York: Longman.

Gleason, S. (2016, April 1). Digital First closes purchase of Orange County Register publisher. *Wall Stree Journal*. Online: http://www.wsj.com/articles/digital-first-closes-purchase-of-orange-county-register-publisher-1459532486

Grant, A. E. (1990, April). The "pre-diffusion of HDTV: Organizational factors and the "diffusion threshold. Paper presented to the Annual Convention of the Broadcast Education Association, Atlanta.

Grant, A. E. & Wilkinson, J. S. (2007, February). Lessons for communication technologies from Web advertising. Paper presented to the Mid-Winter Conference of the Association of Educators in Journalism and Mass Communication, Reno.

Hush-A-Phone Corp. v. AT&T, et al. (1955). FCC Docket No. 9189. Decision and order (1955). 20 FCC 391.

Hush-A-Phone Corp. v. United States. (1956). 238 F. 2d 266 (D.C. Cir.). Decision and order on remand (1957). 22 FCC 112.

Kottler, P. & Keller, K. L. (2005). *Marketing management, 12th ed.* Englewood Cliffs, NJ: Prentice-Hall.

McClatchy. (2008). *About the McClatchy Company*. Retrieved from http://www.mcclatchy.com/100/story/ 179.html.

Microwave Communications, Inc. (1969). FCC Docket No. 16509. Decision, 18 FCC 2d 953.

Papper, R. E., Holmes, M. A. & Popovich, M. N. (2009). Middletown media studies II: Observing consumer interactions with media. In A. E. Grant & J. S. Wilkinson (Eds.) *Understanding media convergence: The state of the field*. New York: Oxford.

Rogers, E. M. (2003). *Diffusion of innovations, 5th ed.* New York: Free Press.

Wilkinson, J.S., Grant, A. E. & Fisher, D. J. (2012). *Principles of convergent journalism (2nd ed.).* New York: Oxford University Press.

5

Communication Policy & Technology

Lon Berquist, M.A.*

Throughout its history, U.S. communication policy has been shaped by evolving communication technologies. As a new communication technology is introduced into society, it is often preceded by an idealized vision, or Blue Sky perspective, of how the technology will positively impact economic opportunities, democratic participation, and social inclusion. Due, in part, to this perspective, government policy makers traditionally look for policies and regulations that will foster the wide diffusion of the emerging technology. At the same time, however, U.S. policy typically displays a light regulatory touch, promoting a free-market approach that attempts to balance the economic interests of media and communication industries, the First Amendment, and the rights of citizens.

Indeed, much of the recent impetus for media deregulation was directly related to communication technologies as "technological plenty is forcing a widespread reconsideration of the role competition can play in broadcast regulation" (Fowler & Brenner, 1982, p. 209). From a theoretical perspective, some see new communication technologies as technologies of freedom where "freedom is fostered when the means of communication are dispersed, decentralized, and easily available" (Pool, 1983, p. 5). Others fear technologies favor government and private interests and become technologies of control (Gandy, 1989). Still others argue that technologies are merely neutral in how they shape society. No matter the perspective, the purpose of policy and regulation is to allow society to shape the use of communication technologies to best serve the citizenry.

Background

The First Amendment is a particularly important component of U.S. communication policy, balancing freedom of the press with the free speech rights of citizens. The First Amendment was created at a time when the most sophisticated communication technology was the printing press. Over time, the notion of "press" has evolved with the introduction of new communication technologies. The First Amendment has evolved as well, with varying degrees of protection for the traditional press, broadcasting, cable television, and the Internet.

Communication policy is essentially the balancing of national interests and the interests of the communications industry (van Cuilenburg & McQuail, 2003). In the United States, communication policy is often shaped in reaction to the development of a new technology. As a result, policies vary according to the

---

* Senior IT Policy Analyst, Texas A&M University (College Station, Texas)

particular communication policy regime: press, common carrier, broadcasting, cable TV, and the Internet. Napoli (2001) characterizes this policy tendency as a "technologically particularistic" approach leading to distinct policy and regulatory structures for each new technology. Thus, the result is differing First Amendment protections for the printed press, broadcasting, cable television, and the Internet (Pool, 1983).

In addition to distinct policy regimes based on technology, scholars have recognized differing types of regulation that impact programming, the industry market and economics, and the transmission and delivery of programming and information. These include content regulation, structural regulation, and technical regulation.

Content regulation refers to the degree to which a particular industry enjoys First Amendment protection. For example, in the United States, the press is acknowledged as having the most First Amendment protection, and there certainly is no regulatory agency to oversee printing. Cable television has limited First Amendment protection, while broadcasting has the most limitations on its First Amendment rights. This regulation is apparent in the type of programming rules and regulations imposed by the Federal Communications Commission (FCC) on broadcast programming that is not imposed on cable television programming.

Structural regulation addresses market power within (horizontal integration) and across (vertical integration) media industries. Federal media policy has long established the need to promote diversity of programming by promoting diversity of ownership. The *Telecommunications Act of 1996* changed media ownership limits for the national and local market power of radio, television, and cable television industries; however, the FCC is given the authority to review and revise these rules. Structural regulation includes limitations or permissions to enter communication markets. For example, the *Telecommunications Act of 1996* opened up the video distribution and telephony markets by allowing telephone companies to provide cable television service and for cable television systems to offer telephone service (Parsons & Frieden, 1998).

Technical regulation needs prompted the initial development of U.S. communication regulation in the 1920s, as the fledgling radio industry suffered from signal interference while numerous stations transmitted without any government referee (Starr, 2004). Under FCC regulation, broadcast licensees are allowed to transmit at a certain power, or wattage, on a precise frequency within a particular market. Cable television systems and satellite transmission also follow some technical regulation to prevent signal interference.

Finally, in addition to technology-based policy regimes and regulation types, communication policy is guided by varying jurisdictional regulatory bodies. Given the global nature of satellites, both international (International Telecommunications Union) and national (FCC) regulatory commissions have a vested interest in satellite transmission.

Regulation of U.S. broadcasting is exclusively the domain of the federal government through the FCC. The telephone industry is regulated primarily at the federal level through the FCC, but also with regulations imposed by state public utility commissions. Cable television, initially regulated through local municipal franchises, is now regulated at the federal level, the local municipal level, and, for some, at the state level (Parsons & Frieden, 1998).

The FCC was created by Congress in 1934 to be the expert agency on communication matters. It is composed of five members appointed by the President, subject to confirmation by the Senate. The President selects one of the commissioners to serve as chairman. No more than three commission members can be of the same political party. Although the FCC is considered an independent agency, this structure leads to the Chairman and majority members affiliated with the same political party of the President, resulting in FCC policy reflecting the policy goals of the administration (Feld & Forscey, 2015).

The Evolution of Communication Technologies

Telegraph

Although the evolution of technologies has influenced the policymaking process in the United States, many of the fundamental characteristics of U.S. communication policy were established early in the history of communication technology deployment, starting with the telegraph. There was much debate on how best to develop the telegraph. For many lawmakers and industry observers, the telegraph was viewed as a natural extension of the Post Office, while others favored government ownership based on the successful European model as the only way to counter the power of a private monopoly (DuBoff, 1984).

In a prelude to the implementation of universal service for the telephone (and the current discussion of a "digital divide"), Congress decreed that, "Where the rates are high and facilities poor, as in this country, the number of persons who use the telegraph freely, is limited. Where the rates are low and the facilities are great, as in Europe, the telegraph is extensively used by all classes for all kinds of business" (Lubrano, 1997, p. 102).

Despite the initial dominance of Western Union, there were over 50 separate telegraph companies operating in the United States in 1851. Interconnecting telegraph lines throughout the nation became a significant policy goal of federal, state, and local governments. No geographic area wanted to be disconnected from the telegraph network and its promise of enhanced communication and commerce.

Eventually, in 1887, the *Interstate Commerce Act* was enacted, and the policy model of a regulated, privately-owned communication system was initiated and formal federal regulation began. Early in the development of communication policy, the tradition of creating communications infrastructure through government aid to private profit-making entities was established (Winston, 1998).

Telephone

Similar to the development of the telegraph, the diffusion of the telephone was slowed by competing, unconnected companies serving their own interests. Although AT&T dominated most urban markets, many independent telephone operators and rural cooperatives provided service in smaller towns and rural areas. Since there was no interconnection among the various networks, some households and businesses were forced to have dual service in order to communicate (Starr, 2004).

As telephone use spread in the early 1900s, states and municipalities began regulating and licensing operators as public utilities, although Congress authorized the Interstate Commerce Commission (ICC) to regulate interstate telephone service in 1910. Primarily an agency devoted to transportation issues, the ICC never became a major historical player in communication policy. However, two important policy concepts originated with the commission and the related *Transportation Act of 1920*. The term "common carrier," originally used to describe railroad transportation, was used to classify the telegraph and eventually the telephone (Pool, 1983). Common carriage law required carriers to serve their customers without discrimination. The other notable phrase utilized in transportation regulation was a requirement to serve the "public interest, convenience, or necessity" (Napoli, 2001). This nebulous term was adopted in subsequent broadcast legislation and continues to guide the FCC even today.

As telephone use increased, it became apparent that there was a need for greater interconnection among competing operators, or the development of some national unifying agreement. In 1907, AT&T President Theodore Vail promoted a policy with the slogan, "One system, one policy, universal service" (Starr, 2004, p. 207). There are conflicting accounts of Vail's motivation: whether it was a sincere call for a national network available to all, or merely a ploy to protect AT&T's growing power in the telephone industry (Napoli, 2001). Eventually, the national network envisioned by Vail became a reality, as AT&T was given the monopoly power, under strict regulatory control, to build and maintain local and long-distance telephone service throughout the nation. Of course, this regulated monopoly was ended decades ago, but the concept of universal service as a significant component of communication policy remains today.

Broadcasting

While U.S. policymakers pursued an efficient national network for telephone operations, they developed radio broadcasting to primarily serve local markets. Before the federal government imposed regulatory control over radio broadcasting in 1927, the industry suffered from signal interference and an uncertain financial future. The *Federal Radio Act* created the Federal Radio Commission and imposed technical regulation on use of spectrum and power, allowing stations to develop a stable local presence.

Despite First Amendment concerns about government regulation of radio, the scarcity of spectrum was considered an adequate rationale for licensing stations. In response to concerns about freedom of the press, the *Radio Act* prohibited censorship by the Radio Commission, but the stations understood that the power of the commission to license implied inherent censorship (Pool, 1983). In 1934, Congress passed the *Communication Act of 1934*, combining regulation of telecommunications and broadcasting by instituting a new Federal Communications Commission.

The *Communication Act* essentially reiterated the regulatory thrust of the 1927 *Radio Act*, maintaining that broadcasters serve the public interest. This broad concept of "public interest" has stood as the guiding force in developing communication policy principles of competition, diversity, and localism (Napoli, 2001; Alexander & Brown, 2007).

Rules and regulations established to serve the public interest for radio transferred to television when it entered the scene. Structural regulation limited ownership of stations, technical regulation required tight control of broadcast transmission, and indirect content regulation led to limitations on station broadcast of network programming and even fines for broadcast of indecent material (Pool, 1983).

One of the most controversial content regulations was the vague Fairness Doctrine, established in 1949, that required broadcasters to present varying viewpoints on issues of public importance (Napoli, 2001). Despite broadcasters' challenges to FCC content regulation on First Amendment grounds, the courts defended the commission's Fairness Doctrine (*Red Lion Broadcasting v. FCC*, 1969) and its ability to limit network control over programming (*NBC v. United States*, 1943). In 1985, the FCC argued the Fairness Doctrine was no longer necessary given the increased media market competition, due in part to the emergence of new communication technologies (Napoli, 2001).

Historically, as technology advanced, the FCC sought ways to increase competition and diversity in broadcasting with AM radio, UHF television, low-power TV, low-power FM, and more recently, digital television.

Cable Television and Direct Broadcast Satellite

Since cable television began simply as a technology to retransmit distant broadcast signals to rural or remote locations, early systems sought permission or franchises from the local authorities to lay cable to reach homes. As cable grew, broadcasters became alarmed with companies making revenue off their programming, and they lobbied against the new technology.

Early on, copyright became the major issue, as broadcasters complained that retransmission of their signals violated their copyrights. The courts sided with cable operators, but Congress passed compulsory license legislation that forced cable operators to pay royalty fees to broadcasters (Pool, 1983). Because cable television did not utilize the public airwaves, courts rebuffed the FCC's attempt to regulate cable.

In the 1980s, the number of cable systems exploded and the practice of franchising cable systems increasingly was criticized by the cable industry as cities demanded more concessions in return for granting rights-of-way access and exclusive multi-year franchises. The *Cable Communications Act of 1984* was passed to formalize the municipal franchising process while limiting some of their rate regulation authority. The act also authorized the FCC to evaluate cable competition within markets (Parsons & Frieden, 1998).

After implementation of the 1984 Cable Act, cable rates increased dramatically. Congress reacted with the *Cable Television Consumer Protection and Competition Act of 1992*. With the 1992 Cable Act, rate regulation returned with the FCC given authority to regulate basic cable rates. To protect broadcasters and localism principles, the act included "must carry" and "retransmission consent" rules that allowed broadcasters to negotiate with cable systems for carriage (discussed in more detail in Chapter 7). Although challenged on First Amendment grounds, the courts eventually found that the FCC had a legitimate interest in protecting local broadcasters (*Turner Broadcasting v. FCC*, 1997). To support the development of direct broadcast satellites (DBS), the 1992 act prohibited cable television programmers from withholding channels from DBS and other prospective competitors. As with cable television, DBS operators have been subject to must-carry and retransmission consent rules.

The *1999 Satellite Home Viewers Improvement Act* (SHIVA) required, and the *Satellite Home Viewer Extension and Reauthorization Act of 2004* (SHVER) reconfirmed, that DBS operators must carry all local broadcast signals within a local market if they choose to carry one (FCC, 2005). DBS operators challenged this in court, but as in *Turner Broadcasting v. FCC*, the courts upheld the government's mandatory carriage requirements (Frieden, 2005). Congress continued these requirements in 2010 with the *Satellite Television Extension and Localism Act* (STELA), and more recently in 2014 with the *STELA Reauthorization Act* (STELAR) (FCC, 2016c).

Policies to promote the development of cable television and direct broadcast satellites have become important components of the desire to enhance media competition and video program diversity, while at the same time preserving localism principles within media markets.

Convergence and the Internet

The *Telecommunications Act of 1996* was a significant recognition of the impact of technological innovation and convergence occurring within the media and telecommunications industries. Because of this recognition, Congress discontinued many of the cross-ownership and service restrictions that had prevented telephone operators from offering video service and cable systems from providing telephone service (Parsons & Frieden, 1998). The primary purpose of the 1996 Act was to "promote competition and reduce regulation in order to secure lower prices and higher-quality service for American telecommunications consumers and encourage the rapid deployment of new telecommunications technologies" (*Telecommunications Act of 1996*). Competition was expected by opening up local markets to facilities-based competition and deregulating rates for cable television and telephone service to let the market work its magic. The 1996 Act also opened up competition in the local exchange telephone markets and loosened a range of media ownership restrictions.

In 1996, the Internet was a growing technological phenomenon, and some in Congress were concerned with the adult content available online. In response, along with passing the *Telecommunications Act of 1996*, Congress passed the *Communication Decency Act* (CDA) to make it a felony to transmit obscene or indecent material to minors. However, the Supreme Court struck down the indecency provisions of the CDA on First Amendment grounds in *Reno v. ACLU* (Napoli, 2001). Section 230 of the CDA protected websites from liability for content posted by users of their online platform; and at the same time allowed online platforms to self-regulate and "restrict access to or availability of material that the provider or user considers to be obscene, lewd, lascivious, filthy, excessively violent, harassing, or otherwise objectionable" (Caplan, Hanson & Donovan, 2018, p 24).

Congress continued to pursue a law protecting children from harmful material on the Internet with the *Child Online Protection Act* (COPA), passed in 1998; however, federal courts have found it, too, unconstitutional due to First Amendment concerns (McCullagh, 2007). It is noteworthy that the courts consider the Internet's First Amendment protection more similar to the press, rather than broadcasting or telecommunications (Warner, 2008).

Similarly, from a regulatory perspective, the Internet does not fall under any traditional regulatory regime such as telecommunications, broadcasting, or cable television. Instead, the Internet was considered an "information service" and therefore not subject to regulation (Oxman, 1999). There are, however, policies in place that indirectly impact the Internet. For example, section 706 of the *Telecommunications Act of 1996* requires the FCC to "encourage the deployment on a reasonable and timely basis of advanced telecommunications capability to all Americans," with advanced telecommunications essentially referring to broadband Internet connectivity (Grant & Berquist, 2000). In addition, Internet service providers have been subject to FCC network neutrality requirements addressed later in this chapter.

Recent Developments

Broadband

In response to a weakening U.S. economy due to the 2008 recession, Congress passed the *American Recovery and Reinvestment Act* (ARRA) of 2009. Stimulus funds were appropriated for a wide range of infrastructure grants, including broadband, to foster economic development. Congress earmarked $7.2 billion to encourage broadband deployment, particularly in unserved and underserved regions of the country.

The U.S. Department of Agriculture Rural Utility Service (RUS) was provided with $2.5 billion to award Broadband Initiatives Program (BIP) grants, while the National Telecommunications and Information Administration (NTIA) was funded with $4.7 billion to award Broadband Technology Opportunity Program (BTOP) grants. In addition, Congress appropriated funding for the *Broadband Data Improvement Act*, legislation that was approved during the Bush Administration in 2008 but lacked the necessary funding for implementation.

Finally, as part of ARRA, Congress directed the FCC to develop a *National Broadband Plan* that addressed broadband deployment, adoption, affordability, and the use of broadband to advance healthcare, education, civic participation, energy, public safety, job creation, and investment. While developing the comprehensive broadband plan, the FCC released its periodic broadband progress report, required under section 706 of the *Telecommunications Act of 1996*. Departing from the favorable projections in previous progress reports, the FCC conceded that "broadband deployment to *all* Americans is not reasonable and timely" (FCC, 2010c, p. 3).

In addition, the Commission revised the dated broadband benchmark of 200 Kb/s by redefining broadband as having download speeds of 4.0Mb/s and upload speeds of 1.0 Mb/s." (FCC, 2010a, p. xi).

In preparing the *National Broadband Plan*, the FCC commissioned a number of studies to determine the state of broadband deployment in the United States. In analyzing U.S. broadband adoption in 2010, the FCC determined 65% of U.S. adults used broadband (Horrigan, 2010); however, researchers forecast that regions around the country, particularly rural areas, would continue to suffer from poor broadband service due to lack of service options or slow broadband speeds (Atkinson & Schultz, 2009).

The targeted goals of the *National Broadband Plan* through 2020 included:

- At least 100 million homes should have affordable access to download speeds of 100 Mb/s and upload speeds of 50 Mb/s.

- The United States should lead the world in mobile innovation with the fastest and most extensive wireless network in the world.

- Every American should have affordable access to robust broadband service, and the means and skills to subscribe.

- Every American community should have affordable access to at least 1 Gb/s broadband service to schools, hospitals, and government buildings.

- Every first responder should have access to a nationwide, wireless, interoperable broadband public safety network.

- Every American should be able to use broadband to track and manage their real-time energy consumption (FCC, 2010a).

The most far-reaching components of the plan included freeing 500 MHz of wireless spectrum, including existing television frequencies, for wireless broadband use, and reforming the traditional telephone Universal Service Fund to support broadband deployment.

The *Broadband Data Improvement Act* (2008) required the FCC to compare U.S. broadband service capability to at least 25 countries abroad. Recent international comparisons of broadband deployment confirm that the U.S. continues to lag in access to broadband capability (FCC, 2018c); and within the U.S. 25 million Americans lack access to wireline broadband service, with over 31% of rural residents lacking access to broadband (FCC, 2018a).

Broadband data from the Organisation for Economic Co-operation and Development (OECD) reports the United States ranked 15th among developed nations for fixed wireline broadband penetration (see

Table 5.1). Other studies show the United States ranked 12th for average connection download speed (15.3 Mb/s), with top-ranked Korea offering significantly greater broadband speed (29 Mb/s), followed by Norway (21.3 Mb/s), Sweden (20.6 Mb/s), Switzerland (18.7 Mb/s), Latvia (18.3 Mb/s), and Japan (18.2 Mb/s) (OECD, 2017).

Table 5.1
International Fixed Broadband Penetration

Country	Broadband Penetration*
Switzerland	45.8
Denmark	42.9
Netherlands	42.2
France	42.0
Norway	40.9
Korea	40.9
Germany	39.4
Iceland	38.7
United Kingdom	38.6
Belgium	38.2
Sweden	37.4
Canada	37.2
Luxembourg	35.7
Greece	34.2
United States	33.6

* Fixed Broadband access per 100 inhabitants

Source: OECD (2017)

The evolving trends in broadband deployment and the increasing broadband speeds offered by providers, prompted the FCC to reevaluate their benchmark definition of broadband established in 2010 (with download speeds of 4.0Mb/s and upload speeds of 1.0 Mb/s). Because of speeds required to access high definition video, data, and other broadband applications, the FCC determined in their *2015 Broadband Progress Report* that in order to meet Congress' requirement for an "advanced telecommunications capability" the benchmark measure for broadband should be increased to download speeds of 25.0 Mb/s and upload speeds of 3.0 Mb/s (FCC, 2015a).

In January 2017, the newly appointed Trump majority FCC Commission announced the formation of a Broadband Deployment Advisory Committee to make recommendations to the Commission on how best to accelerate deployment of broadband by reducing regulatory barriers to broadband investment (FCC, 2018a).

Universal Service Reform

For 100 years, universal service policies have served the United States well, resulting in significant telephone subscription rates throughout the country. But telephone service, an indispensable technology for 20th century business and citizen use, has been displaced by an even more essential technology in the 21st century—broadband Internet access.

Just as the universal deployment of telephone infrastructure was critical to foster business and citizen communication in the early 1900s, broadband has become a crucial infrastructure for the nation's economic development and civic engagement. However, 24 million Americans lack access to broadband, and broadband deployment gaps in rural areas remain significant (FCC, 2018a).

In response to uneven broadband deployment and adoption, the FCC shifted most of the billions of dollars earmarked for subsidizing voice networks in the Universal Service Fund (USF) to supporting broadband deployment. Two major programs were introduced to modernize universal service in the United States: the Connect America Fund and the Mobility Fund.

The Connect America Fund, initiated in 2014, earmarked $4.5 billion annually over six years to fund broadband and high-quality voice in geographic areas throughout the U.S. where private investment in communication was limited or absent. The funding was released to telecommunications carriers in phases, with Phase I releasing $115 million in 2012 to carriers providing broadband infrastructure to 400,000 homes, businesses, and institutions previously without access to broadband. In March 2014, the second round of Phase I funding provided $324 million for broadband deployment to connect more than 1.2 million Americans (FCC, 2014). In 2015, the FCC awarded $1.7 billion through Connect America so carriers could provide broadband to more than eight million rural residents (FCC, 2015b). The Mobility Fund targets wireless availability in unserved regions of the nation by ensuring all areas of the country achieve 3G service, with enhanced opportunities for 4G data and voice service in the future (Gilroy, 2011). In 2012, the Mobility Fund Phase I awarded close to $300 million to fund advanced voice and broadband service for primarily rural areas in over 30 states (Wallsten, 2013).

In 2017, the FCC announced the Connect America Phase II Auction which will provide an additional $1.98 billion over ten years to provide voice and broadband service in unserved rural areas. The Commission also established a Mobility Fund Phase II Auction to allocate $4.53 billion over the next decade to advance the deployment of 4G Long-Term Evolution (LTE) high speed wireless to unserved areas and to preserve service where it might not otherwise exist (FCC, 2018a).

In 2016, the FCC reformed the Lifeline program—a fund developed in 1985 to make telephone service more affordable for eligible low-income Americans. To modernize Lifeline for the digital age, the program was revised to support broadband service to low-income households by leveraging $2.25 billion to provide the same $9.25 monthly subsidy previously used exclusively for telephone costs. The subsidy could be utilized for fixed broadband, mobile services, or bundled service plans combining broadband and voice (FCC, 2016b).

After a critical General Accountability Office report (2017) on the operation of the Lifeline program, the FCC explored deficiencies in the Lifeline and proposed a number of rules to modify the program including:

- Establishing a budget cap for Lifeline

- Limiting subsidies to facility-based providers

- Reviving the role of states in designating Lifeline eligible telecommunications carriers (ETC)

- Improving eligibility verification and recertification to end fraud and abuse (FCC, 2017d)

The Connect America Fund, the Mobility Fund, and the Lifeline reform program maintain the long tradition of government support for communication development, as intended by Congress in 1934 when they created the FCC to make "available...to all the people of the United States...a rapid, efficient, Nationwide, and world-wide wire and radio communication service with adequate facilities at reasonable charges"

(FCC, 2011a, p. 4). Ultimately the Connect America Fund is expected to connect 18 million unserved Americans to broadband, with the hope of creating 500,000 jobs and generating $50 billion in economic growth (FCC, 2011b).

Spectrum Reallocation

Globally, the United States ranks 5th in wireless broadband penetration (See Table 5.2). Historically, mobile broadband speeds have been much greater in Asian and European countries due to greater spectrum availability and wireless technology (Executive Office of the President, 2012). In response to the U.S. spectrum crunch, the *National Broadband Plan* called for freeing 500 MHz of spectrum for wireless broadband.

Table 5.2

International Wireless Broadband Penetration

Country	Broadband Penetration*
Japan	157.4
Finland	145.4
Australia	132.5
Denmark	129.0
United States	128.6
Estonia	125.6
Sweden	122.4
Korea	111.1
Iceland	110.5
Ireland	102.4
New Zealand	101.3
Switzerland	97.8
Norway	96.4
Austria	94.8
Spain	92.7

* Wireless Broadband access per 100 inhabitants
Measure includes both standard and dedicated mobile broadband subscriptions

Source: OECD (2017)

The FCC has begun the steps to repurpose spectrum for wireless broadband service by making spectrum from the 2.3 GHz, Mobile Satellite Service, and TV bands available for mobile broadband service. The Commission has made additional spectrum available for unlicensed wireless broadband by leveraging unused portions of the TV bands, or "white space" that might offer unique solutions for innovative developers of broadband service. Congress supported the effort with passage of the *Middle Class Tax Relief and Job Creation Act of 2012*, including provisions from the *Jumpstarting Opportunity with Broadband Spectrum (JOBS) Act* of 2011 (Moore, 2013). The spectrum reallocation was accomplished through the FCC's authority to conduct incentive auctions where existing license holders, such as broadcasters, relinquish spectrum in exchange for proceeds shared with the Federal government. The FCC began accepting bids for existing broadcast frequencies in March, 2016 with the potential for some broadcast stations to earn up to $900 million for relinquishing their channel for wireless utilization (Eggerton, 2015).

The spectrum auction closed in March 2017 with wireless providers winning 70 MHz of spectrum. Wireless carriers bid a total of $19.8 billion for mobile broadband spectrum, and over $10 billion was shared by 175 broadcasters that elected to repurpose their channels for mobile use. More than $7 billion was deposited to the U.S. Treasury as result of the auction (FCC, 2017g). Also in 2017, the FCC made available an additional 1700 MHz of spectrum bands above 24 GHz for mobile radio services. The FCC expects making these airwaves available may lay the groundwork for Fifth Generation (5G) wireless networks (FCC, 2017f).

Network Neutrality

In 2005, then-AT&T CEO Edward Whitacre, Jr. created a stir when he suggested Google, Yahoo, and Vonage should not expect to use his network for free (Wu, 2017). Internet purists insist the Internet should remain open and unfettered, as originally designed, and decried the notion that broadband providers might discriminate by the type and amount of data content streaming through their lines. Internet service users are concerned that, as more services become available via the Web such as video streaming and voice over Internet protocol (VoIP), Internet service providers (ISPs) will become gatekeepers limiting content and access to information (Gilroy, 2008).

In 2007, the FCC received complaints accusing Comcast of delaying Web traffic on its cable modem service for the popular file sharing site BitTorrent (Kang, 2008). Because of the uproar among consumer groups, the FCC held hearings on the issue and ordered Comcast to end its discriminatory network management practices (FCC, 2008).

Although Comcast complied with the order and discontinued interfering with peer-to-peer traffic like BitTorrent, it challenged the FCC's authority in court. In April 2010, the U.S. Court of Appeals for the D.C. Circuit determined that the FCC had failed to show it had the statutory authority to regulate an Internet service provider's network practices and vacated the order (*Comcast v. FCC*, 2010).

Because broadband service was classified as an unregulated information service, the FCC argued it had the power to regulate under its broad "ancillary" authority highlighted in Title I of the *Telecommunications Act*. The court's rejection of the FCC argument disheartened network neutrality proponents who feared the court's decision would encourage broadband service providers to restrict network data traffic, undermining the traditional openness of the Internet.

In evaluating the court decision, the FCC revisited net neutrality and determined it could establish rules for an open Internet through a combination of regulatory authority.

The FCC Open Internet Rule was adopted in November 2010 and established three basic rules to ensure Internet providers do not restrict innovation on the Internet:

- Disclose information about their network management practices and commercial terms to consumers and content providers to ensure *transparency*

- Accept the *no blocking* requirements of lawful content and applications

- Prevent *unreasonable discrimination*, broadband Internet providers could not unreasonably discriminate in transmitting lawful network traffic over a consumer's Internet service

In January 2011, both Verizon Communications and Metro PCS filed lawsuits challenging the FCC's Open Internet Rule, and in early 2014, the U.S. Court of Appeals for the D.C. Circuit struck down two vital portions of the rule (*Verizon v. FCC*, 2014).

As in the *Comcast v. FCC* decision, the court determined the FCC lacked legal authority to prevent Internet service providers from blocking traffic and lacked authority to bar wireline broadband providers from discriminating among Internet traffic. However, the court upheld the transparency rule which required Internet service providers to disclose how they manage Internet traffic and affirmed the FCC's general authority to oversee broadband services under Section 706 of the *Telecommunications Act of 1996*.

Rather than appeal the ruling, the FCC set out to adopt new rules that considered the Circuit Court's legal rationale (Mazmanian, 2014). In July 2014, the FCC published a Notice of Proposed Rulemaking highlighting the new rules designed to protect and promote an open Internet for consumers. The FCC received a record 3.7 million public comments before the September deadline, surpassing the 1.4 million complaints received after the infamous 2004 Super Bowl broadcast of Janet Jackson's wardrobe malfunction (GAO, 2015).

In February 2015 the FCC formally adopted the rule for Protecting and Promoting the Open Internet (FCC, 2015d). Most importantly, the rule reclassified broadband Internet access service as a telecommunications service under Title II of the *Communications Act of 1934*, therefore subject to common carrier regulation. In addition, the rule covered mobile broadband Internet access based on the definition of commercial mobile services under Title III.

The three major components of the rule included:

- Broadband providers cannot block access to legal content, applications, services, or devices.

- Broadband providers cannot impair or degrade lawful Internet traffic based on content, applications, services or devices (i.e. no throttling).

- Broadband providers cannot favor some lawful Internet traffic over other lawful traffic in exchange for consideration of any kind.

As with previous FCC network neutrality actions, the industry responded with numerous civil suits. The primary challenge was to the FCC's reclassification of broadband under Title II, the principal component of the Open Internet Order. In addition, industry raised the stakes by arguing the Open Internet Order

violated their First Amendment rights of free speech (Green, 2015). Ultimately, the U.S. District Court of Appeals for the D.C. Circuit consolidated the numerous industry lawsuits into one docket and affirmed the FCC's Open Internet Order. (*United States Telecom Association et al. v. FCC*, 2016).

The election of Donald J. Trump in 2016 resulted in the appointment of Commissioner Ajit Pai as FCC Chairman. As a Commissioner, Pai had previously opposed FCC network neutrality rules. As Chairman, and with a newly appointed Republican majority, he initiated a reexamination of the network neutrality rule and proposed a rule titled Restoring Internet Freedom that rolled back network neutrality requirements.

The FCC received an even larger number of public comments than the record setting comment period for the 2015 Protecting and Promoting the Open Internet rule. Over 21 million comments were received by the FCC in response to the proposed Restoring the Internet rule; however, many submitted comments were determined to be duplicate messages or fraudulent (Hitlin, Olmstead & Toor, 2017).

The FCC adopted the Restoring the Internet Order in December 2017 (FCC, 2017e). The new rule rescinded most of the 2015 Open Internet rule by:

- Restoring the classification of wireline broadband as an information service under Title I of the *Communications Act*

- Reclassifying mobile broadband Internet service as a private mobile service thereby not subject to common carrier regulation

- Allowing broadband providers to utilize paid prioritization of content

- Eliminating blocking and throttling restrictions

- Pre-empting local and state broadband regulations

The FCC contends that the transparency requirements in the new Order will counter concerns about broadband service providers thwarting consumers' access to content. Broadband providers are required to publicly disclose their network management practices, network performance characteristics and commercial terms of the broadband internet access services they offer "sufficient for consumers to make informed choices" (FCC, 2017e, p. 308).

Competition and Technology Transformation

Technological transformation and policy changes traditionally improve video programming capabilities and competition among multichannel video programming distributors (MVPDs) which include cable television, direct broadcast satellite, and cable services offered through telephone systems. However, according to the FCC, in 2015 only 17.9% of U.S. homes had access to at least four MVPDs, down from 38.1% in 2014 (FCC, 2017a).

The FCC's *Eighteenth Report on Video Competition* (2017a) showed that from 2014 to 2015, cable lost video market share to telephone systems (See Table 5.3). More importantly, the report discovered online video distributors (OVDs) such as Netflix, Amazon Prime, and Hulu have significantly increased video programming options for consumers, with more than 59.4 million households watching online video in 2015. It is estimated that 70% of North American Internet wireline traffic during prime time viewing hours is due to streaming video and audio.

Table 5.3

MVPD Market Share

Multichannel Video Programming Distributor	2014	2015
Cable	53.4%	53.1%
DBS	33.3%	33.2%
Telco Cable	12.9%	13.4%

Source: FCC (2017a), p. 7

Although MVPDs and OVDs offer an alternative to traditional broadcast television, in 2015 11% of all U.S. households, or 12.4 million homes, still rely exclusively on over-the-air broadcasting for video news and entertainment. Although the introduction of video competition has gradually decreased broadcast television viewership over time, the FCC competition report shows that the prime-time audience share for network affiliates was 32% for 2013-2014 and increased to 33% in 2014-2015 (FCC, 2017a).

In 2017, the FCC enacted revised broadcast ownership rules that eliminated the previous ban of newspaper-broadcast cross-ownership within a market; a rule dating back to 1975. Additionally, the FCC eliminated rules prohibiting cross-ownership of both a radio and a television station within a single market. Finally, the new ownership rules modified the Local Television Ownership Rule making it easier to own more than one station in a local market (FCC, 2017b).

Factors to Watch

The 2017 Restoring Internet Freedom rule has revived strong divisions among industry and policymakers regarding network neutrality. Utilizing the Congressional Review Act, the Senate has attempted to overrule the FCC's network neutrality rule (Fung, 2018). More than 20 states have sued the FCC in the U.S. District Court of Appeals in Washington, D.C. in an effort to overturn the FCC's network neutrality repeal (Neidig, 2018). Internet titans Amazon, Facebook, Google, and Netflix plan to join lawsuits to preserve net neutrality through the Internet Association, the lobbying group of the Internet companies (Finley, 2018).

During the eight years of the Obama Administration, the Democratic controlled FCC concluded in their annual broadband progress reports that broadband was not being deployed to all Americans in a reasonable and timely fashion. This conclusion was based on continuing disparities in broadband availability and broadband adoption throughout the United States (FCC, 2016a). In 2018, the Republican controlled FCC under Chairman Pai concluded that, through the FCC's revised deregulatory policies (including the Restoring Internet Freedom rule), "the Commission is now encouraging the deployment on a reasonable and timely basis of advanced telecommunications capability to all Americans" (FCC, 2018a, p. 3).

One of the final actions of the Obama era FCC was to implement an Internet privacy rule that established a framework for requiring customer consent before Internet service providers could use and share customers' personal information (2016d). The 115th U.S. Congress eventually leveraged the Congressional Review Act to pass legislation nullifying this FCC privacy rule which was signed by President Trump in 2017 (Fung, 2017). The new legislation allows ISPs to mine consumer's Web browsing history, app usage history, and location data. The FCC is prohibited from passing similar privacy regulations in the future, and the Federal Trade Commission is expected to provide limited consumer protection for ISP customers.

In 2018, the FCC created a new Office of Economics and Analytics charged with ensuring that economic analysis is deeply and consistently incorporated into the agency's regular operations and will support work across the FCC and throughout the decision-making process. Specifically, it will provide stronger economic analysis and more data driven research to develop policies and strategies within the FCC (FCC, 2018b).

In addition to revising broadcast cross-ownership rules, the FCC is considering raising national broadcast ownership limit. As of mid-2018, regulation limits an entity from owning or controlling television stations that reach more than 39 percent of the television households in the country (FCC, 2017c). Critics of media concentration fear the FCC is reconsidering the cap in support of the Sinclair Broadcast Group's $39 billion takeover of Tribune Media, as the proposed merger surpasses the current cap (Reardon, 2017). AT&T's proposed merger with Time Warner has been delayed as the Department of Justice antitrust division insists that the acquisition cannot be approved without divestment of either DirecTV or Turner Networks, which includes CNN (Johnson, 2017).

Controversies surrounding online content including libel (Browne-Barbour, 2015), cyberbullying, and fake news (Hwang, 2017; Stanford Law and Policy Lab, 2017) have led to calls for Congress to revisit Section 230 of the Communications Decency Act that may challenge the strong First Amendment protections of the Internet and social media. Congress is also considering strengthening Federal election law to require greater disclosure for political ads on social media, after discovering Russian efforts to disrupt the 2016 presidential election via social media (Kelly, 2017).

Despite the Trump administration's pro-business and deregulatory focus, the FCC's continuing efforts to promote broadband confirm a strong national commitment to ubiquitous broadband service for all Americans. The FCC's new benchmark of 25 Mb/s for the definition of broadband acknowledges the increasing capabilities of broadband in the U.S. The FCC's efforts to expand the use of wireless spectrum recognizes that mobile Internet has become a vital medium for distributing data, voice, and video—with global mobile data traffic expected to grow sevenfold by 2021 (Cisco, 2017). The Connect America Fund and the Mobility Fund provide significant funding for broadband expansion. President Trump's proposed $1.5 trillion infrastructure plan, includes $50 billion directed at a broad range of rural infrastructure projects, including broadband service (Eggerton, 2018). Observers will watch to see if the FCC and administration policies result in enhanced broadband deployment and faster Internet speeds for all Americans.

It is likely the conflicts among the FCC, Congress, industry, and consumer advocates will continue as ideological and political clashes permeate the current policy environment. As communication policy is reshaped to accommodate new technologies, policymakers must explore new ways to work together to serve the public interest.

Bibliography

Alexander, P. J. & Brown, K. (2007). Policy making and policy tradeoffs: Broadcast media regulation in the United States. In P. Seabright & J. von Hagen (Eds.). *The economic regulation of broadcasting markets: Evolving technology and the challenges for policy.* Cambridge: Cambridge University Press.

Atkinson, R.C. & Schultz, I.E. (2009). Broadband in America: Where it is and where it is going (according to broadband service providers). Retrieved from http://broadband.gov/docs/Broadband_in_America.pdf.

The Broadband Data Improvement Act of 2008, Pub. L. 110-385, 122 Stat. 4096 (2008). Retrieved from http://www.ntia.doc.gov/advisory/onlinesafety/BroadbandData_PublicLaw110-385.pdf.

Browne-Barbour, V.S. (2015). Losing their license to libel. *Berkeley Technology Law Journal 30* (2), 1505-1559.

Caplan, R., Hanson, L, & Donovan J. (2018, Feb.). Dead reckoning: Navigating content moderation after "fake news". *Data & Society.* Retrieved from https://datasociety.net/pubs/oh/DataAndSociety_Dead_Reckoning_2018.pdf.

Comcast v. FCC. No. 08-1291 (D.C. Cir., 2010). Retrieved from http://hraunfoss.fcc.gov/edocs_public/attachmatch/DOC-297356A1.pdf.

Cisco, (2017, Sept. 15). Cisco visual network index: Global mobile data traffic forecast update, 2016-2021. Retrieved from https://www.cisco.com/c/en/us/solutions/collateral/service-provider/visual-networking-index-vni/complete-white-paper-c11-481360.html.

DuBoff, R. B. (1984). The rise of communications regulation: The telegraph industry, 1844-1880. *Journal of Communication, 34* (3), 52-66.

Eggerton, J. (2015). FCC sets up incentive auction opening bid prices. *Broadcasting & Cable.* Retrieved from http://www.broadcastingcable.com/news/washington/fcc-sets-incentive-auction-opening-bid-prices/145025.

Eggerton, J. (2018) President fleshes out infrastructure plan. *Broadcasting & Cable.* Retrieved from http://www.broadcasting-cable.com/news/washington/president-fleshes-out-infrastructure-plan/171737.

Executive Office of the President: Council of Economic Advisors (2012, Feb. 12). *The economic benefits of new spectrum for wireless broadband.* Retrieved from https://obamawhitehouse.archives.gov/sites/default/files/cea_spectrum_report_2-21-2012.pdf.

Federal Communications Commission. (2005, September 8). *Retransmission consent and exclusivity rules: Report to Congress pursuant to section 208 of the Satellite Home Viewer Extension and Reauthorization Act of 2004.* Retrieved from http://hraunfoss.fcc.gov/edocs_public/attachmatch/DOC-260936A1.pdf.

Federal Communications Commission. (2008, August 20). *Memorandum opinion and order* (FCC 08-183). Broadband Industry Practices. Retrieved from http://hraunfoss.fcc.gov/edocs_public/attachmatch/FCC-08-183A1.pdf.

Federal Communications Commission. (2010a). *Connecting America: National broadband plan.* Retrieved from http://hraunfoss.fcc.gov/edocs_public/attachmatch/DOC-296935A1.pdf.

Federal Communications Commission. (2010b). *Mobile broadband: The benefits of additional spectrum.* Retrieved from http://download.broadband.gov/plan/fcc-staff-technical-paper-mobile-broadband-benefits-of-additional-spectrum.pdf.

Federal Communications Commission. (2010c). *Sixth broadband deployment report* (FCC 10-129). Retrieved from http://hraunfoss.fcc.gov/edocs_public/attachmatch/FCC-10-129A1_Rcd.pdf.

Federal Communications Commission. (2011a). *Bringing broadband to rural America: Update to report on rural broadband strategy.* Retrieved from http://hraunfoss.fcc.gov/edocs_public/attachmatch/DOC-307877A1.pdf.

Federal Communications Commission. (2011b). *FCC releases 'Connect America Fund" order to help expand broadband, create jobs, benefit consumers.* Federal Communications Commission News. Retrieved from http://hraunfoss.fcc.gov/edocs_public/attachmatch/DOC-311095A1.pdf.

Federal Communication Commission. (2014). *Universal service implementation progress report.* Retrieved from http://transition.fcc.gov/Daily_Releases/Daily_Business/2014/db0324/DOC-326217A1.pdf.

Federal Communications Commission. (2015a). *2015 Broadband progress report* (FCC 15-10). Retrieved from https://apps.fcc.gov/edocs_public/attachmatch/FCC-15-10A1.pdf.

Federal Communications Commission. (2015b). *Connect America fund offers carriers nearly $1.7 billion to expand broadband to over 8.5 million rural Americans.* Federal Communications Commission News. Retrieved from http://transition.fcc.gov/Daily_Releases/Daily_Business/2015/db0429/DOC-333256A1.pdf.

Federal Communications Commission. (2015c). *Memorandum Opinion and Order* (FCC 15-25). Retrieved from https://apps.fcc.gov/edocs_public/attachmatch/FCC-15-25A1.pdf.

Federal Communications Commission. (2015d). *Protecting and Promoting the Open Internet* (FCC 15-24). Retrieved from https://apps.fcc.gov/edocs_public/attachmatch/FCC-15-24A1.pdf.

Federal Communications Commission. (2016a). *2016 broadband progress report* (FCC 16-6). Retrieved from https://apps.fcc.gov/edocs_public/attachmatch/FCC-16-6A1.pdf.

Federal Communications Commission. (2016b). *FCC modernizes Lifeline program for the digital age.* Federal Communications Commission News. Retrieved from http://transition.fcc.gov/Daily_Releases/Daily_Business/2016/db0404/DOC-338676A1.pdf.

Federal Communications Commission. (2016c). *In the matter of designated market areas: Report to congress pursuant to section 109 of the STELA Reauthorization Act of 2014* (DA 16-613). Retrieved from https://apps.fcc.gov/edocs_public/attachmatch/DA-16-613A1.pdf.

Federal Communications Commission. (2016d). *In the matter of protecting the privacy of customers of broadband and other telecommunications services* (FCC 16-148). Retrieved from https://apps.fcc.gov/edocs_public/attachmatch/FCC-16-148A1.pdf.

Federal Communications Commission. (2017a). *Eighteenth report on video competition* (DA 17-71). Retrieved from https://apps.fcc.gov/edocs_public/attachmatch/DA-17-71A1.pdf.

Federal Communications Commission. (2017b). *In the matter of 2014 quadrennial regulatory review: Review of the commission's broadcast ownership rules and other rules adopted pursuant to section 202 of the Telecommunications Act of 1996* (FCC 17-156). Retrieved from https://apps.fcc.gov/edocs_public/attachmatch/FCC-17-156A1.pdf.

Federal Communications Commission. (2017c). *In the matter of amendment of section 73.3555(e) of the commission's rules: National television multiple ownership rule* (FCC 17-169). Retrieved from https://apps.fcc.gov/edocs_public/attachmatch/FCC-17-169A1.pdf.

Federal Communications Commission. (2017d). *In the matter of bridging the digital divide for low-income consumers: Lifeline and Link Up reform and modernization* (FCC 17-155). Retrieved from https://apps.fcc.gov/edocs_public/attachmatch/FCC-17-155A1.pdf.

Federal Communications Commission. (2017e). *In the matter of restoring Internet freedom: Declaratory ruling, report and order* (FCC 17-166). Retrieved from https://transition.fcc.gov/Daily_Releases/Daily_Business/2018/db0220/FCC-17-166A1.pdf.

Federal Communications Commission. (2017f). *In the matter of use of spectrum bands above 24 GHz for mobile radio services* (FCC 17-152). Retrieved from https://apps.fcc.gov/edocs_public/attachmatch/FCC-17-152A1.pdf.

Federal Communications Commission. (2017g). Incentive auction closing and channel reassignment public *notice* (DA 17-314). Retrieved from https://apps.fcc.gov/edocs_public/attachmatch/DA-17-314A1.pdf.

Federal Communications Commission. (2018a). *2018 broadband deployment report* (FCC 18-10). Retrieved from https://transition.fcc.gov/Daily_Releases/Daily_Business/2018/db0202/FCC-18-10A1.pdf.

Federal Communications Commission. (2018b). In the matter of establishment of the office of economics and analytics (FCC 18-7). Retrieved from https://transition.fcc.gov/Daily_Releases/Daily_Business/2018/db0131/FCC-18-7A1.pdf.

Federal Communications Commission. (2018c). *Sixth international broadband data report* (DA 18-99). Retrieved from https://transition.fcc.gov/Daily_Releases/Daily_Business/2018/db0209/DA-18-99A1.pdf.

Feld, H. & Forscey, K. (2015, Feb. 25). Why presidents advise the FCC. *The Hill*. Retrieved from http://thehill.com/blogs/congress-blog/the-administration/233678-why-presidents-advise-the-fcc.

Finley, K. (2018, Jan. 16). Tech giants to join legal battle over net neutrality. *Wired*. Retrieved from https://www.wired.com/story/tech-giants-to-join-legal-battle-over-net-neutrality/.

Fowler, M. S. & Brenner, D. L. (1982). A marketplace approach to broadcast regulation. *University of Texas Law Review 60* (207), 207-257.

Frieden, R. (2005, April). *Analog and digital must-carry obligations of cable and satellite television operators in the United States.* Retrieved from http://ssrn.com/abstract=704585.

Fung, B. (2017, April 4). Trump has signed repeal of the FCC privacy rules. Here's what happens next. *Washington Post*. Retrieved from https://www.washingtonpost.com/news/the-switch/wp/2017/04/04/trump-has-signed-repeal-of-the-fcc-privacy-rules-heres-what-happens-next/?utm_term=.f3f0a71254e2.

Fung, B. (2018, Jan. 15). The senate's push to overrule the FCC on net neutrality now has 50 votes, Democrats say. *Washington Post*. Retrieved from https://www.washingtonpost.com/news/the-switch/wp/2018/01/15/the-senates-push-to-overrule-the-fcc-on-net-neutrality-now-has-50-votes-democrats-say/?utm_term=.d1deb1002bb7.

Gandy, O. H. (1989). The surveillance society: Information technology and bureaucratic control. *Journal of Communication, 39* (3), 61-76.

General Accountability Office (2015). *Federal Communications Commission: Protecting and promoting the open internet* (GAO-15-560R). Retrieved from http://www.gao.gov/assets/680/670023.pdf.

General Accountability Office (2017). Additional action needed to address significant risks in FCC's Lifeline Program (GAO-170538). Retrieved from https://www.gao.gov/assets/690/684974.pdf.

Gilroy, A. A. (2008, Sept. 16). Net neutrality: Background and issues. Congressional Research Service, RS22444. Retrieved from http://www.fas.org/sgp/crs/misc/RS22444.pdf.

Gilroy, A.A. (2011, June 30). Universal service fund: Background and options for reform. Congressional Research Service, RL33979. Retrieved from http://www.fas.org/sgp/crs/misc/RL33979.pdf.

Grant, A. E. & Berquist, L. (2000). Telecommunications infrastructure and the city: Adapting to the convergence of technology and policy. In J. O. Wheeler, Y. Aoyama & B. Wharf (Eds.). *Cities in the telecommunications age: The fracturing of geographies.* New York: Routledge.

Green, A. (2015, Dec. 3). Fate of FCC's net neutrality in federal court. *Roll Call*. Retrieved from http://www.rollcall.com/news/fate_of_fccs_net_neutrality_in_federal_court-245001-1.html.

Hitlin, P., Olmstead, K., & Toor, S. (2017, Nov. 29). Public comments to Federal Communications Commission about net neutrality contain many inaccuracies and duplicates. Pew Research Center. Retrieved from http://www.pewinternet.org/2017/11/29/public-comments-to-the-federal-communications-commission-about-net-neutrality-contain-many-inaccuracies-and-duplicates/.

Horrigan, J. (2010). *Broadband adoption and use in America* (OBI Working Paper No. 1). Retrieved from http://hraunfoss.fcc.gov/edocs_public/attachmatch/DOC-296442A1.pdf.

Hwang, T. (2017, Dec.). *Dealing with disinformation: Evaluating the case for CDA 230 amendment*. Retrieved from https://ssrn.com/abstract=3089442.

Johnson, T. (2017, Nov. 15). AT&T-Time Warner, Sinclair-Tribune test Trump era appetite for big media mergers. *Variety*. Retrieved from http://variety.com/2017/biz/news/att-time-warner-sinclair-tribune-1202614806/.

Kang, C. (2008, March 28). Net neutrality's quiet crusader. *Washington Post*, D01.

Kelly, E. (2017, Nov. 1). Senators threaten new rules if social media firms can't stop Russian manipulation themselves. *USA Today*. Retrieved from https://www.usatoday.com/story/news/politics/2017/11/01/senators-say-social-media-companies-must-do-more-stop-russias-election-meddling/820526001/.

Lubrano, A. (1997). *The telegraph: How technology innovation caused social change*. New York: Garland Publishing.

Mazmanian, A. (2014, Feb. 19). FCC plans new basis for old network neutrality rules. *Federal Computer Week*. Retrieved from http://fcw.com/articles/2014/02/19/fcc-plans-new-basis-for-old-network-neutrality-rules.aspx.

McCullagh, D. (2007, March 22). Net porn ban faces another legal setback. *C/NET News*. Retrieved from http://www.news.com/Net-porn-ban-faces-another-legal-setback/2100-1030_3-6169621.html.

Moore, L.K. (2013). Spectrum policy in the age of broadband: Issues for Congress. *CRS Reports to Congress*. CRS Report R40674. Retrieved from http://www.fas.org/sgp/crs/misc/R40674.pdf.

Napoli, P. M. (2001). *Foundations of communications policy: Principles and process in the regulation of electronic media*. Cresskill, NJ: Hampton Press.

National Broadcasting Co. v. United States, 319 U.S. 190 (1949).

Neidig, H. (2018, Jan. 16). States sue over net neutrality. *The Hill*. Retrieved from http://thehill.com/policy/technology/369203-states-sue-fcc-over-net-neutrality-repeal.

Organisation for Economic Co-operation and Development. (2017). *OECD broadband portal*. Retrieved from http://www.oecd.org/sti/broadband/oecdbroadbandportal.htm.

Oxman, J. (1999). *The FCC and the unregulation of the Internet*. OPP Working Paper No. 31. Retrieved from http://www.fcc.gov/Bureaus/OPP/working_papers/oppwp31.pdf.

Parsons, P. R. & Frieden, R. M. (1998). *The cable and satellite television industries*. Needham Heights, MA: Allyn & Bacon.

Pool, I. S. (1983). *Technologies of freedom*. Cambridge, MA: Harvard University Press.

Reardon, M. (2017, Dec. 29). In 2017, the FCC made life easier for your internet provider. *CNET*. Retrieved from https://www.cnet.com/news/fcc-net-neutrality-repeal-easier-life-for-your-internet-provider/.

Red Lion Broadcasting Co. v. Federal Communications Commission, 395 U.S. 367 (1969).

Stanford Law and Policy Lab. (2017). Fake news and disinformation: The roles of the nation's digital newsstands, Facebook, Google, Twitter and Reddit. Retrieved from https://law.stanford.edu/wp-content/uploads/2017/10/Fake-News-Misinformation-FINAL-PDF.pdf.

Starr, P. (2004). *The creation of the media*. New York: Basic Books.

Telecommunications Act of 1996, Pub. L. No. 104-104, 110 Stat. 56 (1996). Retrieved from http://transition.fcc.gov/Reports/tcom1996.pdf.

Turner Broadcasting System, Inc. v. FCC, 512 U.S. 622 (1997).

United States Telecom Association et al. v. FCC, 825 F.3d 674, 689 (D.C. Cir., 2016).

van Cuilenburg, J. & McQuail, D. (2003). Media policy paradigm shifts: Toward a new communications policy paradigm. *European Journal of Communication, 18* (2), 181-207.

Verizon v. FCC. No. 11-1355 (D.C. Cir., 2014). Retrieved from http://www.cadc.uscourts.gov/internet/opinions.nsf/3AF8B4D938CDEEA685257C6000532062/$file/11-1355-1474943.pdf.

Wallsten, S. (2013, April 5). Two cheers for the FCC's mobility fund reverse auction. Technology Policy Institute. Retrieved from https://techpolicyinstitute.org/policy_paper/two-cheers-for-the-fccs-mobility-fund-reverse-auction/.

Warner, W. B. (2008). Networking and broadcasting in crisis. In R. E. Rice (Ed.), *Media ownership research and regulation*. Creskill, NJ: Hampton Press.

Winston, B. (1998). *Media technology and society: A history from the telegraph to the Internet*. New York: Routledge.

Wu, T. (2017, Dec. 6). How the FCC's net neutrality plan breaks with 50 years of history. *Wired*. Retrieved from https://www.wired.com/story/how-the-fccs-net-neutrality-plan-breaks-with-50-years-of-history/.

Section **11**

Electronic Mass Media

6

Digital Television & Video

Peter B. Seel, Ph.D.*

Overview

Digital television and video technology continues to evolve globally at an accelerated pace driven by Moore's Law, with improved display technologies and continued improvements in digital cameras. Digital displays are getting larger, with 65-inch UDTV sets now selling for half ($900) of what they cost in 2016—and massive 88-inch home OLED displays are now on sale. Image enhancements such as High-Dynamic Range (HDR) and QLED quantum dots are driving the sales of premium displays.

The global transition to digital television transmission is now largely complete. The inclusion of high-quality 4K video cameras in the next generation of 5G mobile phones will improve their images and increase the number of online videos posted by connected global citizens. The daily posting of millions of digital videos is a driving force for social media and legacy media sites worldwide.

The United States government concluded its digital television spectrum auction in 2017, generating $20 billion in proceeds, which it will split with the 175 stations which either sold their former spectrum or will share it with other television stations. The bidders will use the former television spectrum for mobile telecommunication services, and 957 TV stations will need to shift their transmission frequencies to new assignments. The newly adopted ATSC 3.0 digital television standard will enhance the diffusion of Internet-delivered Over-The-Top video programming that will enable viewers to bypass cable and satellite delivery services in the United States.

Introduction

Digital television and video displays are simultaneously getting larger and smaller (in home flat screens that exceed seven feet in diagonal size and, at the other extreme, tiny screens in head-mounted digital displays such as Microsoft's Holo-Lens). There is also improved image resolution at both ends of this viewing spectrum. The improvement in video camera quality in mobile phones has led to a global explosion in the number of videos posted online and live online chat sessions. Program delivery is also shifting from over-the-air (OTA) broadcasting and cable television where viewers have content "pushed" to them to an expanding online environment where they "pull" what they want to watch on demand. This global transformation in delivery is affecting every aspect of motion media viewing.

---

*Professor, Department of Journalism and Media Communication, Colorado State University (Fort Collins, Colorado).

In the United States, this online delivery process is known as *Over-The-Top* television—with OTT as the obligatory acronym now used in advertising these services. It should more accurately be called over-the-top *video* as the term refers to on-demand, non-broadcast, Internet-streamed video content viewed on a digital display, but the confusion reflects the blurring of the line between broadcast, cable, and DBS-delivered "linear" television and on-demand streamed content (Seel, 2012). "Over-The-Top" indicates that there is an increasingly popular means of accessing television and video content that is independent of traditional linear OTA television broadcasters and multichannel video programming distributors (MVPDs, which include satellite, cable, and telco content providers). However, much of the content consists of television programs previously broadcast (or cablecast) accessed using services such as Hulu.com—in addition to streamed movies and programs from providers such as Netflix and Amazon, many now available in the immersive 4K Ultra-High-Definition format.

The term "Over-The-Top" is unique to the U.S.—in other regions of the world it is called "Connected" video or television. To move past this confusion, we suggest the introduction of an all-encompassing term: **Internet-Delivered Television** or **I-DTV**. The use of this meta-term would subsume all OTT, Connected, and IPTV (Internet Protocol television) streamed programming, and create a clear demarcation between this online delivery and broadcast-cable-DBS transmission.

In the United States, a massive retooling is occurring in broadcast and video production operations as the nation undergoes a second digital television transition. The first ended on June 12, 2009 with the cessation of analog television transmissions and the completion of the *first* national transition to digital OTA broadcasting using the Advanced Television Systems Committee (ATSC 1.0) DTV standard. (*DTV Delay Act*, 2009). Development work on improved DTV technology for the U.S. has been underway since 2010, and the Federal Communication Commission voted in November of 2017 for a national *voluntary* adoption of a new standard known as "ATSC 3.0" over the next decade (2018-2028) (FCC, 2017; Eggerton, 2018). Details about the U.S. ATSC 3.0 standard are outlined in the Recent Developments section.

Most flat-screen displays larger than 42-inches sold globally in 2018 and beyond will be "smart" televisions that can easily display these motion-media sources and online sites such as Facebook and Pinterest. An example is Samsung's "Tizen" display that offers voice control in accessing a panoply of Internet-delivered and MVPD-provided content. Remote controls for these displays can easily toggle between broadcast, cable, or DBS content and Internet-streamed OTT videos (see Figure 6.1)

Figure 6.1

Samsung's "Tizen" Interface Screen for its "Smart" Digital Television Displays

Source: Samsung

Digital television displays and high-definition video recording are increasingly common features in mobile phones used by 6.8 billion people of the world's population of 7.6 billion. Digital video cameras in mobile phones have also become more powerful and less expensive. An example is the $700 Apple iPhone 8 (see Figure 6.3) which records 4K video at 24, 30, and 60 frames/second and has optical image stabilization for tracking shots.

The democratization of "television" production generated by the explosion in the number of devices that can record digital video has created a world where over one billion users watch more than one billion videos each day at YouTube.com. Local YouTube sites are available in more than 88 countries, with content available in 76 languages, and half of all global video viewing is on mobile devices (YouTube Statistics, 2018). The online distribution of digital video and television programming is an increasingly disruptive force to established broadcasters and program producers. They have responded by making their content available online for free (or via subscription) as increasing numbers of viewers seek to "pull" digital television content on request rather than watch at times when it is "pushed" as broadcast programming. Television news programs routinely feature video captured by bystanders, such as the horrific aftermath of the terrorist attacks in Barcelona, Spain in August of 2017. The increasing ubiquity of digital video recording capability also bodes well for the global free expression and exchange of ideas via the Internet. However, it also makes possible the posting of hateful videos from groups seeking to incite fear in online audiences. The expanding "universe" of digital television and video is driven by improvements in high-definition video cameras for professional production and the simultaneous inclusion of higher-quality video capture capability in mobile phones.

Figures 6.2 and 6.3
Two Contemporary High-Definition Digital Cameras

On the left, the $49,000 Arriflex Alexa XT camera is used by professionals to shoot theatrical motion pictures and high-end television programs with 4K image resolution (2160 X 4096 pixels). On the right, the $700 Apple iPhone 8 records 4K video at 24, 30, and 60 frames/second with optical image stabilization—it can also multi-task as a mobile phone.

Sources: Arriflex and Apple

Another key trend is the recently completed global conversion from analog to digital television (DTV) technology. The United States completed its national conversion to digital broadcasting in 2009, Japan completed its transition in 2012, India in 2014, and most European nations did so in 2015. At the outset of high-definition television (HDTV) development in the 1980s, there was hope that one global television standard might emerge, easing the need to perform format conversions for international program distribution. There are now multiple competing DTV standards

based on regional affiliations and national political orientation. In many respects, global television has reverted to a "Babel" of competing digital formats reminiscent of the advent of analog color television. However, DTV programming in the widescreen 16:9 aspect ratio is now a commonplace sight in all nations that have made the conversion. The good news for consumers is that digital television displays have become commodity products with prices dropping rapidly each year. In the United States, a consumer can purchase a quality 40-inch LCD digital television for $300—far below the $10 per diagonal inch benchmark that was crossed in 2010 (see Table 6.1).

Background

The global conversion from analog to digital television technology is the most significant change in television broadcast standards since color images were added in the 1960s. Digital television combines higher-resolution image quality with improved multichannel audio, and new "smart" models include the ability to seamlessly integrate Internet-delivered "television" programming into these displays. In the United States, the Federal Communications Commission (FCC, 1998) defines DTV as "any technology that uses digital techniques to provide advanced television services such as high definition TV (HDTV), multiple standard definition TV (SDTV) and other advanced features and services" (p. 7420).

Digital television programming can be accessed via linear over-the-air (OTA) fixed and mobile transmissions, through cable/telco/satellite multichannel video program distributors MVPDs), and through Internet-delivered I-DTV sites. I-DTV is an on-demand "pull' technology in that viewers seek out a certain program and watch it in a video stream or as a downloaded file. OTA linear broadcasting is a "push" technology that transmits a digital program to millions of viewers at once. I-DTV, like other forms of digital television, is a scalable technology that can be viewed as lower quality, highly compressed content or in HDTV-quality on sites such as Vimeo.com or YouTube.com.

Table 6.1

Average U.S. Retail Prices of LCD, LCD-LED, OLED-HD, 4K, OLED 4K, and OLED 8K Television Models from 2013-2017

Display Sizes (diagonal)	Average Retail Price in 2013	Average Retail Price in 2015	Average Retail Price in 2017
40-42 inch LED HDTV	$ 450 (1080p)	$ 330 (1080p)	$ 265 (1080p)
46-47 inch LED HDTV	$ 590 (1080p)	$ 583 (1080p)	$ 440 (1080p)
55-inch 4K Ultra-HD TV	n.a.	$ 957 (2160p)	$ 580 (2160p)
55-inch 4K HDR UHD TV	n.a.	$ 3,000 (2160p)	$ 800 (2160p)
65-inch 4K Ultra-HD TV	n.a.	$ 1,862 (2160p)	$ 865 (2160p)
65-inch 4K HDR UHD TV	n.a.	$ 5,000 (2160p)	$ 1,400 (2160p)
86-inch 4K HDR Super-UHD TV	n.a.	$ 10,000 (2160p)	$ 4,500 (2160p)
88-inch 4K QLED TV	n.a.	n.a.	$ 18,000 (2160p)
88-inch 8K OLED TV	n.a.	n.a.	$ 20,000 (4320p)

Sources: U.S. retail surveys by P.B. Seel for all data. n.a. = display was not available

In the 1970s and 1980s, Japanese researchers at NHK (Japan Broadcasting Corporation) developed two related analog HDTV systems: an analog "Hi-Vision" *production* standard with 1125 scanning lines and 60 fields (30 frames) per second; and an analog "MUSE" *transmission* system with an original bandwidth of 9 MHz designed for satellite distribution throughout Japan. The decade between 1986 and 1996 was a significant era in the diffusion of HDTV technology in Japan, Europe, and the United States. There were a number of key events during this period that shaped advanced television technology and related industrial policies:

- In 1986, the Japanese Hi-Vision system was rejected as a world HDTV production standard by the CCIR, a subgroup of the International Telecommunication Union (ITU). By 1988, a European research and development consortium, EUREKA EU-95, had created a competing system known as HD-MAC that featured 1250 widescreen scanning lines and 50 fields (25 frames) displayed per second (Dupagne & Seel, 1998).

- In 1987, the FCC in the United States created the Advisory Committee on Advanced Television Service (ACATS). This committee was charged with investigating the policies, standards, and regulations that would facilitate the introduction

of advanced television (ATV) services in the United States (FCC, 1987).

- U.S. testing of analog ATV systems by ACATS was about to begin in 1990 when the General Instrument Corporation announced that it had perfected a method of digitally transmitting a high-definition signal. Ultimately, the three competitors (AT&T/Zenith, General Instrument/MIT, and Philips/Thomson/Sarnoff) merged into a consortium known as the Grand Alliance and developed a single digital broadcast system for ACATS evaluation (Brinkley, 1997).

The FCC adopted a number of key decisions during the ATV testing process that defined a national transition process from analog NTSC to an advanced digital television broadcast system:

- In 1990, the Commission outlined a simulcast strategy for the transition to an ATV standard (FCC, 1990). This strategy required that U.S. broadcasters transmit both the new ATV signal and the existing NTSC signal concurrently for a period of time, at the end of which all NTSC transmitters would be turned off.

- The Grand Alliance system was successfully tested in the summer of 1995, and a U.S. digital television standard based on that technology was recommended to the FCC by the Advisory Committee (Advisory Committee on Advanced Television Service, 1995).

- In May 1996, the FCC proposed the adoption of the ATSC Digital Television (DTV) Standard that specified 18 digital transmission variations in HD and SD formats.

- In April 1997, the FCC defined how the United States would make the transition to DTV broadcasting and set December 31, 2006 as the target date for the phase-out of NTSC broadcasting (FCC, 1997). In 2005, after it became clear that this deadline was unrealistic due to the slow consumer adoption of DTV sets, it was reset at February 17, 2009 for the cessation of analog full-power television broadcasting (*Deficit Reduction Act*, 2005).

However, as the February 17, 2009 analog shutdown deadline approached, it was apparent that millions of over-the-air households with analog televisions had not purchased the converter boxes needed to continue watching broadcast programs. Cable and satellite customers were not affected, as provisions were made for the digital conversion at the cable headend or with a satellite set-top box. Neither the newly inaugurated Obama administration nor members of Congress wanted to invite the wrath of millions of disenfranchised analog television viewers, so the shut-off deadline was delayed by an act of Congress 116 days to June 12, 2009 (*DTV Delay Act*, 2009). Between 2009 and 2016, the proportion of U.S. television households using digital HDTV displays climbed from 24 percent to over 90 percent, a significant adoption rate over seven years (DTV Household Penetration, 2016).

Recent Developments

There are two primary areas affecting the diffusion of digital television and video in the United States: an ongoing battle over the digital television spectrum between broadcasters, regulators, and mobile telecommunication providers; and the diffusion of new digital display technologies. The latter include now ubiquitous 4K-resolution Ultra-High-Definition (UHD) sets with display enhancements such as

improved High-Dynamic Range (HDR) and the evolution of QLED quantum dot technology. The spectrum battle is significant in the context of what is known as the *Negroponte Switch*, which describes the conversion of "broadcasting" from a predominantly over-the-air service to one that is now wired and the simultaneous transition of telephony from a traditional wired service to a wireless one for an increasing number of users (Negroponte, 1995). The growth of wireless broadband services has placed increasing demands on the available radio spectrum—and broadcast television is a significant user of that spectrum. The advent of digital television in the United States made this conflict possible in that the assignment of new DTV channels demonstrated that spectrum assigned to television could be "repacked" at will without the adjacent-channel interference problems presented by analog transmission (Seel, 2011).

The DTV Spectrum Auction in the United States

Some context is necessary to understand why the federal government was seeking the return of digital television spectrum that it had allocated in the 2009 U.S. DTV switchover. As a key element of the transition, television spectrum between former channels 52-70 was auctioned off for $20 billion to telecommunication service providers, and the existing 1,500 broadcast stations were "repacked" into lower terrestrial broadcast frequencies (Seel & Dupagne, 2010). The substantial auction revenue that was returned to the U.S. Treasury was not overlooked by federal officials and, as the U.S. suffered massive budget deficits in the global recession of 2008-2010, they proposed auctioning more of the television spectrum as a way to increase revenue without raising taxes in a recession. The U.S. Congress passed the *Middle Class Tax Relief and Job Creation Act of 2012* (*Middle Class*, 2012) and it was signed into law by President Barack Obama. The *Act* authorized the Federal Communication Commission to reclaim DTV spectrum assigned to television broadcasters and auction it off to the highest bidder (most likely wireless broadband providers).

The success of this unique auction depended on several factors. Television spectrum holders had to weigh the value of their surrendered spectrum against three possible options: taking their OTA signal off the air and returning their assigned TV channel, moving to a lower VHF (Very High Frequency) channel in a local repack, or perhaps moving to a higher VHF channel in their market. Stations also had the option to share an OTA channel with another broadcaster, which DTV technology permits. The auction value to the broadcaster declined with each option. Broadcasters seeking to participate in the auction filed applications with the FCC by January 12, 2016 specifying which of the three options they selected (FCC, 2016).

The auction concluded in April 2017 with bids of nearly $20 billion for the TV spectrum sold by 175 broadcast licensees to telecommunication service providers (Pressman, 2017). This amount is similar to that generated in the repacking of the television spectrum in the original U.S. DTV transition, but was much less than the $30-$60 billion predicted by some analysts before the bidding (Irwin, 2018). A significant task is ahead as 957 television stations will have three years (2018-2021) to shift their signals into lower VHF bands to clear the former UHF spectrum for the telecommunication companies which bid on it. Of the 175 stations that sold their former UHF spectrum, 12 are expected to go off the air, 30 will shift their channel assignments to lower VHF frequencies, and the balance will share another station's spectrum (Pressman, 2017).

How will local viewers be affected? In the United States, a majority of television viewers watch local channels via a cable company or from a satellite provider, so the number of OTA viewers is often comparatively small. Cable or satellite viewers would likely *not* notice a change after the channel repack, as their providers assign their own channels on their respective systems. OTA viewers in each market will have to rescan their digital television channels after the repack, as they will be on different frequencies. Viewers may need to purchase a new digital television set, as the new "Next Gen TV" system is based on the *incompatible* ATSC 3.0 standard (Eggerton, 2018). The U.S. incentive auction may be seen in the future as a fruitless exercise, as many telecommunication providers failed to make bids on the vacated television spectrum and many U.S. television stations have to endure yet another channel repacking process. However, the auction did compensate 175 television stations with half of the $20 billion proceeds from the auction for giving up all or part of their spectrum and the U.S. treasury will net $10 billion in income from it.

The ATSC 3.0 DTV Standard

The new ATSC 3.0 system was adopted by the FCC on November 16, 2017 as a *voluntary* digital television transmission standard for the United States (FCC, 2017). Broadcast television companies wanted to be more competitive with Internet-based video content providers. These companies lobbied Congress and the Commission to create a new system that would combine OTA broadcasting with related OTT programming which would allow broadcasters to more precisely geo-target advertising to consumers in ways that resemble Internet-based marketing. It is a convergent technology where broadcast MPEG video compression meets Web-delivered content for simulcast delivery to millions of viewers (Eggerton, 2017, Eggerton, 2018).

There are multiple benefits of the transition to ATSC 3.0 standard for broadcasters: Over-the-Air digital data throughput increases from 19.3 Mbps to 25 Mbps or better in a 6 MHz OTA broadcast channel. This increased data rate would facilitate the OTA transmission of 4K programming, especially if improved High-Efficiency Video Coding (HEVC) is used in this process, and broadcast programming could contain 22.2 channel audio (Baumgartner, 2016). Consumers could use flash drives (much like those used for Chromecast reception) to upgrade older DTV sets for the display of program content that might be a blend of broadcast programs with related content delivered via a broadband connection to a smart display (see figure 6.5 that illustrates the future of television). The improved emergency alert system can also be geo-targeted to a specific area or neighborhood in a broadcast region (Eggerton, 2018).

It is estimated that the average conversion cost per television station will be $450,000, and that the national conversion to the ATSC 3.0 "Next Gen TV" (NGTV) standard may take up to 10 years—three years for the channel repacking process to be completed and then up to seven additional years for the complete national transition. Phoenix, Arizona will provide a test market in 2018 for NGTV with ten stations collaborating on ATSC 3.0 transmissions and OTT feature development (Amy, 2018).

Current Status

Display Features. Two key developments are the arrival of Ultra-High-Definition (UHD) televisions that have 4K screen resolution and the diffusion of "premium" 4K displays that offer High-Dynamic Range (HDR) and/or QLED "quantum dot" image enhancements. Note that a 55-inch 4K UHD television that cost an average price of $957 in 2015 could be purchased for about half that amount (averaging $580) in 2017. Most higher-end 4K displays are "smart" televisions that can easily display any desired programming source, whether it be broadcast, cable or satellite-cast, or Internet-delivered content. The advent of the *incompatible* ATSC 3.0 digital television standard in the U.S. will require the purchase of new displays with the required chipset for accessing the features of the new standard, but most cable and DBS subscribers will have this conversion accomplished by the provider.

DTV adoption and viewing. DTV penetration in the United States in July 2016 was estimated at 91 percent of 118.4 million U.S. television households or 107.74 million homes (DTV Household Penetration, 2016). This figure has steadily increased from 38 percent since the U.S. switchover in 2009 (Leitchman Research Group, 2009). The conversion to HDTVs and now UHD TVs is being propelled by the rapidly falling prices of DTV displays as shown in Table 6.1.

Digital Television Technology

Display Options

Consumers have many technological options for digital television displays, and most sets sold today include higher-resolution screens at 4K (2160p) that feature interactive (smart) features. The dominant display technologies are:

- *4K Television/Video Production and Display.* These Ultra-High-Definition (UHD) televisions have 8,294,400 pixels (3840 pixels wide by 2160 high)—four times that of a conventional HDTV display. One problem with 4K technology is that the data processing and storage requirements are four times greater than with 2K HDTV. Just as this appeared to be a significant problem for 4K program production and distribution, a new compression scheme has emerged to deal with

this issue: High-Efficiency Video Coding (HEVC). It uses advanced compression algorithms in a new global H.265 DTV standard adopted in January 2013 by the International Telecommunications Union. It superseded their H.264 AVC standard and uses half the bit rate of MPEG-4 coding making it ideal for 4K production and transmission (Butts, 2013).

- *Organic Light Emitting Diode (OLED)*—Sony introduced remarkably bright and sharp OLED televisions in 2008 that had a display depth of 3 mm—about the thickness of three credit cards. Mobile phone manufacturers such as Apple and Samsung have adopted AMOLED versions of these displays as they are very sharp, colorful, and bright. Korean manufacturer LG Electronics has decided to stake its future on this technology and recently introduced an 88-inch 8K ("*triple 8*") OLED model that features a remarkable resolution of 7680 pixels wide X 4320 pixels high (also referenced as 4320p) (see Figure 6.4).

- *High Dynamic Range (HDR) Imaging.* Another recent development in television technology has been the inclusion of HDR enhancements to HD and UHD displays. Improving the dynamic range between the light and dark areas of an image adds to its visual impact. Greater detail can be seen in the shadow areas of a displayed scene and also in the bright highlights. The technology was developed by Dolby Laboratories in its plan known as *Dolby Vision*. While also increasing the range of contrast in televised images, HDR-enabled displays have much higher screen brightness than those without it (Waniata, 2015).

- *Quantum Dots.* Consumers considering purchasing a UHD display have the *quantum dots* option. A familiarity with theoretical physics is not required to understand that this term refers to the addition of a thin layer of nano-crystals between the LCD backlight and the display screen. This unique application of nanotechnology increases image color depth by 30 percent without adding extra pixels to the display (Pino, 2016).

Programming

With the advent of 4K UHD televisions, consumers are now demanding more 4K content. Netflix has obliged by producing and distributing multiple original series and motion pictures in 4K resolution. Amazon has emerged as another provider of 4K content and the company is starting to produce their own content in 4K, much like Netflix (Pino, 2016). Apple plans to enter the TV/film production business as a content provider with over $1 billion allocated for production in 2018. Netflix will spend over $7 billion on original programming during that same period, and Amazon spent $4.5 billion on its "Prime" video productions in 2017 (Spangler, 2017). Facebook is actively promoting its "Watch" video delivery service that it introduced in 2017 and, with over two billion global users, it could emerge as a significant provider of OTT video content worldwide in the next decade (Gahan, 2017).

Figure 6.4

LG Electronics Introduced a New 88-Inch 8K Display at the 2018 CES Show

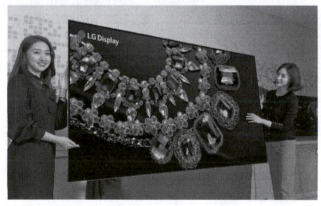

Source: LG Electronics.

Factors to Watch

The global diffusion of DTV technology will evolve over the second and third decades of the 21st century as prices drop for UHD and 4K displays. In the United States, the future inclusion of ATSC 3.0 DTV receivers and the inclusion of UHD-quality video recording capability in most mobile phones will enhance this diffusion on a personal and familial level. In the process, digital television and video will influence what people watch, especially as smart televisions, note-sized phones, tablet computers, and the arrival of 5G

mobile phones between 2018 and 2020 will offer easy access to on-demand, Internet-delivered content that is more interactive than traditional television.

These trends will be key areas to watch between 2018 and 2020:

- *Evolving 4K and 8K DTV technologies*. The continued development of higher-resolution video production and display technologies will enhance the viewing experience of television audiences around the world. While there are substantial data processing/storage and transmission issues with 4K (and certainly 8K) DTV technologies, these will be resolved in the coming decade. The roll-out of fiber-to-the-home (FTTH) with 1 Gb/s transmission speeds in many communities will enhance the online delivery of these high-resolution programs. When combined with ever-larger, high-resolution displays in homes, audiences can have a more immersive viewing experience (with a greater sense of telepresence) by sitting closer to their large screens.

- *Cord-Shavers and Cord-Nevers*. The number of younger television viewers seeking out alternative viewing options via OTT-TV is steadily increasing. Expect to see more television advertisers placing ads on popular Internet multimedia sites such as YouTube to reach this key demographic. With a smart television, a fast Internet connection, and a small antenna for receiving over-the-air digital signals, viewers can see significant amounts of streamed and broadcast content without paying monthly cable or satellite fees. MVPDs such as cable and satellite companies are concerned about this trend as it enables these viewers to "cut the cable" or "toss the dish," so these companies are seeking unique programming content such as premium sports events (e.g., the World Cup and the Super Bowl) to hold on to their customer base.

- *New OTT Programming Providers*. New players Apple and Facebook will join an increasingly crowded field of OTT companies, such as Netflix, CBS AllAccess, and Amazon. This is a golden era for video, television, and "film" program producers as these new channels are competing for a limited number of quality program producers.

Already fragmented audiences for these services will become even more so, as the number of companies compete for their attention via cable, streamed OTT, and Wi-Fi access.

Figure 6.5

The Future of Television

The future of television is a blended display of broadcast content—in this case, a severe weather alert—combined with an Internet-delivered run-down of the local newscast for random access and other streamed media content around the broadcast image.

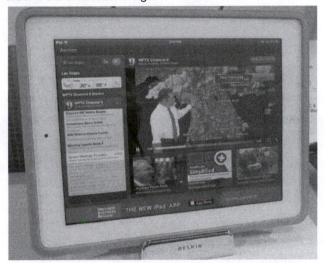

Source: P.B. Seel

- **U.S. Adoption of the ATSC 3.0 Standard.** The FCC has approved the voluntary national adoption of the new, incompatible ATSC 3.0 digital television standard, and broadcasters are making plans to add OTT content to their OTA programming. The ATSC 3.0 standard will facilitate the merger of broadcast television and streamed on-demand programing in a hybrid model. For example, a large-screen smart television could display conventional broadcast programming surrounded by additional windows of related Internet-delivered content (see Figure 6.5).

The era of the blended television-computer or "tele-computer" has arrived. Most large-screen televisions are now "smart" models capable of displaying multimedia content from multiple sources: broadcast, cablecast, DBS, and all Internet-delivered motion media. As the typical home screen has evolved toward larger (and smarter) models, it is also getting thinner and larger with the arrival of new LCD, QLED, and OLED models. The image quality of new high-definition displays is startling and will improve further with the present delivery of 4K Ultra-HDTV sets and the arrival of new 8K OLED models in the near future. Television, which has been called a "window on the world," will literally resemble a large window in the home for viewing live sports, IMAX-quality motion pictures in 2D and 3D, and any video content that can be steamed over the Internet. On the other end of the viewing scale, thanks to new wireless technology, mobile customers with 5G phones will have access to this same diverse content on their high-definition AMOLED screens. Human beings are inherently visual creatures and this is an exciting era for those who like to watch motion media programs of all types in ultra-high-definition at home and away.

Careers in Digital Television & Video Production

There are numerous ways to get started in video and television production or distribution. Working at a campus television station provides the essential on-the-job production experience required to seek an entry-level professional position. In the U.S., there are over 1,500 television stations that provide hundreds of entry-level production or sales positions each year, most of them in smaller markets. Similar entry-level positions are available in developed nations around the world that have local and national television services. The rapid expansion of Internet-delivered videos has also created many opportunities for producers of this type of online content.

As newspapers, magazines and other publications add streamed video to their online sites, there are entry-level production opportunities with these companies. Another entry-level option is to seek employment with a video production company or a related organization which would provide useful professional experience shooting, editing, and posting videos.

Bibliography

Advisory Committee on Advanced Television Service. (1995). *Advisory Committee final report and recommendation.* Washington, DC: Author.

Amy, D. B. (2018, January 18). *Making the most out of ATSC 3.0: How Next Gen TV can drive a broadcast revenue renaissance.* Panel presentation, National Association of Television Program Executives 2018 Conference, Miami Beach.

Baumgartner, F. (2016, 4 January). April SBE Ennis workshop to focus on ATSC 3.0. *TV Technology.* Retrieved from, http://www.tvtechnology.com/broadcast-engineering/0029/april-sbe-ennes-workshop-to-focus-on-atsc-30/277680

Brinkley, J. (1997). *Defining vision: The battle for the future of television.* New York: Harcourt Brace & Company.

Butts, T. (2013, January 25). Next-gen HEVC video standard approved. *TV Technology.* Retrieved from, http://www.tvtechnology.com/article/nexgen-hevc-video-standard-approved/217438.

Deficit Reduction Act of 2005. Pub. L. No. 109-171, § 3001-§ 3013, 120 Stat. 4, 21 (2006).

DTV Delay Act of 2009. Pub. L. No. 111-4, 123 Stat. 112 (2009).

DTV Household Penetration in the United States from 2012-2016. (2016). Statista.com. Retrieved from, https://www.statista.com/statistics/736150/dtv-us-household-penetration/.

Dupagne, M., & Seel, P. B. (1998). *High-definition television: A global perspective.* Ames: Iowa State University Press.

Eggerton, J. (2017, November 16). FCC launches next-gen broadcast TV standard. *Broadcasting & Cable.* Retrieved from, http://www.broadcastingcable.com/news/washington/fcc-launches-next-gen-broadcast-tv-standard/170165.

Eggerton, J. (2018, February 1). ATSC 3.0 rollout can begin next month. *Broadcasting & Cable.* Retrieved from, http://www.broadcastingcable.com/news/washington/atsc-30-rollout-can-begin-next-month/171487.

Federal Communications Commission. (1987). Formation of Advisory Committee on Advanced Television Service and Announcement of First Meeting, 52 Fed. Reg. 38523.

Federal Communications Commission. (1990). Advanced Television Systems and Their Impact Upon the Existing Television Broadcast Service (*First Report and Order*), 5 FCC Rcd. 5627.

Federal Communications Commission. (1997). Advanced Television Systems and Their Impact Upon the Existing Television Broadcast Service (*Fifth Report and Order*), 12 FCC Rcd. 12809.

Federal Communications Commission. (1998). Advanced Television Systems and Their Impact Upon the Existing Television Broadcast Service (*Memorandum Opinion and Order on Reconsideration of the Sixth Report and Order*), 13 FCC Rcd. 7418.

Federal Communications Commission. (2016, March 23). *Broadcast Incentive Auction.* Retrieved from, https://www.fcc.gov/about-fcc/fcc-initiatives/incentive-auctions.

Federal Communications Commission. (2017, November 16). *Authorizing Permissive Use of the "Next Generation" Broadcast Television Standard.* Retrieved from, https://apps.fcc.gov/edocs_public/attachmatch/FCC-17-158A1.docx.

Gahan, B. (2017, December 5). Facebook Watch will overtake YouTube as the biggest video platform. *Mashable.* Retrieved from, https://mashable.com/2017/12/05/how-facebook-watch-wil-overtake-youtube-as-biggest-video-platform/

Irwin, D. (2018, January 16). Was the incentive auction necessary? *TV Technology.* Retrieved from, http://www.tvtechnology.com/opinions/0004/was-the-incentive-auction-necessary/282577.

Leichtman Research Group. (2009, November 30). *Nearly half of U.S. households have an HDTV set* (Press release). Retrieved from, http://www.leichtmanresearch.com/press/113009release.html.

Middle Class Tax Relief and Job Creation Act of 2012. Pub. L. No. 112-96.

Negroponte, N. (1995). *Being Digital.* New York: Alfred A. Knopf.

Pino, N. (2016, February 16). 4K TV and UHD: Everything you need to know about Ultra HD. *TechRadar.* Retrieved from, http://www.techradar.com/us/news/television/ultra-hd-everything-you-need-to-know-about-4k-tv-1048954.

Pressman, A. (2017, April 19). Why almost 1,000 TV stations are about to shift channels. *Fortune,* Retrieved from http://fortune.com/2017/04/19/tv-stations-channels-faq/.

Seel. P. B. (2011). Report from NAB 2011: Future DTV spectrum battles and new 3D, mobile,and IDTV technology. *International Journal of Digital Television.* 2, 3. 371-377.

Seel. P. B. (2012). The 2012 NAB Show: Digital TV goes over the top. *International Journal of Digital Television,* 3, 3. 357-360.

Seel, P. B., & Dupagne, M. (2010). Advanced television and video. In A. E. Grant & J. H. Meadows (Eds.), *Communication Technology Update* (12th ed., pp. 82-100). Boston: Focal Press.

Spangler, T. (2017, August 16). Apple sets $1 billion budget for original TV shows, movies. *Variety.* Retrieved from, http://variety.com/2017/digital/news/apple-1-billion-original-tv-shows-movies-budget-1202529421/.

Waniata, R. (2015, May 3). What are HDR TVs and why is every manufacturer selling them now? *DigitalTrends.* Retrieved from, http://www.digitaltrends.com/home-theater/hdr-for-tvs-explained.

YouTube Statistics. (2018). YouTube Inc. Retrieved from, https://www.youtube.com/intl/en-GB/yt/about/press/.

7

Multichannel Television Services

Paul Driscoll, Ph.D. & Michel Dupagne, Ph.D.[*]

Overview

Consumers are using four methods to receive packages of television channels and/or programming, including cable, satellite, telephone lines, and Internet delivery (OTT—Over-the-Top Television). This multichannel video programming distributor (MVPD) market is more competitive than ever, with consumers actively switching from one service to others, as well as subscribing to multiple services at the same time. Overall the MVPD industry continued to face substantial subscriber losses in 2016 and 2017 due to the growing cord-cutting trend. Cable operators, which still dominate the MVPD market with a 55% penetration rate, shed 2% of their video subscriber base between 2016 and 2017; direct broadcast satellite (DBS) providers with a market share of 34% lost 5%; and telephone companies with an 11% share saw their subscriber count drop by 8%. In all, the traditional MVPDs lost about 4% of their video subscribers between 2016 and 2017 and the same downward trend is expected to continue. The top four U.S. pay-TV providers (AT&T, Comcast, Charter, and DISH) served 79% of the mainstream video subscribers in 2017. The top virtual MVPDs, such as Sling TV and DirecTV Now, added more than an estimated 1.5 million subscribers in 2017. It is unclear whether the cord-cutting phenomenon will stabilize in the near future.

Introduction

Until the early 1990s, most consumers who sought to receive multichannel television service enjoyed few options other than subscribing to their local cable television operator. Satellite television reception through large dishes was available nationwide in the early 1980s, but this technology lost its luster soon after popular networks decided to scramble their signals. Multichannel multipoint distribution service (MMDS), using microwave technology and dubbed wireless cable, existed in limited areas of the country in the 1980s, but has never become a major market player. So, for all intents and purposes, the cable industry operated as a de facto monopoly with little or no competition for multichannel video service during the 1980s.

The market structure of subscription-based multichannel television began to change when DirecTV launched its direct broadcast satellite (DBS) service in 1994 and when DISH delivered satellite signals to its first customers in 1996. Another watershed moment occurred when Verizon introduced its fiber-to-the-home FiOS service in 2005 and AT&T started deploying U-verse in 2006. While few cable overbuilders

---

[*] Driscoll is Associate Professor and Vice Dean for Academic Affairs, Department of Journalism and Media Management, University of Miami (Coral Gables, Florida) and Dupagne is Professor in the same department.

(i.e., two wired cable systems overlapping and competing with one another for the same video subscribers in the same area) have ever existed, the multichannel video programming distributor (MVPD) marketplace became increasingly competitive during the second decade of the 2000s, especially with the explosive growth of hundreds of over-the-top (OTT) programming services. The Federal Communications Commission (FCC) reported that 35% of U.S. homes were able to subscribe to at least four MVPDs in 2013 (FCC, 2015a), but this statistic plunged to only 17.9% in 2015, largely attributable to the acquisition of DirecTV by AT&T. Still, 99% of housing units continued to have access to three competing MVPD services—a cable system and two direct broadcast satellite providers—in 2015 (FCC, 2017). In 2018, the MVPD industry faced a very different landscape, from an 86% near market saturation in earlier decades to 77% in 2017, with an expectation for further decline. So-called cord-cutting activity and cord-never inactivity pose a possibly existential threat to the traditional MVPD business model from over-the-top (OTT) providers like Netflix, Amazon Prime Video, Hulu, and many others. Although consumption and distribution patterns are changing quickly, MVPDs remain powerful businesses, capturing some advantage from their vertical integration as program content creators with the ability to sell their own TV subscriptions over-the-top, often in bundles containing fewer channels.

This chapter will first provide background information about multichannel television services to situate the historical, regulatory, and technological context of the industry. Special emphasis will be given to the cable industry. The other sections of the chapter will describe and discuss major issues and trends affecting the MVPD industry.

Definitions

Before we delve into the content proper, it is important to delineate the boundaries of this chapter and define key terms that are relevant to multichannel television services. Because definitions often vary from source to source, thereby creating further confusion, we will rely as much as possible on the legal framework of U.S. Federal Regulations and laws to define these terms. *The Cable Television Consumer Protection and Competition Act of 1992* defines a multichannel video programming distributor (MVPD) as "a person such as, but not limited to, a cable operator, a multichannel multipoint distribution service, a direct broadcast satellite service, or a television receive-only satellite program distributor, who makes available for purchase, by subscribers or customers, multiple channels of video programming" (p. 244). For the most part, MVPD service refers to multichannel video service offered by cable operators (e.g., Comcast, Charter), DBS providers (DirecTV, DISH), and telephone companies (e.g., AT&T's U-verse, AT&T Fiber, Verizon's FiOS). This chapter will focus on these distributors.

Pay television designates a category of TV services that offer programs uninterrupted by commercials for an additional fee on top of the basic MVPD service (FCC, 2013a). These services primarily consist of premium channels, pay-per-view (PPV), and video on demand (VOD). Subscribers pay a monthly fee to receive such premium channels as HBO and Starz. In the case of PPV, they pay a per-unit or transactional charge for ordering a movie or another program that is scheduled at a *specified time*. VOD, on the other hand, allows viewers to order a program from a video library at *any given time* in return for an individual charge or a monthly subscription fee. The latter type of VOD service is called subscription VOD or SVOD. We should note that the terms "pay television" and "MVPD" are often used interchangeably to denote a TV service for which consumers pay a fee, as opposed to "free over-the-air television," which is available to viewers at no cost.

But perhaps the most challenging definitional issue that confronts regulators is whether over-the-top (OTT) providers could qualify as MVPDs if they offer multiple channels of video programming for purchase, like traditional MVPDs do. In December 2014, the FCC (2014a) proposed to redefine an MVPD as a provider that makes available for purchase "multiple linear streams of video programming, regardless of the technology used to distribute the programming" (p. 15996). In so doing, the Commission sought to create a more flexible and more competitive platform-neutral framework for MVPD operations. This reinterpretation would cover Internet-based distributors as long as they provide multiple linear programming streams. A "stream" would be analogous to a channel, and "linear programming" would refer to prescheduled video programming. Specifically, the redefinition in the FCC's (2014a) *Notice of Proposed Rulemaking* (NPRM) would apply to subscription linear providers (also called linear OTTs), such as Sky Angel, which delivered religious and family-friendly programming over IP (Internet Protocol) until 2014 (see FCC, 2010). We should note that the new MVPD status, if implemented, would carry both regulatory benefits (e.g., program access rules) and obligations (e.g., retransmission consent) for these Internet-based MVPDs. In December 2015, then-FCC Chairman Tom Wheeler indicated that the MVPD redefinition proceeding would be placed on hiatus given the numerous innovative developments influencing the video space (Eggerton, 2015).

In summary, while the FCC is considering expanding the meaning of the term "MVPD," it is likely that this NPRM will progress slowly, due in part to the objections set forth by cable operators. In addition, such a reinterpretation would be limited in scope because it would only concern linear OTT providers—MVPDs that offer multiple programming streams at prescheduled times. Thus, linear OTT would be regulated quite differently from VOD OTT services, such as Netflix and Hulu. It is unclear, though, how the FCC would treat an OTT provider that offers both linear and non-linear video content. Presumably, the FCC (2015a) would continue to view a VOD OTT as an online video distributor (OVD), which "offers video content by means of the Internet or other Internet Protocol (IP)-based transmission path provided by a person or entity other than the OVD" (p. 3255). Even though it supplies more than 300 paid channels, YouTube would still be classified as an OVD because these channels are not programmed according to a regular broadcast or cable schedule. So, as of early 2018, all OTT providers were considered as OVDs according to the FCC's classification.

Interestingly, the FCC (2014a) chose to define OTT far more narrowly than the industry does—as "linear [emphasis added] video services that travel over the public Internet and that cable operators do not treat as managed video services on any cable system" (p. 16026). More than ever, understanding the evolving increasingly complex terminology of the video space is critically important for any student of communication technologies. While OTT providers were addressed in

detail in Chapter 6, the impact of OTT on multichannel video services is inescapable and will be mentioned in this chapter as warranted.

Background
The Early Years

While a thorough review of the history, regulation, and technology of the multichannel video industry is beyond this chapter (see Baldwin & McVoy [1988], Parsons [2008], Parsons & Frieden [1998] for more information), a brief overview is necessary to understand the broad context of this technology.

As TV broadcasting grew into a new industry in the late 1940s and early 1950s, many households were unable to access programming because they lived too far from local stations' transmitter sites or because geographic obstacles blocked reception of terrestrial electromagnetic signals. Without access to programming, consumers were not going to purchase TV receivers, a problem for local stations seeking viewers and appliance stores eager to profit from set sales.

The solution was to erect a central antenna capable of capturing the signals of local market stations, amplify, and distribute them through wires to prospective viewers for a fee. Thus, cable extended local stations' reach and provided an incentive to purchase a set. The first non-commercial Community Antenna TV (CATV) service was established in 1949 in Astoria, Oregon, but multiple communities claim to have pioneered this retransmission technology, including a commercial system launched in Lansford, Pennsylvania, in 1950 (Besen & Crandall, 1981).

Local TV broadcasters initially valued cable's ability to extend their household reach, but tensions arose when cable operators began using terrestrial microwave links to import programming from stations located in "distant markets," increasing the competition for audiences that local stations rely on to sell advertising. Once regarded as a welcome extension to their over-the-air TV signals, broadcasters increasingly viewed cable as a threat and sought regulatory protection from the government.

Evolution of Federal Communications Commission Regulations

At first, the FCC showed little interest in regulating cable TV. But given cable's growth in the mid-1960s, the FCC, sensitive to TV broadcasters' statutory responsibilities to serve their local communities and promote local self-expression, adopted rules designed to protect over-the-air TV broadcasting (FCC, 1965). Viewing cable as only a supplementary service to over-the-air broadcasting, regulators mandated that cable systems carry local TV station signals and placed limits on the duplication of local programming by distant station imports (FCC, 1966). The Commission saw such rules as critical to protecting TV broadcasters, especially struggling UHF-TV stations, although actual evidence of such harm was largely nonexistent. Additional cable regulations followed in subsequent years, including a program origination requirement and restrictions on pay channels' carriage of movies, sporting events, and series programming (FCC, 1969, 1970).

The FCC's robust protection of over-the-air broadcasting began a minor thaw in 1972, allowing an increased number of distant station imports depending on market size, with fewer imports allowed in smaller TV markets (FCC, 1972a). The 1970 pay cable rules were also partially relaxed. Throughout the rest of the decade, the FCC continued its deregulation of cable, sometimes in response to court decisions.

Although the U.S. Supreme Court held in 1968 that the FCC had the power to regulate cable television as reasonably ancillary to its responsibility to regulate broadcast television, the scope of the FCC's jurisdiction under the *Communications Act of 1934* remained murky (*U.S. v. Southwestern Cable Co.*, 1968). The FCC regulations came under attack in a number of legal challenges. For instance, in *Home Box Office v. FCC* (1977), the Court of Appeals for the D.C. Circuit questioned the scope of the FCC's jurisdiction over cable and rejected the Commission's 1975 pay cable rules, designed to protect broadcasters against possible siphoning of movies and sporting events by cable operators.

In the mid-1970s, cable began to move beyond its community antenna roots with the use of domestic satellites to provide additional programming in the form of superstations, cable networks, and pay channels. These innovations were made possible by the FCC's 1972 *Open Skies Policy*, which allowed qualified companies to launch and operate domestic satellites (FCC, 1972b). Not only did the advent of satellite television increase the amount of programming available to households well beyond the coverage area of local TV broadcast channels, but it also provided cable operators with an opportunity to fill their mostly 12-channel cable systems.

Early satellite-delivered programming in 1976 included Ted Turner's "Superstation" WTGC (later WTBS) and the Christian Broadcasting Network, both initially operating from local UHF-TV channels. In 1975, Home Box Office (HBO), which had begun as a pay programming service delivered to cable systems via terrestrial microwave links, kicked off its satellite distribution offering the "Thrilla in Manila" heavyweight title fight featuring Mohammad Ali against Joe Frasier.

Demand for additional cable programming, offering many more choices than local television stations and a limited number of distant imports, stimulated cable's growth, especially in larger cities. According to the National Cable & Telecommunications Association (2014), there were 28 national cable networks by 1980 and 79 by 1990. This growth started an alternating cycle where the increase in the number of channels led local cable systems to increase their capacity, which in turn stimulated creation of new channels, leading to even higher channel capacity, etc.

Formal congressional involvement in cable TV regulation began with the *Cable Communications Policy Act of 1984*. Among its provisions, the *Act* set national standards for franchising and franchise renewals, clarified the roles of Federal, state, and local governments, and freed cable rates except for systems that operated without any "effective competition," defined as TV markets having fewer than three over-the-air local TV stations. Cable rates soared, triggering an outcry by subscribers.

The quickly expanding cable industry was increasingly characterized by a growing concentration of ownership. Congress reacted to this situation in 1992 by subjecting more cable systems to rate regulation of their basic and expanded tiers, and instituting retransmission consent options (requiring cable systems to give something of value to local TV stations not content with simple must-carry status).

The Cable Television Consumer Protection and *Competition Act of 1992* also required that MVPDs make their programming available at comparable terms to satellite and other services. The passage of the *Telecommunications Act of 1996* also reflected a clear preference for competition over regulation. It rolled back much of the rate regulation put in place in 1992 and opened the door for telephone companies to enter the video distribution business.

Distribution of TV Programming by Satellite

By the late 1970s and early 1980s, satellites were being used to distribute TV programming to cable systems, expanding the line-up of channels available to subscribers. But in 1981, about 11 million people lived in rural areas where it was not economical to provide cable; five million residents had no TV at all, and the rest had access to only a few over-the-air stations (FCC, 1982).

One solution for rural dwellers was to install a Television Receive Only (TVRO) satellite dish. Sometimes called "BUGS" (Big Ugly Dishes) because of the 8-to-12-foot diameter dishes needed to capture signals from low-power satellites, TVROs were initially the purview of hobbyists and engineers. In 1976, H. Taylor Howard, a professor of electrical engineering at Stanford University, built the first homemade TVRO and was able to view HBO programming for free (his letter of notice to HBO having gone unanswered) (Owen, 1985).

Interest in accessing free satellite TV grew in the first half of the 1980s, and numerous companies began to market satellite dish kits. From 1985 to 1995, two to three million dishes were purchased (Dulac & Godwin, 2006). However, consumer interest in home satellite TV stalled in the late-1980s when TV programmers began to scramble their satellite feeds.

In 1980, over the strenuous objections of over-the-air broadcasters, the FCC began to plan for direct broadcast satellite (DBS) TV service (FCC, 1980). In September 1982, the Commission authorized the first commercial operation. At the 1983 Regional Administrative Radio Conference, the International Telecommunication Union (ITU), the organization that coordinates spectrum use internationally, awarded the United States eight orbital positions of 32 channels each (Duverney, 1985). The first commercial attempt at DBS service began in Indianapolis in 1983, offering five channels of programming, but ultimately all of the early efforts failed, given the high cost of operations, technical challenges, and the limited number of desirable channels available to subscribers. (Cable operators also used their leverage to dissuade programmers from licensing their products to DBS services.) Still, the early effort demonstrated the technical feasibility of using medium-power Ku-band satellites (12.2 to 12.7 GHz downlink) and foreshadowed a viable DBS service model.

In 1994, the first successful DBS providers were DirecTV, a subsidiary of Hughes Corporation, and U.S. Satellite Broadcasting (USSB) offered by satellite TV pioneer Stanley Hubbard. Technically competitors, the two services offered largely complementary programming and together launched the first high-power digital DBS satellite capable of delivering over 200 channels of programming to a much smaller receiving dish (Crowley, 2013).

DirecTV bought USSB in 1998. EchoStar (now the DISH Network) was established in the United States in 1996 and became DirecTV's primary competitor. In 2005, the FCC allowed EchoStar to take over some of the satellite channel assignments of Rainbow DBS, a failed 2003 attempt by Cablevision Systems Corporation to offer its Voom package of 21 high-definition (HD) TV channels via satellite. In 2003, the U.S. Department of Justice blocked an attempted merger between DirecTV and EchoStar. DirecTV later became a successful acquisition target.

Direct Broadcast Satellite Expansion

DBS systems proved popular with the public but faced a number of obstacles that allowed cable to remain the dominant provider of multichannel television service. One competitive disadvantage of DBS operators over their cable counterparts was that subscribers were unable to view their local-market TV channels without either disconnecting the TV set from the satellite receiver or installing a so-called A/B switch that made it possible to flip between over-the-air and satellite signals.

In 1988, Congress passed the *Satellite Home Viewer Act* that allowed the importation of distant network programming (usually network stations in New York or Los Angeles) to subscribers unable to receive the networks' signals from their over-the-air local network affiliates. Congress followed up in 1999 with the *Satellite Home Viewer Improvement Act* (SHVIA) that afforded satellite companies the opportunity (although not a requirement) to carry the signals of local market TV stations to all subscribers living in that market. SHIVA also mandated that DBS operators carrying one local station carry all other local TV stations requesting carriage in a local market, a requirement known as "carry one, carry all."

Today, almost all of U.S. TV households subscribing to DBS service are able to receive their local stations via satellite. DISH Network supplies local broadcast channels in HD to all 210 U.S. TV markets and DirecTV covers 198 markets (FCC, 2016). Local station signals are delivered from the satellite using spot beam technology that transmits signals into specific local TV markets. Satellites have dozens of feedhorns used to send their signals to earth, from those that cover the entire continental United States (CONUS) to individual spot beams serving local markets.

Cable System Architecture

Early cable TV operators built their systems based on a "tree and branch" design using coaxial cable, a type of copper wire suitable for transmission of radio frequencies (RF). As shown in Figure 7.1, a variety of video signals are received at the cable system's "headend," including satellite transmissions of cable networks, local TV station signals (delivered either by wire or from over-the-air signals received by headend antennas), and feeds from any public, educational, and governmental (PEG) access channels the operator may provide. Equipment at the headend processes these incoming signals and translates them to frequencies used by the cable system.

Figure 7.1
Tree-and-Branch Design

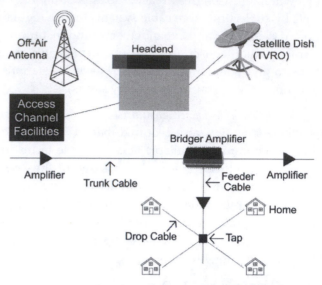

Source: P. Driscoll and M. Dupagne

Most cable companies use a modulation process called quadrature amplitude modulation (QAM) to encode and deliver TV channels. The signals are sent to subscribers through a trunk cable that, in turn, branches out to thinner feeder cables, which terminate at "taps" located in subscribers' neighborhoods. Drop cables run from the taps to individual subscriber homes. Because electromagnetic waves attenuate (lose power) as they travel, the cable operator deploys amplifiers along the path of the coaxial cable to boost signal strength. Most modern TV receivers sold in North America contain QAM tuners, although set-top boxes provided by the cable system are usually used because of the need to descramble encrypted signals.

Over the years, the channel capacity of cable systems has increased dramatically, from 12-22 channel systems through the mid-1970s to today's systems capable of carrying more than 300 TV channels and bi-directional communication. Modern cable systems have 750 MHz or more of radio frequency capacity, up to 1GHz in the most advanced systems (Afflerbach, DeHaven, Schulhuf, & Wirth, 2015). This capacity allows not only plentiful television channel choices, but also enables cable companies to offer a rich selection of pay-per-view (PPV) and video-on-demand (VOD) offerings. It also makes possible high-speed Internet and voice over Internet protocol (VoIP) telephone services to businesses and consumers. Most cable operators have moved to all-digital transmission, although a few still provide some analog channels, usually on the basic service tier.

As shown in Figure 7.2, modern cable systems are hybrid fiber-coaxial (HFC) networks. Transmitting signals using modulated light over glass or plastic, fiber optic cables can carry information over long distances with much less attenuation than copper coaxial, greatly reducing the need for expensive amplifiers and offering almost unlimited bandwidth with high reliability. A ring of fiber optic cables is built out from the headend, and fiber optic strands are dropped off at each distribution hub. From the hubs, the fiber optic strands are connected to optical nodes, a design known as fiber-to-the-node or fiber-to-the-neighborhood (FTTN). From the nodes, the optical signals are transduced back to electronic signals and delivered to subscriber homes using coaxial cable or thin twisted-pair copper wires originally installed for telephone service. The typical fiber node today serves about 500 homes, but additional nodes can be added as data-intensive services like high speed broadband, video on demand, and Internet Protocol (IP) video are added. Depending on the number of homes served and distance from the optical node, a coaxial trunk, feeder, and drop arrangement may be deployed.

Rollouts of HFC cable networks and advances in compression have added enormous information capacity compared to systems constructed with only coaxial copper cable, boosting the number of TV channels available for live viewing, and allowing expansive VOD offerings and digital video recorder (DVR) service. Additionally, given the HFC architecture's bi-directional capabilities, cable companies are able to offer consumers both telephone and broadband data transmission services, generating additional revenues.

Some MVPDs, such as Verizon's FiOS service, Google Fiber, and AT&T Fiber, extend fiber optic cable directly to the consumer's residence, a design known as fiber-to-the-home (FTTH) or fiber-to-the-premises (FTTP). FiOS uses a passive optical network (PON), a point-to-multipoint technology that optically splits a single fiber optic beam sent to the node to multiple fibers that run to individual homes, ending at an optical network terminal (ONT) attached to

the exterior or interior of the residence (Corning, 2005). The ONT is a transceiver that converts light waves from the fiber optic strand to electrical signals carried through the home using appropriate cabling for video, broadband (high-speed Internet), and telephony services.

Figure 7.2
Hybrid Fiber-Coaxial Design

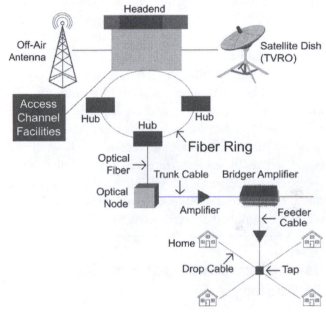

Source: P. Driscoll and M. Dupagne

Although the actual design of FTTH systems may vary, fiber has the ability to carry data at higher speeds compared to copper networks. Fiber broadband services can deliver up to a blazing one gigabit (one billion bits) of data per second. Fiber-to-the-home may be the ultimate in digital abundance, but it is expensive to deploy and may actually provide more speed than most consumers currently need. In addition, technological advances applied to existing HFC systems have significantly increased available bandwidth, making a transition to all-fiber networks a lower priority for existing cable systems in the United States. However, with ever-increasing demand, fiber-to-the-home may ultimately prevail. QAM- modulated signals can be transcoded into Internet Protocol (IP) signals, allowing program distribution via apps, and allows integration of MVPD programming on a variety of third-party devices such as a Roku box, Apple TV or a properly equipped "smart" TV receiver.

Direct Broadcast Satellite System Architecture

As shown in Figure 7.3, DBS systems utilize satellites in geostationary orbit, located at 22,236 miles (35,785 km) above the equator, as communication relays. At that point in space, the satellite's orbit matches the earth's rotation (although they are not travelling at the same speed.) As a result, the satellite's coverage area or "footprint" remains constant over a geographic area. In general, footprints are divided into those that cover the entire continental United States, certain portions of the United States (because of the satellite's orbital position), or more highly focused spot beams that send signals into individual TV markets. DBS operators send or "uplink" data from the ground to a satellite at frequencies in the gigahertz band, or billions of cycles per second. Each satellite is equipped with receive/transmit units called transponders that capture the uplink signals and convert them to different frequencies to avoid interference when the signal is sent back to earth or "downlinked."

Figure 7.3

Basic Direct Broadcast Satellite Technology

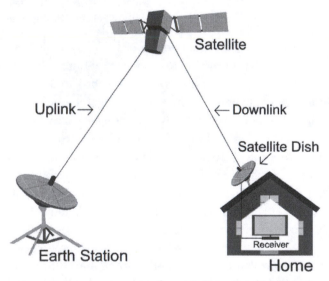

Source: P. Driscoll and M. Dupagne

A small satellite dish attached to the consumer's home collects the signals from the satellite. A low-noise block converter (LBN) amplifies the signals by as much as a million times and translate the high frequency satellite signal to the lower frequencies used by TV receivers. The very high radio frequencies used in satellite communications generate extremely short wavelengths, sometimes leading to temporary signal interference in rainy weather when waves sent from the satellite are absorbed or scattered by raindrops.

Unlike cable service that features bi-directional data flows, commercial DBS is a one-way service that relies on broadband connections to offer most of its VOD services. But with its nationwide footprint, DBS service is available in rural areas where the infrastructure cost of building a cable system would be prohibitive.

Video Compression

The large increase in the number of available programming channels over cable and satellite services is possible due to advances in digital video compression. Developed by the ITU and the MPEG group, video compression dramatically reduces the number of bits needed to store or transmit a video signal by removing redundant or irrelevant information from the data stream. Satellite signals compressed using the MPEG-2 format allowed up to eight channels of digital programming to be squeezed into bandwidth previously used for one analog channel. Video compression yields increased capacity and allows satellite carriage of high-definition TV channels and even some 4K Ultra High Definition (UHD) channels.

Impressive advances in compression efficiency have continued, including the adoption by DBS and some cable operators of MPEG-4 Advanced Video Coding (AVC) that reduces by half the bit rate needed compared with MPEG-2 (Crowley, 2013). Interest is now focused on High Efficiency Video Coding (HEVC), approved in 2013 by the ITU and MPEG, which doubles the program carrying capacity compared with MPEG-4 compression. Even more powerful codecs like HEVC/H.265 and VP9 will enable MVPDs and virtual MVPDs (vMVPDs) to carry additional high-definition and 4K UHD channels (Baumgartner, 2017b).

Satellite, cable, streaming services and even broadcast station operators will continue to deploy advances in compression techniques to provide consumers with picture quality that exploits improvements in TV set display technology, including features like 4K UHD, high contrast resolution (HCR)/ high dynamic range (HDR), and higher frame rates. (For more on home video, see Chapter 16.)

Recent Developments
Statewide Video Franchising

Prior to 2005, the required process of cable franchising (i.e., authorizing an entity to construct and operate a cable television system in an area) was generally handled by local governments. The *Cable Act of 1984* did not mandate that the franchising authority be a local body, but subjecting cable operators to local jurisdiction made sense since they used public rights-of-way to deploy their infrastructure.

On the other hand, cable operators have long argued that local franchising requirements are onerous and complex, which often involve negotiations and agreements with multiple local governments (e.g., county and city).

In 2005, Texas became the first state to enact a law shifting the responsibilities of cable/video franchise requirements from the local to the state government (see Parker, 2011). More than 20 other states followed suit. In those states, it is typically the department of state or the public utilities commission that issues cable franchises (sometimes called certificates) to wireline MVPDs (National Conference of State Legislatures [NCSL], 2014). Other state agencies may share other cable television service duties (e.g., quality complaints) that previously fell within the purview of local governments. As of November 2014 (latest statistics), at least 25 states awarded cable/video franchising agreements (NCSL, 2014).

Few studies have investigated the impact of statewide cable franchising reforms on competition. Concerns remain about the amount of franchise fees, funding and regulatory oversight of public, educational, and governmental (PEG) access channels, and the possibility and effectiveness of anti-redlining provisions to prevent service discrimination in low-income areas. The FCC (2015a) reported that some states did away with or curtailed the requirements for PEG channels whose funding was often a key provision in local franchise agreements.

A la Carte Pay-TV Pricing

A la carte MVPD service would allow subscribers to order individual channels as a customized programming package instead of relying on standard bundled packages or tiers. This issue has attracted considerable policy interest, especially in the mid-2000s, because consumers have often complained to their elected officials or policymakers about the high cost of cable service in relation to other services. Indeed, the FCC (2013b) has documented for nearly two decades that cable rate increases outpace the general inflation rate. It reported that the average price of expanded basic cable service (cable programming service tier) grew by a compound annual growth rate of 6.1% from 1995 to 2012. In contrast, the consumer price index for all items rose by only 2.4% per year during those 17 years.

The Commission also found that the number of channels available in the expanded package soared from 44 in 1995 to 150 in 2012, or by an annual rate of 5.8% (FCC, 2013b). But the per-channel price hardly changed during that time (0.2% annually), which bolstered the case of the cable industry for legitimate price hikes. In absolute terms, though, cable and other MVPD bills continue to increase far above inflation, due largely to climbing license fees (paid by distributors to non-broadcast programmers for carriage) and retransmission fees (paid by operators to broadcasters for carriage).

According to Lendu, a personal finance website, the monthly MVPD subscription fee averaged $116.93 in December 2017 (Lafayette, 2018a). By the end of 2017, 36% of respondents surveyed by TiVo (2018) reported spending more than $100 a month on their MVPD bill (excluding video-on-demand/movie buys, phone, or Internet service), and 10% indicated paying a monthly bill exceeding $150.

On paper, a bundling strategy makes more economic sense than a pure components strategy (a la carte) when the marginal cost of channel delivery is low, which is often the case with popular channels, and when consumers price channel packages similarly (Hoskins, McFadyen, & Finn, 2004). But studies have shown that consumers would be receptive to a la carte selection if afforded the opportunity. For instance, TiVo (2018) reported that about 81% of the respondents in late 2017 favored the option of customizing their own MVPD package. They ranked ABC, Fox, CBS, Discovery Channel, and A&E as the top five most desirable channels to have in a self-selected package. Respondents also indicated that they would spend an average of $35.87 per month for their a la carte pay-TV package and would select an average of 24 channels. TiVo (2018) pointed out that "the ideal price of $35.87 (for U.S. respondents) for an a la carte package compares to the price of many vMVPD services offered today, such as YouTube TV and Sling TV" (p. 9). Of course, there is no guarantee that a virtual MVPD would match a consumer's ideal self-selection of programming.

In fact, while viewers continue to favor self-selected packages, a process that may involve scrutinizing the channel line-ups of multiple vMVPDs and be time-consuming, they also "crave simplicity" when "faced with overload" (Baumgartner, 2018d). These two goals may well conflict with each other. In addition, it is unclear whether consumers are fully cognizant of the economic implications that a la carte programming would entail. Critics have pointed out that a la carte programming would not necessarily reduce cable bills because programmers would have to raise their license fees substantially to compensate for declining viewership and advertising revenue. Furthermore, channel diversity would likely be reduced (Nocera, 2007).

Bundling is hardly an insignificant pricing matter for MVPDs. Not only does bundling play a key role in monetizing MVPDs' video content, but it can also influence the entire revenue stream of cable operators that offer triple-play packages. Triple play, which refers to the combination of video, voice, and data products, is beneficial to both customers who can save money and some MVPDs, particularly cable operators, who can secure a competitive advantage over other MVPDs and counter the effect of cord-cutting or slow growth in the video segment. Media research group Kagan estimated that, as of December 2017, 65% of the customers served by the top five public multiple system operators (MSO) subscribed to more than one service, and 36% opted for the triple-play bundle (Lenoir, 2018).

In summary, availability of a la carte programming, while still being discussed, may never materialize in the United States, especially since some U.S. operators, perhaps as a preemptive strategy, began introducing "skinny" packages, or bundles with pared-down and cheaper channel offerings in 2015. In contrast, a la carte became a reality in Canada when one MVPD, Bell, started offering individual channels for CAN $4 or CAN $7 per channel per month in March 2016, ahead of the December 2016 regulatory deadline (Bradshaw, 2016).

In March 2015, the Canadian Radio-television and Telecommunications Commission (CRTC) (2015) had mandated all MVPDs to provide Canadian viewers with (1) a CAN $25-a-month entry-level television bundle by March 2016 and (2) the option to subscribe to individual channels beyond the basic service on a "pick-and-pay" basis by December 2016. Similar to the U.S. basic service tier, the Canadian first-tier offering includes all Canadian local and regional television stations, public interest channels, provincial educational channels, and community channels if available. Canadian MVPD subscribers can order supplemental (discretionary) channels beyond the first-level tier on a pick-and-pay basis or in small packages. Because of its importance to MVPD marketing and economics, the issue of skinny bundles in the United States will be addressed further.

Adapting to Changes in the Linear Viewing Model

MVPDs continue to adapt their technology and content distribution offerings to accommodate subscribers' shifting patterns of TV consumption. Today, the average adult American watches three hours and 42 minutes per week of time-shifted TV, and seven out of 10 identify themselves as binge viewers (McCord, 2014; Nielsen, 2017).

The trend toward delayed viewing is one factor disrupting the TV ecosystem, especially for advertising revenues and program ratings. The digital video recorder (DVR), first introduced in 1999, is now present in 64% of television households that subscribe to an MVPD, and 81% of U.S. TV households have a DVR, Netflix, or use VOD (Leichtman Research Group, 2016).

Although available as stand-alone units, DVRs are usually integrated into powerful set-top boxes provided by cable and DBS operators. Increasingly, these set-top boxes also provide access to subscription over-the-top programming like Netflix.

The versatility of DVRs has continued to expand, and the available features are sometimes used by MVPDs to distinguish themselves from competitors. For example, the DISH Network's high-end DVR service, the "Hopper3," boasts simultaneous recording of up to 16 program channels to its 2TB hard drive (500 hours of HD programming) and allows the viewing of four channels simultaneously on one TV receiver. It also features: automatic ad skipping on playback of some network TV shows; voice integration with Amazon's Alexa voice assistant; and built-in apps. It can stream and record 4K video from Netflix and provide remote access on mobile devices for out-of-home viewing of live and recorded TV (using Slingbox™ technology) (Prospero, 2017).

Altice USA (formerly Cablevision Systems) offers a remote-storage DVR (RS-DVR) multi-room service that allows subscribers to record up to 15 programs simultaneously and store them in the "cloud" (remote servers at the system's headend) eliminating the expense of providing hard drives in DVR units. (Cablevision Systems Corp., 2014). Even without a DVR unit, about 70% of MVPD subscribers have access to a huge inventory of VOD and SVOD offerings, including previously aired episodes of programs from channels subscribed to by the consumer (Holloway, 2014).

Another sign of possible upheaval for MVPDs is the growing success of over-the-top (OTT) video services available through a host of streaming media player dongles, boxes, and smart TVs, including Chromecast, Roku, Apple TV, and Sony PlayStation. The rising popularity of broadband-delivered video subscription services, such as Netflix, Hulu, and Amazon Prime Video, have led to concerns that customers, especially younger customers, will be "cutting the cord" on MVPD subscriptions (or never begin subscribing) in favor of more affordable services (Lafayette, 2014). However, for those satisfied with their existing MVPD service, Netflix has announced that it will make its service available for carriage on cable TV systems (Bray, 2014). (For more on OTT, see Chapter 6.)

Retransmission Consent

Few other issues in the MVPD business have created more tensions between distributors and broadcasters (and angered more subscribers) than retransmission consent negotiations going awry.

As noted above, the *Cable Act of 1992* allowed local broadcasters to seek financial compensation for program carriage from MVPDs, a process known as retransmission consent. While most of these agreements are concluded with little fanfare (Lenoir, 2014), some negotiations between parties can degenerate into protracted, mercurial, "who-will-blink-first" disputes, which sometimes lead to high-profile programming blackouts when the broadcaster forces the removal of its signal from the MVPD's line-up. In 2014, there were 94 retransmission blackouts of varying length; that number jumped to 193 in 2015 and to 293 in 2017 (American Television Alliance, 2018).

The economic stakes are high. Broadcasters claim that, like cable networks, they deserve fair rates for their popular programs and increasingly consider retransmission fees as a second revenue stream. Television stations expect to collect more than $10 billion in broadcast retransmission fees in 2018 (see Table 7.1). Few industry observers would disagree that "Retrans saved TV broadcasting, but it cannot save it forever" (Jessell, 2017). On the other hand, the MVPDs, fully aware that excessive customer bills could intensify cord-cutting, have attempted to rein in broadcasters' demands for higher retransmission fees, generally with little success, reflecting consumers' robust appetite for broadcast network programming (Baumgartner, 2016).

The Communications Act (47 U.S.C. § 325(b)(3) (C)) imposes a statutory duty on both MVPDs and broadcasters to exercise "good faith" in retransmission consent negotiations, and regulators have increasingly taken note of clashes that too often leave consumers in the dark. In 2014, the FCC barred the "top-4" broadcast stations in a TV market (that are not commonly owned) from banding together to leverage their bargaining power in retransmission consent negotiations and proposed rules to curtail program blackouts (FCC, 2014b).

Table 7.1

Annual Broadcast Retransmission Fee Projections by Medium, 2016-2018

Medium	2016	2017	2018
Cable ($ millions)	$4,211.10	$4,999.30	$5,482.30
Average cable fee/sub/month ($)	$6.61	$7.94	$8.84
DBS ($ millions)	$2,683.60	$3,210.50	$3,530.70
Average DBS fee/sub/month ($)	$6.70	$8.03	$8.89
Telco ($ millions)	$1,073.70	$1,120.70	$1,073.40
Average telco fee/sub/month ($)	$7.28	$8.73	$9.84
Total retransmission fees ($ millions)	$7,968.40	$9,330.50	$10,086.40
Average fee/sub/month ($)	$6.79	$8.18	$9.07

Note. All figures are estimates as of June 2017. The average fee per month per subscriber for each video medium is obtained by dividing the amount of annual retransmission fee for the medium by the average number of subscribers for that medium. The average number of subscribers, which is calculated by taking the average of the subscriber count from the previous year and the subscriber count of the current year, is meant to estimate the number of subscribers at any given time during the year and reflect better the retransmission fee charged during the year instead of at the year-end. The average estimates refer to the fees paid to television stations on behalf of each subscriber, not to the payment per station.

Source: Kagan. Reprinted with permission

As a result of Congressional action in December 2014, the prohibition on broadcaster coordination in retransmission consent negotiations was applied to all same-market stations not under common legal control (STELA, 2014). The FCC also proposed rules that would label certain broadcaster retransmission negotiation tactics (such as blocking consumers access to a broadcaster's online programming that duplicates its broadcast delivery during a blackout) as evidence of not negotiating in good faith (FCC, 2015b).

AT&T Changes the U.S. MVPD Market

As discussed earlier, AT&T added video programming to its telephone and cellular services with the fiber-optic based U-verse service. Because that service was distributed through AT&T's local telephone network (which had long ago replaced most of its 20th century copper network with a fiber-based network designed to deliver data and video on top of telephone conversations), U-verse was limited to areas in which AT&T provided local telephone service. Not satisfied with the steady growth of U-verse, AT&T bought DirecTV in 2015 for $48.5 billion. Perhaps illustrating the economics of satellite-delivery of video, AT&T immediately began shifting video subscribers to DirecTV. It is unknown how long AT&T intends to keep using the DirecTV name, but the company has been aggressive at crossmarketing its multichannel

video services with telephone, cellular phone, and broadband data services (Huddleston, 2015).

Current Status

TV Everywhere

According to the FCC (2015a), TV Everywhere (TVE) "allows MVPD subscribers to access both linear and video-on-demand ("VOD") programming on a variety of in-home and mobile Internet-connected devices" (p. 3256). Ideally, as its name indicates, TVE should provide these subscribers the ability to watch programs, live or not, at any location as long as they have access to a high-speed Internet connection.

Launched in 2009, TVE has grown into a major marketing strategy of multichannel video operators to retain current subscribers, undercut cord-cutting trends, and attract Millennials to pay-TV subscription by offering channel line-ups on mobile devices (Winslow, 2014). According to a survey by market research company GfK (2015), Generation Y (ages 13-38) respondents in MVPD households used TVE sites and apps more frequently than their Generation X (ages 39-52) and Baby Boomer Generation (ages 53-69) counterparts. Another study reported that among non-subscribers "54% of younger Millennials [ages 18-24] and 47% of older Millennials (ages 25-34) said that they were more likely to subscribe to a service if it offered TV Everywhere capabilities" (Baar, 2015).

Table 7.2
Pay-TV Subscribers' Awareness and Adoption of TV Everywhere, 2012-2017

Date	Awareness	Adoption
Q4 2012	26.4%	26.0%
Q4 2013	36.3%	21.6%
Q4 2014	42.3%	25.2%
Q4 2015	40.0%	21.5%
Q4 2016	49.1%	30.9%
Q2 2017	50.5%	34.7%

Source: TiVo's *Video Trends Report* surveys

By the end of 2014, Comcast and other major MVPDs offered their subscribers numerous channels for in-home TVE viewing and even for out-home viewing, though in a smaller quantity (see Winslow, 2014). But even though TVE awareness has nearly doubled from 2012 to 2017, only about a third of respondents in the latest TiVo survey reported using these TVE apps (see Table 7.2). In its own study, GfK (2015) concluded that "[c]onsumer education continues to be a critical missing piece of the puzzle for TV Everywhere." Interestingly, the authentication process, whose user-friendliness has been questioned (e.g., Dreier, 2014; Winslow, 2014), is perceived as easy to use by a large majority (68%) of surveyed users (TDG, 2014).

Skinny Bundles

"Skinny" bundles or packages refer to "smaller programming packages with fewer channels at potentially lower costs to consumers (and, less altruistically, also allow providers to save on the cost of programming)" (Leichtman Research Group, 2015, p. 1). Even though the basic service tier can be technically considered as a slimmed-down bundle (Hagey & Ramachandran, 2014), the term really applies to selected groups of channels from the (expanded) programming service tier. A skinny bundle does not have to be quantitatively tiny. For instance, DISH Network's original Sling TV streaming service, introduced in 2015 and perhaps the best-known example of a skinny package that also happens to be an OTT service, contains a dozen channels for $20 a month (Fowler, 2015). By the

end of 2017, more than 2.2 million viewers subscribed to Sling TV (Farrell, 2018a).

At least on the surface, consumers seem to prefer smaller TV packages. According to TiVo (2018), the ideal number of channels selected by respondents averaged 24 in late 2017. In addition, the ideal cost that these respondents were willing to pay averaged $35.97 per month. By comparison, MVPD subscribers received an average of 194 channels in 2015 (James, 2015).

Are skinny packages economically and mutually beneficial for consumers and distributors? It is too early to make such assessment, but we must recognize that pared-down bundles may have some drawbacks for both constituencies and may not become the cost-saving panacea that many hope. While expecting lower monthly bills, consumers may be disappointed at the limited menu of channels and frustrated by the extra charges they will have to pay to obtain additional channels. As for the distributors, they may embrace skinny packages as a marketing strategy to stabilize cord-cutting losses, shed less popular channels from their line-ups, and court "cord-nevers" (i.e., never subscribing to an MVPD) with cheaper alternatives (see Hagey & Ramachandran, 2014). But they also risk cord shaving (i.e., downgrading video services) from their traditional pay-TV subscribers and higher churn (i.e., subscriber turnover). Charles Ergen, CEO of DISH, acknowledged that the churn for low-cost Sling TV was higher than for other services, but he also added that subscribers "tend to come back over a period of time" (Williams, 2016).

Decline of MVPD Subscription

As shown in Table 7.3, the MVPD penetration (MVPD subscribers/TV households) was 77% in 2017. The number of basic cable subscribers decreased by 4% from 2014 (54.0 million) to 2017 (51.9 million). For the DBS and telco (telephone company) providers, the subscriber decline was even worse (6% and 18%, respectively). Interestingly, the subscriber gap between these providers and their cable rivals has been widening in recent years (see Table 7.3). Kagan projected that this gradual decline in MVPD subscription would continue in 2018. It expected that the (traditional) MVPD penetration would fall to 73% in 2018, down from 81% in 2016 and 85% in 2014.

Table 7.3

U.S. Multichannel Video Industry Benchmarks (in millions), 2014-2018

Category	2014	2015	2016	2017	2018
Basic cable subscribers	54.0	53.3	52.8	51.9	51.3
Digital service cable subscribers	48.2	49.4	50.6	50.3	50.2
High-speed data cable subscribers	55.8	59.6	63.4	66.4	68.2
Voice service cable subscribers	27.3	28.7	29.3	29.4	29.3
DBS subscribers	33.5	33.1	33.2	31.5	30.4
Telco video subscribers	13.0	13.0	11.5	10.6	8.3
All multichannel video subscribers	100.9	99.7	97.7	94.0	90.0
TV households	119.4	120.0	121.1	122.4	123.7
Occupied households	120.6	121.3	122.4	123.7	125.1

Note. All figures are estimates. These counts exclude virtual MVPD subscribers.   *Source*: Kagan. Reprinted with permission

The media research firm also noted that cable networks continue to lose subscribers, due to carriage cancellation by MVPDs and cord-cutting or shaving by consumers. Nielsen's universe estimates for 99 cable networks dropped by an average compound annual growth rate (CAGR) of 1.4% for each year from January 2014 to January 2018 (Robson, 2018). It is unclear whether the subscriber losses in the multichannel video sector will slow down or stabilize through a combination of TVE and skinny bundle marketing strategies.

Table 7.4

Top Ten MVPDs in the United States, as of December 2017 (in thousands)

Company	Subscribers
AT&T (DirecTV/U-verse)	24,116
Comcast	22,357
Charter Communications	16,997
DISH Network	11,030
Verizon Communications (FiOS)	4,619
Cox Communications	3,852
Cablevision Systems (Altice-owned)	2,470
Suddenlink Communications (Altice-owned)	1,112
Frontier Communications	961
Mediacom Communications	821
Estimated total U.S. MVPD market	93,973

Note. These counts exclude virtual MVPD subscribers.

Source: Kagan. Reprinted with permission

The ranking of the top MVPDs changed dramatically in 2015 with AT&T's acquisition of DirecTV. As indicated in Table 7.4, the top four providers served 79% of all MVPD subscribers at the end of 2017, generating a high degree of concentration in that industry (see Hoskins et al., 2004). All 10 MVPDs listed accounted for a 94% share of the estimated total MVPD market. We should note that this table reports the number of subscribers for Cablevision and Suddenlink separately, even though both companies are owned by European-based Altice.

Factors to Watch

MVPD Consolidation

Scale is perhaps the most important economic strategy in the media business because it allows a larger media firm to reduce input costs (e.g., programming costs) and derive competitive advantages against rivals (e.g., channel revenues and advertising opportunities). AT&T Entertainment Group CEO John Stankey pointed out that "[w]e didn't buy DirecTV because we love satellite exclusively as a distribution medium; we bought it because it gave us scale in entertainment and scale in distribution of entertainment" (*Broadcasting & Cable*, 2016, p. 19).

Major consolidation of distribution and programming sectors continues against a backdrop of substantial existing horizontal ownership concentrations. For example, after completing the acquisition of Suddenlink in December 2015, European-based Altice continued to grow with the purchase of Cablevision Systems for $18 billion. In June 2016, Altice became the fourth largest cable operator in the nation, serving

4.6 million video and broadband subscribers. The company announced that it would "squeeze some $900 million in cost savings out of the consolidation of Cablevision and Suddenlink within three years of the deal's closing" (Littleton, 2016).

In October 2016, AT&T announced its intention to buy the programming conglomerate Time Warner for $85 billion ($108 billion including debt). In November 2017, the U.S. Department of Justice filed suit against the transaction, arguing that the vertical integration of AT&T's distribution networks with Time Warner's programming networks could lead to anti-competitive behavior against rival MVPDs and slow the growth of emerging online distributors (*U.S. v. AT&T*, 2017). Time Warner's holdings include Turner Networks (with its sports holdings), HBO, CNN, music interests, and more. In December 2017, Disney, the world's biggest content creator, paid over $52 billion to acquire key programming assets from 20th Century Fox (Farrell, 2017).

Government has a number of tools to protect consumers from potentially anticompetitive consequences from completed mergers, including divestiture, imposition of behavioral conditions, or both, before granting approval. For example, in April 2016, the U.S. Department of Justice approved Charter Communications' $65.5 billion acquisition of two major MSOs, Time Warner Cable and Bright House Networks, subject to pro-competition and consumer-friendly restrictions. Charter agreed to comply with the following key conditions (Eggerton, 2016; Kang & Steel, 2016):

- Not to strike any agreement with video programmers that would limit programming availability on online video distributors (OVDs) like Netflix and Amazon Prime Video for seven years

- Not to use usage-based pricing and enforce data caps on broadband users for seven years

- Not to charge OVDs interconnection fees for seven years

- To expand availability of high-speed Internet service to another two million homes

- To offer a low-cost broadband service option to low-income households

The seismic consolidations are likely to continue, but companies ripe for takeover are shrinking. Some companies are exploring more expansion abroad, with Comcast offering $31 billion for satellite service Sky (Farrell, 2018b). Normal consolidation will likely continue between smaller-size operators, but only once the giants sort it out.

Cord-Cutting

By 2018, there was no doubt that cord-cutting (i.e., cancelling MVPD service in favor of alternatives—often OTT services) has adversely affected the number of MPVP subscriptions and associated revenues. A 2015 online survey from Forrester revealed that 24% of surveyed U.S. adults did not pay for cable, which included 18% of cord-nevers and 6% of cord-cutters (Lynch, 2015). More worrisome for the cable industry was the projection that only 50% of consumers aged 18 to 31 would subscribe to cable by 2025. In the more recent Lendu poll conducted in December 2017, 56% of the respondents reported that they did not think they would still be cable subscribers in five years (Lafayette, 2018a).

From Table 7.5. it is clear that the annual subscriber losses due to cord-cutting are substantial for traditional MVPDs. These operators shed more than an estimated 6 million subscribers in the last five years, and there are few signs of abatement for 2018. One writer even compared the evolution of cord-cutting to the continuous gradual decline of broadcast network audiences: "Each year the networks argued to advertisers, that at some point, the ratings decline would begin to stabilize" (Adgate, 2017).

The main reasons for cutting the cord have remained consistently the same over time: cost of MVPD service (factor mentioned by 87% of respondents without pay-TV service in late 2017); use of streaming services (40%) and use of an over-the-air antenna for reception of basic channels (23%; TiVo, 2018).

Table 7.5

Net Subscriber Adds and Losses for the Top Pay-TV Providers/MVPDs, 2010-2017[*]

Year	All Top MVPDs	Top Traditional MVPDs	Top Virtual MVPDs
2017	(1,493,245)	(3,092,245)	1,599,000
2016	(758,804)	(1,903,804)	1,145,000
2015	(398,067)	(924,067)	526,000
2014	(120,383)	(163,383)	43,000[**]
2013	(39,615)	(81,615)	42,000[**]
2012	173,192	161,192	12,000[**]
2011	400,310	400,310	NA
2010	569,227	569,227	NA

Note. NA = not available. [*]The top pay-TV providers or multichannel video programming distributors (MVPDs) represent about 95% of the market. [**]Subscriber data for the top Internet-delivered pay-TV providers or virtual multichannel video programming distributors (vMVPDs) from 2012 to 2014 originated from DISH's Sling TV International service.

Source: Leichtman Research Group

The open question remains whether cord-cutting will eventually stop, even heralding a return to traditional MVPD subscription, or whether MVPD service will follow the path of the music or newspaper industries. For pay-TV providers, the options range from bad to catastrophic and may require them to rethink their video revenue model. On the very negative side, The Diffusion Group predicted that 30 million homes—about one third of the 94 million MVPD subscribers in 2017—would go without pay-TV service by 2030 (Baumgartner, 2017d).

It is also possible that viewers decide to re-subscribe to traditional MVPD service after missing the breadth of channels available in the MVPD line-up or being disappointed by the technical quality and content diversity of virtual MVPD service (see Baumgartner, 2018c; Lafayette, 2018b). For the long term, Nielsen has argued that the decline in pay-TV subscription may reverse itself by the changing viewing habits of older Millennials who have children; these cord-never Millennials may decide to become subscribers once they start families (Lafayette, 2016).

Virtual MVPD Service

Virtual multichannel video programming distributors (MVPD) are over-the-top (OTT) providers that offer low-cost and limited channel packages via streaming technology. The most popular vMVPD services include: Sling TV, DirecTV Now, PlayStation Vue,

Hulu with Live TV, and YouTube TV. Monthly prices for packages range from $20 to $75 (Pierce, 2018).

As shown in Table 7.5, vMVPD service has grown to the detriment of traditional MVPD service. For instance, according to Kagan data, DISH Network lost more than 1 million subscribers in 2017 while Sling TV added more than 700,000 subscribers during the same year. But evaluating precisely this degree of cannibalization is nearly an impossible task without questioning the vMVPD subscribers. The Diffusion Group found that 54% of vMVPD subscribers were cord-cutters and only 9% were cord-nevers (Baumgartner, 2018a). UBS Securities predicted that the number of vMVPD subscribers could reach 15 million by 2020 (Farrell, 2016). Some critics, however, have argued that "[b]eing a vMVPD will remain a truly lousy business" (Baumgartner, 2018c) because its average revenue per user (ARPU) is lower and its churn is higher than that of traditional MVPD services (see Farrell, 2018c).

Smart Technology

In 2017, Comcast CEO Brian Roberts proclaimed that broadband was increasingly "the epicenter" of the company's customer relationships, driving bundle strategies and generating more than $20 billion in annual revenue (James, 2017). In recent years, more cable operators have deployed gigabit Internet service using DOCSIS 3.1 and 3.0 technology in their footprints (Asaf, 2018). But while average Internet speeds have continued to increase, it is becoming apparent that residential broadband penetration is nearing its saturation point. According to Leichtman Research Group, 82% of U.S. homes subscribe to broadband service (Frankel, 2017). Thus, the growth of broadband service will slow down in the years to come with direct economic implications for cable operators (Baumgartner, 2017c).

Given this situation, it is not surprising that cable operators are considering alternative revenue streams, especially in the area of smart technology or Internet of Things (IoT). Smart home technology refers to Internet-connected devices used in the home, such as home security systems, thermostats, and lights. Cable operators are beginning to develop strategies to monetize these smart devices. For instance, according to Xfinity Home Senior Vice President and General Manager Daniel Herscovici, "Building partnerships with device makers and stitching them into an integrated ecosystem will help smart home products and services break into the consumer mainstream and create scale" (Baumgartner,

2017a). Using artificial intelligence techniques, Xfinity Home is working on voice-recognition applications to command home devices (Baumgartner, 2018b). (For more on IoT, see Chapter 12.)

Getting a Job

The dramatically changing MVPD marketplace offers an abundance of entry-level jobs and career paths for quick-learning, adaptable applicants. On the technology side, there is a growing demand for engineers and computer programmers working across dozens of initiatives from system security and cloud computing to network design and facilities management. As major mass media companies, MVPDs also offer a multifaceted set of employment opportunities on the business side, including management, marketing, public relations, business analysis, audience measurement, and much more. Opportunities abound for those with entrepreneurial spirit and being part of the television business often supplies a touch of glamour.

Bibliography

Adgate, B. (2017, December 7). Cord cutting is not stopping any time soon. *Forbes*. Retrieved from https://www.forbes.com.

Afflerbach, A, DeHaven, M., Schulhuf, M., & Wirth, E. (2015, January/February). Comparing cable and fiber networks. *Broadband Communities*. Retrieved from http://www.bbcmag.com.

American Television Alliance. (2018, January 9). *Broadcasters shatter TV blackout record in 2017*. Retrieved from http://www.americantelevisionalliance.org.

Asaf, K. (2018, February 20). DOCSIS drives gigabit internet closer to full availability in cable footprints. *Multichannel Trends*. Retrieved from https://platform.mi.spglobal.com.

Baar, A. (2015, October 19). TV Everywhere key to attracting Millennials. *MarketingDaily*. Retrieved from http://www.mediapost.com.

Baldwin, T. F., & McVoy, D. S. (1988). *Cable communication* (2nd ed.). Englewood Cliffs, NJ: Prentice Hall.

Baumgartner, J. (2016, April 27). TiVo: Big four broadcasters top 'must keep' list. *Multichannel News*. Retrieved from http://www.multichannel.com.

Baumgartner, J. (2017a, January 9). Home, where the smart is. *Multichannel News*. Retrieved from http://www.multichannel.com.

Baumgartner, J. (2017b, August 28). Direct-to-consumer models lead multiscreen push. *Broadcasting & Cable*. Retrieved from http://www.broadcastingcable.com

Baumgartner, J. (2017c, November 7). Broadband 'inching toward saturation,' analyst says. *Multichannel News*. Retrieved from http://www.multichannel.com.

Baumgartner, J. (2017d, November 29). 'Legacy' pay TV market to fall to 60% by 2030: Forecast. *Multichannel News*. Retrieved from http://www.multichannel.com.

Baumgartner, J. (2018a, February 12). Virtual MVPDs ended 2017 with 5.3M subs: Study. *Multichannel News*. Retrieved from http://www.multichannel.com.

Baumgartner, J. (2018b, February 26). Comcast builds dedicated voice team as functionality extends beyond TV. *Multichannel News*. Retrieved from http://www.multichannel.com.

Baumgartner, J. (2018c, March 1). Virtual MVPDs growing 'like weeds': Analyst. *Multichannel News*. Retrieved from http://www.multichannel.com.

Baumgartner, J. (2018d, March 12). Study: No 'magic bullet' for the new bundle. *Broadcasting & Cable*. Retrieved from http://www.broadcastingcable.com.

Besen, S. M., & Crandall, R.W. (1981). The deregulation of cable television. *Journal of Law and Contemporary Problems, 44,* 77-124.

Bradshaw, J. (2016, March 1). The skinny on skinny basic TV. *The Globe and Mail*. Retrieved from http://www. theglobeandmail.com.

Bray, H. (2014, April 28). Amid new technologies, TV is at a turning point. *Boston Globe*. Retrieved from http://www. bostonglobe.com.

Broadcasting & Cable. (2016, March 7). *146*(9), 19.

Cable Communications Policy Act of 1984, 47 U.S.C. §551 (2011).

Cable Television Consumer Protection and Competition Act of 1992, 47 U.S.C. §§521-522 (2011).

Cablevision Systems Corp. (2014). *About multi-room DVR*. Retrieved from http://optimum.custhelp.com/app/answers/detail/a_id/2580/kw/dvr/related/1.

Canadian Radio-television and Telecommunications Commission. (2015). *A world of choice* (Broadcasting Regulatory Policy CRTC 2015-96). Retrieved from http://crtc.gc.ca.

Corning, Inc. (2005). *Broadband technology overview white paper: Optical fiber*. Retrieved from http://www.corning.com/docs/opticalfiber/wp6321.pdf.

Crowley, S. J. (2013, October). *Capacity trends in direct broadcast satellite and cable television services*. Paper prepared for the National Association of Broadcasters. Retrieved from http://www.nab.org.

Dreier, T. (2014, March 10). SXSW '14: ESPN 'frustrated and disappointed' by TV Everywhere. *Streaming Media*. Retrieved from http://www.streamingmedia.com.

Dulac, S., & Godwin, J. (2006). Satellite direct-to-home. *Proceedings of the IEEE, 94,* 158-172. doi: 10.1109/JPROC.2006.861026.

Duverney, D. D. (1985). Implications of the 1983 regional administrative radio conference on direct broadcast satellite services: A building block for WARC-85. *Maryland Journal of International Law & Trade, 9,* 117-134.

Eggerton, J. (2015, December 17). FCC's Wheeler: MVPD redefinition still on 'pause.' *Broadcasting & Cable*. Retrieved from http://www.broadcastingcable.com.

Eggerton, J. (2016, April 25). FCC proposes Charter-Time Warner Cable merger conditions. *Broadcasting & Cable*. Retrieved from http://www.broadcastingcable.com.

Farrell, M. (2016, September 20). Study: Virtual MVPDs could lure 15M subs by 2020. *Multichannel News*. Retrieved from http://www.multichannel. com.

Farrell, M. (2017, December 14). Disney pulls Fox trigger 20th Century Fox studios, FX, NatGeo, regional sports nets and more enter Magic Kingdom. *Multichannel News*. Retrieved from http://www.multichannel.com.

Farrell, M. (2018a, February 21). Sling TV ends year with 2.2M subscribers. *Multichannel News*. Retrieved from http://www.multichannel. com.

Farrell, M. (2018b, March 1). Burke: Comcast has long had its eye on Sky. *Broadcasting & Cable*. Retrieved from http://www.broadcastingcable.com.

Farrell, M. (2018c, March 5). Small dish, deep decline. *Multichannel News*. Retrieved from http://www.multichannel. com.

Federal Communications Commission. (1965). Rules re microwave-served CATV (*First Report and Order*), 38 FCC 683.

Federal Communications Commission. (1966). CATV (*Second Report and Order*), 2 FCC2d 725.

Federal Communications Commission. (1969). Commission's rules and regulations relative to community antenna television systems (*First Report and Order*), 20 FCC2d 201.

Federal Communications Commission. (1970). CATV (*Memorandum Opinion and Order*), 23 FCC2d 825.

Federal Communications Commission. (1972a). Commission's rules and regulations relative to community antenna television systems (*Cable Television Report and Order*), 36 FCC2d 143.

Federal Communications Commission. (1972b). Establishment of domestic communications- satellite facilities by non-governmental entities (*Second Report and Order*), 35 FCC2d 844.

Federal Communications Commission. (1980). *Notice of Inquiry*, 45 F.R. 72719.

Federal Communications Commission. (1982). Inquiry into the development of regulatory policy in regard to direct broadcast satellites for the period following the 1983 Regional Administrative Radio Conference (*Report and Order*), 90 FCC2d 676.

Federal Communications Commission. (2010). Sky Angel U.S., LLC emergency petition for temporary standstill (*Order*), 25 FCCR 3879.

Federal Communications Commission. (2013a). *Definitions*, 47 CFR 76.1902.

Federal Communications Commission. (2013b). *Report on Cable Industry Prices*, 28 FCCR 9857.

Federal Communications Commission. (2014a). Promoting innovation and competition in the provision of multichannel video programming distribution services (*Notice of Proposed Rulemaking*), 29 FCCR 15995.

Federal Communications Commission. (2014b). Amendment of the Commission rules related to retransmission consent (*Report and Order and Further Notice of Proposed Rulemaking*), 29 FCCR 3351.

Federal Communications Commission. (2015a). Annual assessment of the status of competition in the market for the delivery of video programming (*Sixteenth Report*), 30 FCCR 3253.

Federal Communications Commission. (2015b). Implementation of Section 103 of the STELA Reauthorization Act of 2014: Totality of the circumstances test (*Notice of Proposed Rulemaking*), 30 FCCR 10327.

Federal Communications Commission. (2016). *Television broadcast stations on satellite*. Retrieved from https://www.fcc.gov.

Federal Communications Commission. (2017). Annual assessment of the status of competition in the market for the delivery of video programming (*Eighteenth Report*), 32 FCCR 568.

Fowler, G. A. (2015, January 26). Sling TV: A giant step from cable. *The Wall Street Journal*. Retrieved from http://online.wsj.com.

Frankel, D. (2017, December 13). Residential internet penetration has grown only 1% since 2012, LRG says. *FierceCable*. Retrieved from https://www.fiercecable.com.

GfK. (2015, December 16). Over half of viewers in pay TV homes have used "TV Everywhere" services—up from 2012 (*Press Release*). Retrieved from http://www.gfk.com.

Hagey, K., & Ramachandran, S. (2014, October 9). Pay TV's new worry: 'Shaving' the cord. *The Wall Street Journal*. Retrieved from http://www.wsj.com.

Holloway, D. (2014, March 19). Next TV: Hulu sale uncertainty swayed CBS' SVOD deals. *Broadcasting & Cable*. Retrieved from http://www.broadcastingcable.com.

Home Box Office v. FCC, 567 F.2d 9 (D.C. Cir. 1977).

Hoskins, C., McFadyen, S., & Finn, A. (2004). *Media economics: Applying economics to new and traditional media*. Thousand Oaks, CA: Sage.

Huddleston, T. (2015, August 3). Here's AT&T's first TV plan after buying DirecTV. *Fortune*. Retrieved from http://fortune.com.

James, M. (2015, August 14). Consumers want fewer TV channels and lower monthly bills – will 'skinny' packages work? *Los Angeles Times*. Retrieved from http://www.latimes.com.

James, S. B. (2017, October 26). Broadband the 'epicenter' of Comcast customer relationships, CEO says. *S&P Global Market Intelligence*. Retrieved from https://platform.mi.spglobal.com.

Jessell, H. A. (2017, September 29). Retrans saved local TV, now what? *TVNewsCheck*. Retrieved from http://www.tvnewscheck.com.

Kang, C., & Steel, E. (2016, April 25). Regulators approve Charter Communications deal for Time Warner Cable. *The New York Times*. Retrieved from http://www.nytimes.com.

Lafayette, J. (2014, April 28). Young viewers streaming more, pivot study says. *Broadcasting and Cable*. Retrieved from http://www.broadcastingcable.com.

Lafayette, J. (2016, March 28). Nielsen: Pay TV subs could stabilize. *Broadcasting and Cable*, 146(12), 4.

Lafayette, J. (2018a, January 29). Cord cutters say they're saving money, poll says. *Broadcasting and Cable*. Retrieved from http://www.broadcastingcable.com.

Lafayette, J. (2018b, March 14). Cord cutters not returning to pay TV, TiVo Q4 study finds. *Broadcasting and Cable*. Retrieved from http://www.broadcastingcable.com.

Leichtman Research Group. (2015, 3Q). 83% of U.S. households subscribe to a pay-TV service. *Research Notes*. Retrieved from http://www.leichtmanresearch.com.

Leichtman Research Group. (2016, 1Q). DVRs leveling off at about half of all TV households. *Research Notes*. Retrieved from http://www.leichtmanresearch.com.

Lenoir, T. (2014, January 14). High retrans stakes for multichannel operators in 2014. *Multichannel Market Trends*. Retrieved from https://platform.mi.spglobal.com.

Lenoir, T. (2018, March 15). Triple-play cable subs rebound in Q4'17, softening video decline. *Multichannel Trends*. Retrieved from https://platform.mi.spglobal.com.

Littleton, C. (2016, June 21). Altice completes Cablevision acquisition, creating no. 4 U.S. cable operator. *Variety*. Retrieved from http://variety.com.

Lynch, J. (2015, October 6). New study says by 2025, half of consumers under 32 won't pay for cable. *Adweek*. Retrieved from http://www.adweek.com.

McCord, L. (2014, April 29). Study: 61% of frequent binge-viewers millennials. *Broadcasting & Cable*. Retrieved from http://www.broadcastingcable.com.

National Cable and Telecommunications Association. (2014). *Cable's story*. Retrieved from https://www.ncta.com.

National Conference of State Legislatures. (2014). *Statewide video franchising statutes*. Retrieved from http://www.ncsl.org/ research/ telecommunications-and-information-technology/statewide-video-franchising-statutes.aspx.

Nielsen. (2017, 2Q). *The Nielsen total audience report*. Retrieved from http://www.nielsen.com.

Nocera, J. (2007, November 24). Bland menu if cable goes a la carte. *The New York Times*. Retrieved from http://www. nytimes.com.

Owen, D. (1985, June). Satellite television. *The Atlantic Monthly*, 45-62.

Parker, J. G. (2011). Statewide cable franchising: Expand nationwide or cut the cord? *Federal Communications Law Journal, 64*, 199-222.

Parsons, P. (2008). *Blue skies: A history of cable television*. Philadelphia: Temple University Press.

Parsons, P. R., & Frieden, R. M. (1998). *The cable and satellite television industries*. Boston: Allyn and Bacon.

Pierce, D. (2018, February 14). Why you should cut cable—and what you'll miss. *The Wall Street Journal*. Retrieved from http://online.wsj.com.

Prospero, M. (2017, May 26). Dish Hopper 3 review: The best just keeps getting better. *Tom's Guide*. Retrieved from https://www.tomsguide.com/us/dish-hopper-3,review-3544.html.

Robson, S. (2018, January 12). Cable networks are losing subscribers as viewers change habits. *Economics of Networks*. Retrieved from https://platform.mi.spglobal.com.

Satellite Home Viewer Act. (1988). Pub. L. No. 100-667, 102 Stat. 3949 (codified at scattered sections of 17 U.S.C.).

Satellite Home Viewer Improvement Act. (1999). Pub. L. No. 106-113, 113 Stat. 1501 (codified at scattered sections in 17 and 47 U.S.C.).

STELA Reauthorization Act of 2014 (STELAR). 47 U.S.C. § 325(b)(3)(C).

TDG. (2014). How would you rank your experience with the 'TV Everywhere' authentication process? *Statista*. Retrieved http://www.statista.com.

Telecommunications Act. (1996). Pub. L. No. 104-104, 110 Stat. 56 (codified at scattered sections in 15 and 47 U.S.C.).

TiVo. (2018). *Q4 2017 Video trends report*. Retrieved from http://blog.tivo.com.

U.S. v. AT&T, Inc. (2017). Case 1:17-cv-02511. Retrieved from http://www.justice.gov.

U.S. v. Southwestern Cable Co., 392 U.S. 157 (1968).

Williams, J. (2016, March 1). Feeling the churn: A new model for OTT. *SNL Financial*. Retrieved from https://plat-form.mi.spglobal.com.

Winslow, G. (2014, December 8). Operators look for TV Everywhere to live up to its name. *Broadcasting & Cable, 144*(44), 8-9.

8

Radio & Digital Audio

Heidi D. Blossom, Ph.D.*

Overview

Radio is more significant today than it was one hundred years ago. More people are reached every week through radio than any other medium and new forms of radio distribution are making this medium relevant to different generations through streaming technologies. Digital Music is a growing industry as the music industry taps into the digital revolution with subscription-based streaming services. Services such as Spotify and Apple Music are growing exponentially by offering users personalized on-demand and live-streaming experiences. Ultimately, the radio industry is still going strong and the technological advances in digital audio make listening to music and other programming even more convenient and enjoyable.

Introduction

Radio reaches more Americans each week than any other media. In the U.S. alone, there are 11,383 commercial AM/FM stations that reach 93% of the U.S. population every week compared to 85% who watch TV each week. Add to that the 61% of Americans who listen to the tens of thousands of online radio stations available and you have a radio and digital audio powerhouse (Nielsen, 2017). Radio's relationship with its audience has evolved as radio reinvents itself every couple decades to respond to technological advances. New forms of digital media delivery have disrupted the radio industry, further fragmenting the audience.

A major jump occurred in the later part of the 20th century with the introduction of Internet streaming. The number of people tuning in to hear their favorite radio programming online has exponentially increased, forcing the radio industry to once again look at a makeover. Monthly online radio listening went from 5% of the U.S. population in 2000 to 61% in 2017 (Statistia, 2018). The jump occurred as more people began listening on their smart phones rather than through desktop computers. In 2018, 73% of American adults 18+ listened to streaming radio via smartphones. That was an increase of 7% in only one year, while laptop and desktop listening decreased by 6% over the same period (Vogt, 2017). In the 1950's radio went mobile with the invention of the transistor, which shrunk the size of radios allowing people to carry them in their pockets. Digital radio is analogous in that streaming technology is delivered through your mobile device that also fits right into your pocket.

Digital audio is growing exponentially as consumers subscribe to streaming music services such as Spotify, Apple Music, and Pandora. Streaming services have immense brand awareness and usage among a

---

* Chair, Dept. of Mass Communication, North Greenville University (Greenville, South Carolina).
The author would like to thank Rick Sparks, Ph.D. for his contribution to the copyright and legal issues provided for in this chapter.

younger audience, with 85% of 13-15-year-olds streaming music on mobile devices. The radio industry is taking notice as it fights to maintain audience control of the lucrative younger demographics (IFPI, 2017).

Figure 8.1
Radio Studio

Source: HisRadio

Given the 76% market penetration of smartphones among all U.S. cellphone users under the age of 55, it is no wonder that radio listeners are using their mobile devices to expand their choices of radio beyond traditional terrestrial radio (Edison, 2017). While streaming services continue to draw a younger population of listeners, both traditional and online radio remain relevant and vibrant.

Background

"Mary had a little lamb, its fleece was white as snow"

These were the first words ever recorded, launching a global audio recording revolution. Little did Thomas Alva Edison realize what he had invented when he discovered that sound could be recorded and played back. While the audio quality on the first phonograph was barely recognizable, the phonograph was a true media innovation. Recording technology brought music, theater, politics and education into the living rooms of thousands, but by the 1920s the recorded music industry was in decline, mostly due to WWI and the invention of radio, which had hit its stride by the end of the 1920s (Lule, 2014).

Radio

No one expected radio to revolutionize the world, but more than 100 years ago that is exactly what happened. On March 1, 1893 Nikola Tesla, the father of modern communication technology, gave his first public demonstration of radio at the National Electric Light Association in St. Louis, paving the way to further innovations and uses of his new technology (Cheney, 2011). Tesla held more than 700 patents, including wireless communications, cellular communications, robotics, remote control, radar, and many other communication technologies that we still use today. Although many have credited Guglielmo Marconi with the invention of radio, Marconi combined the inventions of Tesla and others to advance already established radio technologies. In 1943 the U.S. Patent office restored credit for the original radio patents back to Nikola Tesla (Coe, 1996).

Figure 8.2
Nikola Tesla

Source: Public Domain

The first radio transmission of actual sound came on December 24, 1906 from Ocean Bluff-Brant Rock, Massachusetts when Reginald Fessenden played *O Holy Night* on his violin and then read a passage from the Bible to wireless operators on ships at sea. The ships had been advised to be listening at 9pm on that Christmas Eve and they were amazed to hear music and voices—something that had never been heard before (O'Neal, 2006). That first live broadcast laid the foundation of the live music and programming format. The first radio news program began in 1920 in Detroit at WWJ, which remains an all-news radio station to this day (Douglas, 2004).

After a ban on radio during World War I, department stores, universities, churches, and entrepreneurs started their own radio stations in the early 1920s. In just four years' time the number of commercial radio stations in the U.S. rose from 34 to 378 stations (Scott, 2008). By the 1930s, radio was in its Golden Age, exerting influence around the world and marking its prominence in mass media. Radio captivated America with live drama, comedy, and music shows that challenged the imagination. By the end of the 1930s, more than 80% of Americans owned a radio (Craig, 2006). Radio connected individuals to the world and allowed listeners to have a front-row seat to the boxing match, hear first-hand the news from around the world, and experience politics and culture like never before—live. It was a powerful mass medium and the only communication technology that was accessible to most, which continues today with the vast majority of Americans listening to the radio each week (Nielsen, 2017).

One of the main reasons radio was so successful through the Great Depression of the 1930s was because there was no competition that compared to radio until the introduction of television. By 1949, television's popularity had risen to a level that caused many to predict the demise of radio. Television eventually replaced radio's prime-time dominance as many popular radio shows moved to television; regardless, radio penetrated places of isolation that television could not go, namely, the automobile. Nevertheless, radio was able to maintain a captive audience as the industry adjusted by shifting its programming to what it still does best—music (Cox, 2002). By the 1950s, the radio DJ was born, and America's Top 40 format ruled the airways (Brewster, 2014).

During the counterculture of the 1960s, young people looked to music to speak to the culture of the day. They began to reject the over-commercialization of the AM dial and turned to FM radio for better sound quality and less commercialization. Radio was

their source for new music. By the 1970s, FM technology was widespread as audiences sought out better reception and high-fidelity stereo sound (Lule, 2014). It was that shift of purpose for radio that redefined it. Consumers wanted quality-sounding music, and radio was their free, go-to source.

At the same time that FM radio was surging, digital audio technology was being developed. By the late 1980s and 1990s, Compact Discs (CD) came into mainstream use, ending the long standing analog era of music recordings. In the meantime, radio technology languished, with no real innovations until Hybrid Digital (HD) radio nearly two decades later. To understand this part of radio's history, it is important to know the difference between analog and digital signals.

Analog vs. Digital

The difference between analog and digital audio is in how the audio is reproduced. Sound is represented as a wave. A radio wave is an electromagnetic wave transmitted by an antenna. Radio waves have different frequencies; by tuning a radio receiver to a specific frequency, listeners can pick up that distinct signal.

Analog recordings make an equivalent physical representation of sound waves on a physical object such as a disc (record) or magnetic tape (cassette tape) (Elsea, 1996). When played back, the audio player reads the physical information on the disc or tape and converts the patterns to sound we can hear. For radio, similar principles apply. Sound wave energy is converted into electrical patterns that imitate a sound wave. These electrical patterns are then superimposed on a carrier wave and transmitted to receivers (radios), which convert the electrical patterns into music, speech or other sounds we can enjoy. While the quality of sound has been refined over the years, one of the marked innovations in the transmission of radio waves was in the split of the waves to create a stereo sound (McGregor, Driscoll, & McDowell, 2010). For the physical radio (hardware), shrinking the size of radios allowed portability, and related developments refined the sound quality (Scannell, 2010).

Digital recordings, on the other hand, are made by taking thousands of "samples" of an analog signal, with a number for each sample representing the size or strength of the signal at that moment. The thousands of samples can be put together to approximate the original wave with the "stairstep" pattern in Figure 8.3 representing what happens when you put these digital samples together to recreate the analog wave. The more samples you take, the smoother the wave—and the more accurate the sound reproduction. The numbers created for digital audio are created through binary code, or a series of 0s and 1s which creates a code that instructs the digital player on what to play (Hass, 2013).

Figure 8.3
Radio Sound Waves

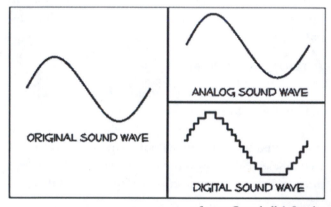

Source: Campbell & Sparks

In the early 1980s, digital audio hit mainstream in the form of the compact disc (CD), which was marketed as small ("compact") and virtually indestructible with near perfect sound. (Sterling & Kittross, 2001). This compared to the cassette tape that disintegrated in heat, or to vinyl records that could scratch, break, or warp. Consumers quickly embraced CD technology as prices dropped and the music industry transitioned to a digital era in CD music production. The public was hooked, and the radio industry had to adjust to higher expectations from consumers who demanded clearer sound. However, converting analog radio to digital radio would be a complex process that would take decades to accomplish.

There were several issues the radio industry had to overcome in the transition from analog to digital. In the 1980s the cost of storing and broadcasting digital signals was exorbitant. In 1983, 1MB of storage cost $319 and would hold about one quarter of one song (Komorowski, 2014). The speed at which audio could

be broadcast was so slow that it was inconceivable to move to a digital broadcast system. In order for digital technology to be practical, the size of the files had to be decreased, and the bandwidth had to be increased.

Digital audio technology advanced with the introduction of codecs that shrunk digital audio into sizes that were more manageable and enabled faster transmission of the audio data. The compression of audio is important for digital radio broadcasting because it affects the quality of the audio delivered on the digital signal. The codec that emerged as a standard was the MPEG-1 Audio Layer 3, more commonly known as MP3. MP3 compressed the audio to about one-tenth the size without perceptibly altering the quality of the sound for FM listeners (Sellars, 2000). While the MP3 standard compresses audio data to a lower audio standard than was being used before digital audio was standardized, the basic technology of binary code and audio compression were being developed for digital radio broadcasting.

The Shift to Digital Audio Broadcasting

European broadcasters beat American broadcasters in adopting digital transmission with their development and adoption of the Eureka 147 DAB codec, otherwise known as MP2 (O'Neill & Shaw, 2010). The National Association of Broadcasters (NAB) was very interested in Eureka 147 DAB in 1991; however, broadcasters eventually lobbied for the proprietary iBiquity IBOC codec designed to deliver a high-quality audio with lower bit rates (or bandwidth) than the Eureka system.

Figure 8.4
HD Radio™ Receiver

Source: Manimecker

iBiquity's proprietary In-band On-channel (IBOC) digital radio technology layers the digital and analog signals, allowing radio broadcasters to keep their assigned frequencies and simply add the digital conversion hardware to upgrade their current broadcast systems (FCC, 2004). Known as HD Radio (Hybrid Digital), this technology also allows for metadata transmission including station identification, song titles and artist information, as well as advanced broadcasting of services such as breaking news, sports, weather, traffic and emergency alerts. The IBOC technology also allows for tagging services allowing listeners to tag music they want to purchase. Ultimately, this technology gives more control to the consumer and engages them in a way that traditional analog radio could not (Hoeg & Lauterbach, 2004).

The FCC approved iBiquity's HD Radio system of digital radio delivery in 2002, and radio stations began broadcasting the HD signals in 2004, delivering CD quality audio for both AM and FM frequencies (HDRadio, 2004). HD radio is free to listeners but requires users to purchase an HD ready radio.

HD radio also opens up the amount of spectrum available to broadcasters, allowing single radio stations to split their signal into multiple digital channels offering more options in programming, otherwise known as multicasting. For example, a radio station may primarily be a country music formatted station (95.5 FM) but offer a news-talk station on the split frequency (95.5-2) and a Spanish language music station on a third channel (95.5-3) (HD Radio, 2004).

While alternative music delivery systems and the lack of consumer knowledge about HD Radio is a barrier to this technology succeeding, the major barrier is found in the automotive industry. In 2015, the audio technology company DTS acquired iBiquity with promises to expand the HD radio market; however, they had a major setback with General Motors cutting the feature from a number of models. While HD Radio is not expected to fail anytime soon, if HD Radio is not standardized in vehicles, this radio technology will not survive.

Radio Automation

Radio automation is the use of broadcast programming software to automate delivery of pre-recorded programming elements such as music, jingles, commercials, announcer segments, and other programming elements. Radio automation is used in both traditional terrestrial radio broadcasting as well as streaming radio stations on the web.

Radio automation also allows for better management of personnel allowing announcers to pre-record their segments in a way that sounds like they are live on the air when in reality they may have recorded the segment hours or even days before. Radio personalities such as John Tesh or Delilah, whose nationally syndicated radio shows draw in millions of listeners, pre-record their programs in a way that sounds like they are live on the air.

In an effort to better allocate resources, radio stations and networks across the globe have moved to automated operations.

Recent Developments
Revenue Boom: $1.1 Billion

In 2017, digital audio was the fastest growing advertising format, representing a 42% increase in just one year, which surpasses the growth rate of search and digital video advertising. Radio station digital revenues were up 13.4% in 2017. In fact, 85% of radio advertisers bought some form of digital advertising; however, digital audio ad revenues account for only 2% of the $1.1 billion in digital advertising revenue. With more unlimited data plans being offered from cellular companies, revenues are expected to grow as more consumers tune to digital audio on their mobile devices (IAB, 2017).

The Streaming Music Revolution

Top streaming services earn their money through subscriptions and the subscriber numbers are increasing. In 2017, revenues from streaming music services such as Spotify, Apple, Amazon, and Google represented a 48% jump to $2.5 billion in just one year (RIAA, 2017). The leader in streaming music, Spotify, boasted more than 70 million paid subscribers globally in 2017, and Apple hit 30 million subscribers at the same time.

While CD sales plummeted, the music industry saw a 5.9% growth in 2016 due to the continued growth of streaming music. Consumers are buying into streaming to access more than 40 million songs; the widest diversity of music offering in history (IFPI, 2017). As older music technologies die out, digital music technologies will continue to grow and innovate.

Figure 8.5
Ways to Listen to Digital Audio

Source: Spotify

On-Demand vs. Live-Streaming

On-demand music streaming services such as Spotify and Google Play Music allow users to specify a particular song or artist and immediately hear it. Much like the old jukebox, on-demand services allow listeners to build a playlist with the exact songs and artists they would like to hear. On-demand also offers the uploading of owned music to a "locker room" which stores that music online, allowing access to that music even if the service doesn't offer those song choices or artists. On-demand music streaming also offers an offline listening experience, meaning users can tag their favorite songs and listen to them whenever they like. The only limitation is the storage capacity of the smartphone or music playing devices being used.

Live-streaming is the real-time transmission of music or programming over the Internet. This is often done through automation software that creates a stream of pre-recorded music, but there is a trend of transmitting actual live audio or video feeds to users. Streaming companies such as Pandora are music services that allow the user to create a personalized channel of music based on songs, artists, music genres or even moods of music they like.

These streaming services use an algorithm of listeners' indications of what they "like," "don't like," or "skip." The service then selects other songs or artists that are similar in style to the ones specified. Just like on-demand music services, there are no DJs on live-streaming. The only speaking voices users hear is the occasional company identification and advertising announcements. "Premium" users can pay a subscription fee and eliminate advertising on their streaming service for uninterrupted music and can often download music for offline listening.

Podcasting

Podcasting continues to grow in popularity as more than 24% of Americans claimed to have listened to a podcast in the last month (Edison, 2017). Podcasts are digital audio recordings of music or talk, available for download onto a mobile device or computer. Podcasts are typically produced as a series which is available as individual downloads or automatic subscription downloads. Radio broadcasts converted to podcasts are some of the most popular, with radio programs such as *This American Life*, *TED Radio Hour*, and *Fresh Air*, popular NPR radio shows leading the charge of weekly shows converted to podcasts. *This American Life*, an hour long themed journalistic, informational, and sometimes comedic radio program, had more than five million downloads every month of their popular podcast (PODTRAC, 2017).

Podcasts can be listened to when an individual has the time, making podcasting a convenience-medium for those who listen to them. Anyone with a computer or mobile device and an Internet connection can create a Podcast. In 2016, 4.6 billion podcasts were downloaded from the podcast hosting service, Libsyn, which was an increase of 2 billion downloads in just two years. The number of listeners is the key statistic to podcasting, and that number is slowly growing. Pew research reports that in 2016, 40% of Americans had listened to a podcast and that number was up by eleven percent from the year before (Vogt, 2017).

Copyright Royalties

Since 2005, the Copyright Royalty Board (CRB), comprised of three U.S. judges, has set rates and terms for copyright statutory licenses, including music licensing. A significant part of determining webcasting royalties for streaming music was the CRB's designation of record labels as the entity of "willing seller" of licensing, with broadcast and webradio as the "willing buyers" (Cheng, 2007). This designation gives the music industry a powerful role in lobbying for ever-higher royalty rates for both radio broadcasting and online streaming music services. In 2015, the CRB substantially increased streaming music rates for the period 2016-2020, when the *Webcaster Settlement Act* expired. Most small webcasters could not survive, with the outcome being what one veteran webcaster called "Bloody Sunday" with the massive expiration of tens of thousands of webstations (Hill, 2016).

Music Artists Say 'No more!'

While many radio professionals and hobbyists accuse the music industry of a money grab by its aggressive enforcement of increased licensing and royalty rates for web radio, a different perspective is represented by Taylor Swift's famous 2014 action of not allowing Spotify to stream her then-new album, *1989*. Swift's label asked Spotify to add some conditions to the streaming service: 1) to listen for free, the consumer must be outside the U.S., and 2) if the consumer was based in the U.S., the consumer must pay a subscription fee to listen to the music. After Spotify refused to budge on both demands, Swift's music was pulled from the service and wasn't heard (on that platform) until June 2017. Taylor Swift continues to limit releases on streaming services to help drive fans to pay to download her albums on buy them on CD.

In a personal letter to the CEO of Spotify, Swift accused the company of undervaluing the artists they played. According to Swift, while Spotify was reaping huge profits, it gave back relatively little to the musicians who created the music. On average, artists on Spotify earn a fraction of a cent per play, or between $0.006 and $0.0084 (Spotifyartists.com, 2016). But the real amount the artist receives is far less, given that more than 70% of Spotify's revenue goes to the record label, the publishers and distributors (Linshi, 2014).

Swift was emphatic in her opposition: "I'm not willing to contribute my life's work to an experiment that I don't feel fairly compensates the writers, producers, artists, and creators of this music" (Prigg, 2015).

The music industry is finally tapping in to the digital revolution and subscription based streaming services are growing in popularity creating a new track of revenue for the music industry.

Satellite Radio

SiriusXM satellite radio is a subscription-based radio service that uses satellite technology to deliver its programming. Subscribers purchase a radio equipped with a satellite receiver and can listen to their favorite stations wherever they are in the U.S. This service uses geosynchronous satellite technology to deliver its programming, meaning no matter where users are in the U.S., they can hear their favorite SiriusXM station (Keith, 2009). Satellite radio comes standard in most new and many pre-owned cars and trucks but requires a subscription activation. Satellite radio services have been offered since 2001 and have seen slow but steady growth in subscriptions. There were more than 27.5 million satellite radio subscribers in 2010. That figure has steadily grown to more than 32.7 million subscribers in 2017, making satellite radio a real growth industry as radio technology progresses. Most recently, SiriusXM has opened channels in emergency situations for people who do not have subscriptions (SiriusXM, 2018).

Shifts in ownership in the radio industry are as frequent as shifts in music tastes among the public. For example, Sirius and XM started as competing services before merging in 2008. Then in 2017 Sirius/XM purchased a 19% share in Pandora. As this book is going to press, the finances of the two largest radio station owners in the U.S. are in question, and additional ownership changes in radio are likely before you read this paragraph.

Figure 8.6

SiriusXM In-Dash Satellite Radio

Source: SiriusXM

Factors to Watch

FM Radio in Smartphones

Most smartphones today are equipped with an FM radio receiver already built into the phone; however, at the time this was written, most cellphone carriers or manufacturers had not activated the chip. While smartphones have access to Internet radio streaming, using the FM radio receiver in your smartphone uses less battery power than streaming and also does not use cellular data. When cell towers are down, this chip would still receive FM radio signal, making it an important device during emergency situations. The FCC has urged cellphone manufacturers to activate the feature; however, it is not yet mandatory in the United States (FCC, 2017). The National Association of Broadcasters has also lobbied for all cellphone manufacturers and cellular carriers to enable usage of the FM chip in their phones. NAB Labs has created an app that users can download which allows users to tune to local radio stations using their smartphone's unlocked built-in FM radio chip (NextRadio, 2018).

Radio Turns to Video

Most radio stations have an interactive website that connects with social media and includes, of all things, video. Many prime time radio shows offer live video streamed from the radio studio for people to watch online. Formats such as news-talk stations populate their websites with local news videos to attract a video-craved audience, while music driven formats feature their own radio personalities interviewing popular artists or relating video human interest stories which connect with their audience. Ultimately, many stations are trying to brand themselves as an entertainment package complete with video, audio, and social media.

In an effort to engage more with the audience, many radio stations have live video streaming on social media. Streaming on Facebook Live or YouTube Live allows stations to connect with their audiences to enhance their radio programming and visually engage the audience by connecting them with celebrity guest interviews, contests, or connecting with concerts or community events at remote broadcasts away from the radio station (Zarecki, 2017).

Localism

With an overabundance of music and programming choices available to the digital media consumer, there is still one thing that radio offers that digital algorithms cannot satiate, and that is connectedness. While people may gravitate towards digital audio content tailored to their current whim, local radio has the opportunity to draw the audience in with content that connects them with their community.

The radio industry is still going strong with 271 million Americans listening to radio every week (Nielsen, 2017). With the technological advances in digital audio and the many options available to listen to music and programming, radio is still the go-to source for listening pleasure; however, the radio industry cannot be complacent. If the radio industry is to survive, it must continue to innovate and adapt to the technology of today and the innovations of tomorrow.

Career Opportunities in Radio & Digital Audio:

- Announcer
- Audio Producer
- News Reporter
- Programmer
- Promotions/Marketing
- Recording Artist
- Sales Associate
- Social Media Manager
- Sound Engineer
- Videographer

Getting a Job

The radio and digital audio industries remain vibrant, even though consolidation has reduced the number of media entities. Positions are plentiful in local radio and in digital media career paths.

Most people think of the announcer or DJ as the key role at a radio station, but radio has many facets and needs many types of workers. Most radio professionals are happy to talk to someone interested in their line of work, so you might call up your favorite station and ask for an interview with the manager, program director, or on-air talent about how they got into the business.

While a degree in Mass Communication is not necessary to enter the radio industry, having an educational background in communications, broadcasting, or business is helpful for entry to a broadcast career and generally provides more opportunities for advancement.

Figure 8.7
Radio Announcer

Source: Houston Public Media/KUHF-FM

Bibliography

Brewster, B. & Broughton, F. (2014). Last night a DJ saved my life: The history of the disc jockey. New York: Grove/Atlantic, Inc.

Cheney, M. (2011). Tesla: Man out of time. New York: Simon and Schuster.

Cheng, J. (2007, March 20). NPR fights back, seeks rehearing on Internet radio royalty increases. Ars Technica. Retrieved from http://arstechnica.com

Coe, L. (1996). Wireless radio: A brief history. McFarland. Newington, Connecticut: American.

Cox, J. (2002). Say goodnight, Gracie: The last years of network radio. Jefferson, NC: McFarland.

Craig, S. (2006). The More They Listen, the More They Buy: Radio and the Modernizing of Rural America, 1930-1939. Agricultural history, 1-16.

Douglas, S. J. (2004). Listening in: Radio and the American imagination. Minneapolis: U of Minnesota Press.

Edison Research (2017). The infinite dial 2017. Retrieved from http://www.edisonresearch.com

Elsea, P. (1996). University of California Santa Cruz Technical Essays: Analog recording of sound. Retrieved from http://artsites.ucsc.edu/EMS/music/tech_background/te-19/teces_19.html

FCC (2004) Digital Audio Broadcasting Systems and Their Impact on the Terrestrial Radio Broadcast Service: Further Notice of Proposed Rulemaking and Notice of Inquiry. MM Docket No. 99-325. Washington, DC

FCC (2017) Chairman Pai Urges Apple to Activate FM Chips to Promote Public Safety.) Retrieved from https://www.fcc.gov/document/chairman-pai-urges-apple-activate-fm-chips-promote-public-safety

HD Radio (2004) Retrieved from http://hdradio.com/us-regulatory/fcc_approval_process

Hass, J. (2013) Introduction to Computer Music: Volume One. Indiana University.

Hill, B. (2016, February 15). CRB full decision released; small webcasters not mentioned. Radio & Internet News (RAIN). Retrieved from http://www.rainnews.com

Hoeg, W., & Lauterbach, T. (Eds.). (2004). Digital audio broadcasting: principles and applications of digital radio. West Sussex, UK: John Wiley & Sons.

IAB (2017, December). IAB internet advertising revenue report. Retrieved from https://www.iab.com/wp-content/uploads/2017/12/IAB-Internet-Ad-Revenue-Report-Half-Year-2017-REPORT.pdf

IFPI (2017, September) ifpi Connecting With Music: Music Consumers Insight Report. Retrieved from http://www.ifpi.org/downloads/Music-Consumer-Insight-Report-2017.pdf

Keith, M. C. (2009). *The Radio Station: broadcast, satellite & Internet*. Burlington, MA: Focal Press.

Komorowski, M. (2014) A history of storage cost. Retrieved from http://www.mkomo.com/cost-per-gigabyte

Linshi, J. (2014, November 3). Here's why Taylor Swift pulled her music from Spotify. Time. Retrieved from http://time.com

Lule, J. (2014). Understanding Media and Culture: An Introduction to Mass Communication New York: Flat World Education.

McGregor, M., Driscoll, P., & McDowell, W. (2010). Head's Broadcasting in America. Boston: Allyn & Bacon.

NextRadio (2018) What is NextRadio? Retrieved from http://nextradioapp.com/

Nielsen (2017, June). State of the media: Audio today. Retrieved from http://www.nielsen.com/us/en/insights/reports/2017/state-of-the-media-audio-today-2017.html

O'Neal, J. E., (2006) Fessenden: World's First Broadcaster? Radio World. New York, NY Retrieved, from http://www.radioworld.com/article/fessenden-world39s-first-broadcaster/15157

O'Neill, B., & Shaw, H. (2010). Digital Radio in Europe: Technologies, Industries and Cultures. Bristol, U.K.: Intellect.

PODTRAC (2017). Podcast Industry Audience Rankings. Retrieved from http://analytics.podtrac.com/industry-rankings/

Prigg, M. (2015, August 4). Taylor Swift really won't shake it off: Singer hits out at Spotify. Retrieved from http://www.dailymail.co.uk

RIAA (2017). News and Notes on 2017 Mid-Year RIAA Revenue Statistics. Retrieved from https://www.riaa.com/wp-content/uploads/2017/09/RIAA-Mid-Year-2017-News-and-Notes2.pdf

Scannell, P. (2010). The Ontology of Radio. In B. O'Neill, *Digital Radio in Europe*. Chicago: University of Chicago Press.

Scott, C. (2008). "History of the Radio Industry in the United States to 1940". EH.Net, edited by Robert Whaples. March 26, 2008. Retrieved from http://eh.net/encyclopedia/thehistoryoftheradioindustryintheunitedstatesto1940/

Sellars, P. (2000). Perceptual coding: How MP3 compression works. Retrieved from http://www.soundonsound.com/sos/may00/articles/mp3.htm

SiriusXM (2018, January 31) SiriusXM Reports Fourth Quarter and Full-Year 2017 Results, Retrieved from http://investor.siriusxm.com/investor-overview/press-releases/press-release-details/2018/SiriusXM-Reports-Fourth-Quarter-and-Full-Year-2017-Results/default.aspx

Spotifyartists.com (2016, March). How is Spotify contributing to the music business? Retrieved from http://www.spotifyartists.com

Statista (2018) Share of monthly online radio listeners in the United States from 2000 to 2017. Retrieved from https://www.statista.com/statistics/252203/share-of-online-radio-listeners-in-the-us/

Sterling, C. H., & Kittross, J. M. (2001). Stay tuned: A history of American broadcasting. Mahwah, NJ: LEA.

This American Life (2016) About Us. Retrieved from http://www.thisamericanlife.org/about

Vogt, N. (2017, June 16). Audio and Podcasting Fact Sheet. Retrieved from http://www.journalism.org/fact-sheet/audio-and-podcasting/

Zarecki, Tom (2017, March 8). Is Facebook Live Upstaging Your Station? Retrieved from https://radioink.com/2017/03/08/facebook-live-upstaging-station/

9

Digital Signage

Jennifer Meadows, Ph.D.*

Overview

Digital signage is a ubiquitous technology that we seldom notice in our everyday lives. Usually in the form of LCD digital displays with network connectivity, digital signage is used in many market segments including retail, education, banking, corporate, and healthcare. Digital signage has many advantages over traditional print signs including the ability to update quickly, use dayparting, and incorporate multiple media including video, social media, and interactivity. Interactive technologies are increasingly being incorporated into digital signage allowing the signs to deliver customized messages to viewers. New technologies including BLE beacons and LinkRay are being used to communicate with users' smartphones, and digital signs can now collect all sorts of information on those who view them.

Introduction

When thinking about the wide assortment of communication technologies, digital signage probably didn't make your top-10 list. But consider that digital signage is becoming ubiquitous, although many of us go through the day without ever really noticing it. When you hear the term digital signage perhaps the huge outdoor signs that line the Las Vegas Strip or Times Square in New York City come to mind. Digital signage is actually deployed in most fast food restaurants, airports, and on highways. Increasingly it can be found in schools, hospitals, and even vending machines.

The major markets for digital signage are retail, entertainment, transportation, education, hospitality, corporate, and health care. Digital signage is usually used to lower costs, increase sales, get information or wayfinding, merchandising, and to enhance the customer experience (Intel, 2016). Throughout this chapter, examples of these markets and uses will be provided.

The Digital Screenmedia Association defines digital signage as "the use of electronic displays or screens (such as LCD, LED, OLED, plasma or projection, discussed in Chapter 6) to deliver entertainment, information and/or advertisement in public or private spaces, outside of home" (Digital Screenmedia, n.d.). Digital signage is frequently referred to as DOOH or digital outside of the home.

The Digital Signage Federation expands the definition a little to include the way digital signage is networked and defines digital signage as "a network of digital displays that is centrally managed and addressable for targeted information, entertainment, merchandising and advertising" (Digital Signage Federation, n.d.). Either way, the key components to a digital signage system include the display/screen, the content for the display, and some kind of content management system.

---

* Professor and Chair, Department of Media Arts, Design, and Technology, California State University, Chico (Chico, California).

The displays or screens for digital signage can take many forms. LCD including those that use LED technology is the most common type of screen, but you will also see plasma, OLED and projection systems. Screens can be flat or curved, rigid or flexible, and can include touch and gesture interactivity and cameras. Screens can range in size from a tiny digital price tag to a huge stadium scoreboard.

Content for digital signage can also take many forms. Similar to a traditional paper sign, digital signs can contain text and images. This is where the similarity ends, though. Digital signage content can also include video (live and stored, standard, 3D, high definition, 4K, and 8K), animation, augmented reality (AR), RSS feeds, social networking, and interactive features.

Unlike a traditional sign, digital signage content can be changed quickly. If the hours at a university library were changed for spring break, a new traditional sign would have to be created and posted while the old sign was taken down. With a digital sign, the hours can be updated almost instantly.

Another advantage of digital signage is that content can also be delivered in multiple ways besides a static image. The sign can be interactive with a touch screen or gesture control. The interaction can be with a simple touch or multi-touch where the users can swipe, and pinch. For example, restaurants can have digital signage with the menu outside. Users can then scroll through the menu on the large screen just like they would on a smartphone. Digital signs can also include facial recognition technologies that allow the system to recognize the viewer's age and sex, then deliver a custom message based upon who is looking at the sign.

All of these technologies allow digital signage to deliver more targeted messages in a more efficient way, as well as collect data for the entity that deploys the sign. For example, a traditional sign must be reprinted when the message changes, and it can only deliver one message at a time. Digital signage can deliver multiple messages over time, and those messages can be tailored to the audience using techniques such as dayparting.

Dayparting is the practice of dividing the day into segments so a targeted message can be delivered to a target audience at the right time (Dell, 2013). For example, a digital billboard may deliver advertising for coffee during the morning rush hour and messages about prime-time television shows in the evening. Those ads could be longer because drivers are more likely to be stopped or slowed because of traffic. An advertisement for a mobile phone during mid-day could be very short because traffic will be flowing.

Digital signage also allows viewers to interact with signs in new ways such as the touch and gesture examples above and through mobile devices. In some cases, viewers can move content from a sign to their mobile device using near field communication (NFC) technology or Wi-Fi. For example, visitors to a large hospital could interact with a digital wayfinding sign that can send directions via NFC to their phones.

Another advantage with digital signage is social networking integration. For example, retailers can deploy a social media wall within a store that shows feeds from services like Instagram, Twitter, Facebook, and Tumblr.

New communication technologies are bringing change to digital signage. For example, artificial intelligence (AI) technologies might enable a virtual hotel concierge who can answer questions and provide local information. Gamification helps engage customers with brands and IoT (Internet of Things) technology, especially sensors, can be incorporated into digital signage.

All this interactivity brings great benefits to those using digital signs. Audience metrics can be captured such as dwell time (how long a person looks at the sign), where the person looks on the screen (eye tracking), age and sex of the user, what messages were used interactively, and more.

Considering all the advantages of digital signage, why are people still using traditional signs? Although digital signage has many advantages such as being able to deliver multiple forms of content; allowing interactivity, multiple messages, and dayparting; easily changeable content; and viewer metrics; the big disadvantage is cost. Upfront costs for digital signage are much higher than for traditional signs. Over time, the ROI (return on investment) is generally high for digital signage, but for many businesses and organizations this upfront cost is prohibitive. See Table 9.1 for a comparison of traditional and digital signs.

Table 9.1

Paper versus Digital Signage

Traditional/Paper Signs	Digital Signage
● Displays a single message over several weeks	● Displays multiple messages for desired time period
● No audience response available	● Allows audience interactivity
● Changing content requires new sign creation	● Content changed quickly and easily
● Two-dimensional presentation	● Mixed media = text, graphics, video, pictures
● Lower initial costs	● High upfront technology investment costs

Source: Janet Kolodzy

Cost is often a factor in choosing a digital signage system. There are two basic types of systems: premise based and Software as a Service (SaaS). A premise-based system means that the entire digital signage system is in-house. Content creation, delivery, and management are hosted on location. Advantages of premise-based systems include control, customization, and security. In addition, there isn't an ongoing cost for service. Although a doctor's office with one digital sign with a flash drive providing content is an inexpensive option, multiple screen deployments with a premise-based system are generally expensive and are increasingly less popular.

SaaS systems use the Internet to manage and deliver content. Customers subscribe to the service, and the service usually provides a content management system that can be accessed over the Internet. Templates provide customers an easy interface to create sign content, or customers can upload their own content in the many forms described earlier. Users then pay the SaaS provider a regular fee to continue using the service. Both Premises and SaaS systems have advantages and disadvantages and many organizations deploy a mix of both types.

So, to review the technology that makes digital signage work, first there needs to be a screen or display. Next, there should be some kind of media player which connects to the screen. Content can be added to the media player using devices such as flash drives and DVDs, but more commonly the media player resides on a server and content is created, managed, and delivered using a server-on-site or accessed over the Internet.

Digital signage, then, is a growing force in out-of-home signage. Kelsen (2010) describes three major types of digital signage: point of sale, point of wait, and point of transit. These categories are not mutually exclusive or exhaustive but they do provide a good framework for understanding general categories of digital signage use.

Point of Sale digital signage is just what it sounds like—digital signage that sells. The menu board at McDonald's is an example of a point of sale digital sign. A digital sign at a See's Candy store in San Francisco International Airport, shows product images until someone stands in front of it. Then users can interact with the sign to get more information about products and an interactive history of the business (Mottl, 2016). Micro-digital-signage (MDS) allows what is called "Intelligent Shelving." These small digital signs are placed on shelves in retail stores to give price, promotion, and product information using eye catching features such as HDTV (Retail Solution, 2016).

Point of Wait digital signage is placed where people are waiting and there is dwell time, such as a bank, elevator, or doctor's office. One example is the digital signage deployed inside taxis. Point of Wait signage can convey information such a weather, news, or advertising.

Point of Transit digital signs target people on the move. This includes digital signs in transit hubs such as airports and train stations, signage that captures the attention of drivers, like digital billboards, and walkers, such as street facing retail digital signs. For example, Dylan's Candy Bar in New York City used a digital sign that turned people walking by into candy (Weiss. 2017).

Figure 9.1 gives examples of where you might find the three different types of digital signage.

Figure 9.1

Types of Digital Signage

Point of Sale

Restaurant Menu Board
Price Tags
Point of Payment
Mall

Point of Wait

Taxi
Doctor's Office
Elevator
Bank Line

Point of Transit

Airport
Billboard
Bus Shelter
Store Window

Source: J. Meadows

Figure 9.2:

Point of Mind Digital Signage at The Cosmopolitan, Las Vegas, NV

Source: J. Meadows

Arguably there is a fourth category that I'll call Point of Mind. This is digital signage employed to create an environment or state of mind. Another term often used is techorating. Hotel lobbies, restaurants, and offices use digital signage to create an environment for guest/clients. For example, the Cosmopolitan in Las Vegas uses digital signage for environmental design throughout the property. The lobby features digital signage depicting different elevator scenarios moving up and down columns. See Figure 9.2.

Background

The obvious place to begin tracing the development of digital signage is traditional signage. The first illuminated signs used gas, and the P.T. Barnum Museum was the first to use illuminated signs in 1840 (National Park Service, n.d.). The development of electricity and the light bulb allowed signs to be more easily illuminated, and then neon signs emerged in the 1920's (A Brief History, 1976). Electronic scoreboards such as the Houston Astros scoreboard in the new Astrodome in 1965 used 50,000 lights to create large messages and scores (Brannon, 2009).

Billboards have been used in the United States since the early 1800s. Lighting was later added to allow the signs to be visible at night. In the 1900s billboard structure and sizes were standardized to the sizes and types we see today (History of OOH, n.d.).

The move from illuminated signs to digital signs developed along the same path as the important technologies that make up digital signage discussed earlier: screen, content, and networks/content management. The development of the VCR in the 1970's contributed to digital signage as the technology provided a means to deliver a custom flow of content to television screens. These would usually be found indoors because of weather concerns.

A traditional CRT television screen was limited by size and weight. The development of projection and flat screen displays advanced digital signage. James Fergason developed LCD display technology in the early 1970's that led to the first LCD watch screen (Bellis, n.d.).

Plasma display monitors were developed in the mid 1960's but, like the LCD, were also limited in size, resolution, and color. It wasn't until the mid-1990's that high definition flat screen monitors were developed and used in digital signage. Image quality and resolution are particularly important for viewers to be able to read text, which is an important capability for most signage.

As screen technologies developed, so did the content delivery systems. DVDs replaced VCRs, eventually followed by media servers and flash memory devices. Compressed digital content forms such as jpeg, gif, flv, mov, and MP3 allowed signage to be quickly updated over a network.

Interactivity with screens became popular on ATM machines in the 1980's. Touch screen technology was developed in the 1960's but wasn't developed into widely used products until the 1990's when the technology was incorporated in personal digital assistant devices (PDAs) like the Palm Pilot.

Digital billboards first appeared in 2005 and immediately made an impact, but not necessarily a good one. Billboards have always been somewhat contentious. Federal legislation, beginning with the Bonus Act in 1958 and the Highway Beautification Act in 1965, tie Federal highway funding to the placement and regulation of billboards (Federal Highway Administration, n.d.). States also have their own regulations regarding outdoor advertising including billboards.

The development of digital billboards that provide multiple messages and moving images provoked attention because the billboards could be driver distractors. Although Federal Transportation Agency rules state billboards that have flashing, intermittent or moving light or lights are prohibited, the agency determined that digital billboards didn't violate that rule as long as the images are on for at least 4 seconds and are not too bright (Richtel, 2010).

States and municipalities also have their own regulations. For example, a law passed in Michigan in early 2014 limits the number of digital billboards and their brightness, how often the messages can change, and how far apart they can be erected. Companies can erect a digital billboard if they give up three traditional billboards or billboard permits (Eggert, 2014).

While distraction is an issue with billboards, privacy is an overall concern with digital signage, especially as new technologies allow digital signs to collect data from viewers. The Digital Signage Federation adopted its Digital Signage Privacy Standards in 2011. These voluntary standards were developed to "help preserve public trust in digital signage and set the stage for a new era of consumer-friendly interactive marketing" (Digital Signage Federation, 2011).

The standards cover both directly identifiable data such as name, address, date of birth, and images of individuals, as well as pseudonymous data. This type of data refers to that which can be reasonably associated with a particular person or his or her property. Examples include IP addresses, Internet usernames, social networking friend lists, and posts in discussion forums. The standards also state that data should not be knowingly collected on minors under 13.

The standards also recommend that digital signage companies use other standards for complementary technologies, such as the Mobile Marketing Association's Global Code of Conduct and the Code of Conduct from the Point of Purchase Association International.

Other aspects of the standard include fair information practices such as transparency and consent. Companies should have privacy policies and should give viewers notice.

Audience measurement is divided into three levels:

- *Level I*—Audience counting such as technologies that record dwell time but no facial recognition.

- *Level II*—Audience targeting. This is information that is collected and aggregated and used to tailor advertisements in real time. Finally,

- *Level III*—Audience identification and/or profiling. This is information collected on an individual basis. The information is retained and has links to individual identity such as a credit card number.

The standards recommend that Level I and II measurement should have opt-out consent while Level III should have opt-in consent (Digital Signage Federation, 2011).

As the use of digital signage has grown, so too has the number of companies offering digital signage services and technologies. Some of the long-standing companies in digital signage include Scala, BrightSign, Dynamax, Broadsign, and Four Winds Interactive. Other more traditional computer and networking companies including IBM, AT&T, Intel, and Cisco have also begun to offer full service digital signage solutions.

Many digital signage solutions include ways for signs to deliver customized content using short wave wireless technologies such as NFC (near field communication), RFID, and BLE (Bluetooth Low Energy) beacons.

For example, using NFC, a shopper could use their phone to tap a sign to get more information on a product. With RFID, chips placed in objects can then communicate with digital signs. For example, Scala's connected wall allows users to pick up an object such as a shoe with embedded RFID and then see information about that shoe on a digital sign.

BLE Beacon technology works a little differently. Small low power beacons are placed within a space such as a store. Then shoppers can receive customized messages from the beacons. Users will have to have an app for the store for the beacons to work. So, if a user has an app for a restaurant, as they are walking by the restaurant, the beacon can send a pop-up message with specials or a coupon to entice the user to enter the restaurant.

Mobile payment through NFC and digital signage continues to develop. Vending machines that use NFC technology to allow consumers to pay for items are becoming more personalized. For example, Costa Coffee has the Costa Express CEM-200 vending machine. This isn't the old stale coffee in a small paper cup machine of the past. This machine recognizes users' demographics and offers drinks targeted for that group. It also has a touch screen and cashless payment (Getting Started, 2016). You can see the machine in action here: atomhawk.com/case-study/costa-coffee-cem-200.

Panasonic's LinkRay visibile light communication technology is now being used in some digital signage displays. The technology allows users to communicate with the digital signs using their smartphone camera via a LinkRay app. The Peterson Automotive Museum in Orlando Florida is using this technology to engage with visitors (Panasonic LinkRay, 2017).

Recent Developments

Innovations in digital signage continue to emerge at a fast pace making it a must-have technology rather than a nice-to-have technology. Some of the major developments in digital signage are increased integration with augmented reality, Internet of Things, Artificial Intelligence, and E-Ink.

One fun trend in digital signage is augmented reality. "Augmented reality (AR) is a live, direct or indirect, view of a physical, real-world environment whose elements are augmented by computer-generated sensory input such as sound, video, graphics or GPS data" (Augmented Reality, n.d.). So, imagine a store where you can virtually try on any item—even ones not in stock at the time.

AR and digital signage has been used previously in some creative ways. In one innovative application Pepsi used AR technology to create all kinds of unusual surprises on a street in London. People in a bus shelter could look through a seemingly transparent window at the street, and then monsters, wild animals, and aliens would appear, interacting with people and objects on the street. The transparent sign was actually a live video feed of the street (Pepsi Max Excites, 2014). A video of the deployment can be seen at www.youtube.com/watch?v=Go9rf9GmYpM. There was a similar deployment of the technology in Vienna for the Premiere of The Walking Dead. Zombies surprised unsuspecting people. See it here www.youtube.com/watch?v=B7FzWUhgqck.

While those AR installations were pretty specialized, nowadays, AR can be found in retail stores across the country. One of the most common ways that AR is being used is with beauty products. Instead of having to try out makeup with samples that everyone else has touched, AR technology allows customers to "try on" different makeup with an AR mirror. Expect AR applications to increase for retail to allow users to try on clothes, makeup, glasses and even hair colors. Both VR and AR are discussed in more depth in Chapter 15.

Social networking continues to be a popular addition to digital signage. It's fairly common to find digital signage with Twitter feeds, especially with Point of Wait digital signage. News organizations, in particular, use Twitter to keep their digital signage updated with their news feed. Facebook and Instagram posts can also be part of digital signage. For example, corporate digital signage could feature company Instagram posts. In addition, companies can use their own followers' posts and images. Imagine walking into a clothing or makeup store and seeing Instagram posts of customers wearing the clothing for sale or demonstrating how to apply makeup. Universities and schools use social networking and digital signage to connect students to the institution and to recruit new students. It's also important for educational institutions and retailers to use the technologies that their students and customers are using.

"Alexa, what's on sale?" Virtual assistants like the Amazon Echo and Google Home are now popular home technologies, and this technology is now going beyond just home devices to be included in everything from cars to digital signage. For example, MangoSigns has an Alexa app that lets users search for information using voice (MangoSigns, n.d.)

Alexa isn't the only technology; digital signs can now respond to users 'voices or other factors using AI technology including sensors. For example, Apotek Hjartat created an anti-smoking digital billboard. When someone was smoking near the billboard, sensors reacted and played a video of a person coughing and looking disgusted. Bannister (2017) discusses how this billboard could have been even more effective using AI to determine if the person put out the cigarette or where the smoke was coming from.

Another application of AI in digital signage is on the backend. Digital signage can use AI technologies to deliver the strongest personalized message to users. Using factors such as facial analysis and other metrics and big data, AI can determine with most potentially successful advertisement to deliver (Henkel, 2016).

Facial recognition technologies embedded in digital signage are increasing with good customer feedback. For example, Lancôme used a "smart mirror" with Ulta Beauty. Customers would use the mirror to try on makeup and the mirror would collect useful data such as make up and style preferences, social media handles and contact information. Unilever used smile recognition technology and a digital kiosk to dispense ice cream treats to users who smiled, all the while collecting user data (Facial recognition, n.d.)

One major part of a digital sign that hasn't been discussed yet is the actual screen and media players. Advances in OLED, 4K, Ultra HD, and 8K screens (discussed in Chapter 6) and displays have made digital signs clearer even at very large sizes.

OLED screens are almost as thin as the old paper posters. LG is the top manufacturer for OLED screens in the world. At the Consumer Electronics Show in 2018 LG unveiled a 65-inch rollable LED screen as well as "wallpaper" screens (Katzmaier, 2018). At this point the super thin screens aren't used much in digital signage because of cost and fragility but expect their deployment to rise, especially for interior installations. The sturdy LCD remains king of outdoor digital signage.

Most digital signs are LCD and use traditional LCD display technologies. E-Ink has often been seen to be limited to smaller e-readers such as the Amazon Kindle. Easy to read in sunlight and using less energy, E-Ink has its advantages and now it is available for digital signage in the form of e-paper. E-paper is often used with solar energy sources because power is only needed when the display changes. Look for more e-paper display installations in the future (E-Ink, 2018).

Current Status

The digital signage industry is growing steadily. MarketsandMarkets reported that the digital signage market is expected by grow to $32.8 billion by 2023 (Rohan, 2017). Grant View Research reports a market size of $31.7 billion by 2025. Using either report, the digital signage market is expected to growth by almost 8% from 2016 to 2025. The global digital signage market is expected to grow by 5.2% over the same time. (Digital Signage Market Size, 2017).

The top markets in the U.S. for digital signage are retail, corporate, banking, healthcare, education, and transportation. Retail is the largest segment followed by transportation and banking (Digital Signage Market Analysis, 2017).

Digital billboards, despite their relative unpopularity with some states and municipalities continue to grow. According to Statista (2017), in 2017 there were 7000 digital billboards across the USA; that's up from 6400 in 2016.

According to the Nielsen OOH Compilation Report, 95% of U.S. residents over 16 noticed out-of-home advertising, and 83% of billboard views look at the advertising, message all or most of the time. The same study found that one in four OOH viewers have interacted with an advertisement with an NFC sensor or QR code (Williams, 2017).

Factors to Watch

The digital signage industry continues to mature and develop. Trends worth noting include increased data collection, use of data, and attempts to regulate that data. Short-range wireless such as BLE beacons and NFC will continue to increase. Digital signage will grow as implementation costs drop dramatically. In addition, the market is mature enough and adoption has spread enough that companies waiting to adopt are moving forward with deployment.

The use of augmented reality and artificial intelligence will continue to grow. Expect digital signage to deliver more customized and personalized messages.

Costs for screens, even 4K and higher resolution will continue to drop and LCD will see a challenge from OLED displays, especially for indoor applications.

Getting a Job

Digital signage is a large and growing industry that few people seem to notice, which makes it a market full of career opportunities. Whether you are studying computer science, graphic design, advertising, marketing, or business information systems, digital signage involves a wide range of industries from displays to networking and interactive graphics. As mentioned above, many more organizations, institutions, and businesses are adopting digital signage and need people who can design and manage these important systems. The Digital Signage Experts Group has published standards for digital signage education and offers digital signage expert certification (Digital Signage Experts Group, n.d.). You can find this at www.dseg. org /certification-programs/.

Bibliography

A Brief History of the Sign Industry. (1976). American Sign Museum. Retrieved from http://www.signmuseum.org/a-brief-history-of-the-sign-industry/.

Augmented Reality. (n.d.). *Mashable*. Retrieved from http://mashable.com/category/augmented-reality/.

Bannister, D. (2017). AI opens doors for digital signage. Digital Signage Today. Retrieved from https://www.digitalsignagetoday.com/blogs/ai-opens-doors-for-digital-signage/.

Bellis, M. (n.d.). Liquid Crystal Display—LCD. Retrieved from http://inventors.about.com/od/lstartinventions/a/LCD.htm.

Brannon, M. (2009). The Astrodome Scoreboard. *Houston Examiner*. Retrieved from http://www.examiner.com/article/the-astrodome-scoreboard.

Dell, L. (2013). Why Dayparting Must Be Part of Your Mobile Strategy. *Mashable*. Retrieved http://mashable.com /2013/08/14/dayparting-mobile/.

Digital Screenmedia Association. (n.d.). Frequently Asked Questions. Retrieved from http://www.digitalscreenmedia.org/faqs#faq15.

Digital Signage Experts Group (n.d.) Digital Signage Certification Programs. Retrieved from http://www.dseg.org/certification-programs/.

Digital Signage Federation. (n.d.). Digital Signage Glossary of Terms. Retrieved from http://www.digitalsignagefederation.org/glossary#D.

Digital Signage Federation. (2011). Digital Signage Privacy Standards. Retrieved from http://www.digitalsignagefederation.org/Resources/Documents/Articles%20and%20Whitepapers/DSF%20Digital%20Signage%20Privacy%20Standards %2002-2011%20(3).pdf.

Digital Signage Market Analysis By Type, By Component, By Technology, By Application, By Location, By Content Category, By Size, By Region, And Segment Forecasts 2014-2025, (2017). Grand View Research. Retrieved from https://www.grandviewresearch.com/industry-analysis/digital-signage-market

Digital Signage Market Size Worth $31.71 Billion by 2025. (2017). Grant View Research. Retrieved from https://www.grandviewresearch.com/press-release/global-digital-signage-market

Eggert, D. (2014, February 1). Mich law may limit growth of digital billboards. *Washington Post*. Retrieved from http://www.washingtontimes.com/news/2014/feb/1/mich-law-may-limit-growth-of-digital-billboards/?page=all.

E-Ink delivers e-paper display to Japan (2028). Digital Signage Today. Retrieved from https://www.digitalsignagetoday.com/news/e-ink-delivers-e-paper-display-to-japan.

Facial recognition technology is popping up in consumer activations and event interactives. (n.d.). Cramer. Retrieved from http://cramer.com/trent/issue-26/

Federal Highway Administration (n.d.) Outdoor Advertising Control. Retrieved from http://www.fhwa.dot.gov/real_estate/practitioners/oac/oacprog.cfm.

Getting Started in Digital Signage (n.d.). Retrieved from http://www.intel.com/content/dam/www/public/us/en/documents/guides/digital-signage-step-by-step-guide.pdf.

Henkel, C. (2016). Why to Use Artificial Intelligence in your digital signage network? LinkedIn. Retrieved from https://www.linkedin.com/pulse/why-use-artificial-intelligence-your-digital-signage-network-henkel/

History of OOH. (n.d.). Outdoor Advertising Association of America. Retrieved from http://www.oaaa.org/outofhomeadvertising/historyofooh.aspx.

Intel. (2016). Digital Signage At-a—Glance. Retrieved from http://www.intel.com/content/dam/www/public/us/en/documents/guides/digital-signage-guide.pdf.

Katzmaier, D. (2018). LG OLED TVs don't mess with success in 2018. Cnet. Retrieved from https://www.cnet.com/news/lg-oled-tvs-dont-mess-with-success-in-2018/

Kelsen, K. (2010). *Unleashing the Power of Digital Signage*. Burlington, MA: Focal Press.

MangoSigns (n.d.). Amazon Alexa App. Retrieved from https://mangosigns.com/apps/alexa

Mottl, J. (2016). See's Candy offers taste of history with digital signage at SF Int'l Airport, *Digital Signage Today*. Retrieved from http://www.digitalsignagetoday.com/articles/sees-candy-offers-taste-of-history-with-digital-signage-at-sf-intl-airport/.

National Park Service. (n.d.). Preservation Briefs 24-34. Technical Preservation Services. U.S. Department of the Interior.

Panasonic LinkRay technology enhances visitor experience at Petersen Automotive Museum. (2017). Panasonic Retrieved from http://shop.panasonic.com/about-us-latest-news-press-releases/06152017-infocommpetersen.html

Pepsi Max Excites Londoners with Augmented Reality DOOH First. (2014). Grand Visual. Retrieved from http://www.grandvisual.com/pepsi-max-excites-londoners-augmented-reality-digital-home-first/.

Retail Solution: Digital Self Displays (2016). Retrieved from http://www.intel.com/content/www/us/en/retail/solutions/digital-shelf-display.html.

Richtel, M. (2010, March 2). Driven to Distraction. *New York Times*. Retrieved from http://www.ntimes.com/2010/03/02/technology/02billboard.html?pagewanted=all&_r=0.

Rohan (2017). Digital Signage Market worth 32.84 billion USD by 2023. MarketsandMarkets. Retrieved from https://www.marketsandmarkets.com/PressReleases/digital-signage.asp

Statista (2017). Number of digital billboards in the United Statesin 2016 and 2017. Retrieved from https://www.statista.com/statistics/659381/number-billboards-digital-usa/

Weiss, L. (2017). Retailers look to storefront digital ads to target passerby. New York Post. Retrieved from https://nypost.com/2017/09/26/retailers-look-to-storefront-digital-ads-to-target-passersby/

Williams, D. (2075). *NielsenOOH Compilation Report*. Retrieved from https://www.fliphound.com/files/billboard-out-of-home-advertising-resources/nielsen-oaaa-compilation-report-final-digital-billboard-ooh-study-2017.pdf

10

Cinema Technologies

Michael R. Ogden, Ph.D. *

Overview

Film's history is notably marked by the collaborative dynamic between the art of visual storytelling and the technology that makes it possible. Filmmakers have embraced each innovation—be it sound, color, special effects, or widescreen imagery—and pushed the art form in new directions. Digital technologies have accelerated this relationship. As today's filmmakers come to terms with the inexorable transition from film to digital acquisition, they also find new, creative possibilities. Among enabling technologies are cameras equipped with larger image sensors, expanded color gamut and high dynamic range, 2K, 4K, Ultra-high definition (8K plus) image resolutions, and high frame rates for smoother motion. Emerging standards in 2D and 3D acquisition and display are competing with the rise of VR and immersive cinema technologies. Digital Cinema Packages are now the standard for theater distribution while advances in laser projection deliver more vibrant colors and sharper contrast. These developments promise moviegoers a more immersive cinematic experience. It is also hoped they will re-energize moviegoers, lure them back into the movie theaters, and increase profits.

Introduction

Storytelling is a universally human endeavor. In his book, *Tell Me A Story: Narrative and Intelligence* (1995), computer scientist and cognitive psychologist Roger Schank conjectures that not only do humans think in terms of stories; our very understanding of the world is in terms of stories we already understand. "We tell stories to describe ourselves not only so others can understand who we are but also so we can understand ourselves... . We interpret reality through our stories and open our realities up to others when we tell our stories" (Schank, 1995, p. 44). Stories touch all of us, reaching across cultures and generations.

Robert Fulford, in his book, *The Triumph of Narrative: Storytelling in the Age of Mass Culture* (1999), argues that storytelling formed the core of civilized life and was as important to preliterate peoples as it is to us living in the information age. However, with the advent of mass media—and, in particular, modern cinema—the role of storyteller in popular culture shifted from the individual to the "Cultural Industry" (c.f. Horkheimer & Adorno, 1969; Adorno, 1975; Andrae 1979), or what Fulford calls the "industrial narrative" (1999)—an apparatus for the production of meanings and pleasures involving aesthetic strategies and psychological processes (Neale 1985) and bound by its own set of economic and political determinants and made possible by contemporary technical capabilities.

In other words, cinema functions as a social institution providing a form of social contact desired by citizens immersed in a world of "mass culture." It also exists as a psychological institution whose purpose is to encourage the movie-going habit (Belton, 2013).

* Professor & Assistant Dean, College of Communication & Media Sciences, Zayed University (Dubai, United Arab Emirates).

Simultaneously, cinema evolved as a technological institution, its existence premised upon the development of specific technologies, most of which originated during the course of industrialization in Europe and America in the late 19th and early 20th Centuries (*e.g.*, film, the camera, the projector and sound recording). The whole concept and practice of filmmaking evolved into an art form dependent upon the mechanical reproduction and mass distribution of "the story"—refracted through the aesthetics of framing, light and shade, color, texture, sounds, movement, the shot/counter-shot, and the *mise en scène* of cinema.

"[T]here is something remarkable going on in the way our culture now creates and consumes entertainment and media," observes Steven Poster, former National President of the International Cinematographers Guild (2012, p. 6). Through movies, television, and the Internet, contemporary society now absorbs more stories than our ancestors could have ever imagined (Nagy, 1999). We live in a culture of mass storytelling.

However, the name most associated with cinematic storytelling, Kodak, after 131 years in business as one of the largest producers of film stock in the world, filed for Chapter 11 bankruptcy in 2012 (De La Merced, 2012). The company was hit hard by the recession and the advent of the RED Cinema cameras, the ARRI Alexa, and other high-end digital cinema cameras. In the digital world of bits and bytes, Kodak became just one more 20th Century giant to falter in the face of advancing technology.

Perhaps it was inevitable. The digitization of cinema began in the 1980s in the realm of special visual effects. By the early 1990s, digital sound was widely propagated in most theaters, and digital nonlinear editing began to supplant linear editing systems for post-production. "By the end of the 1990s, filmmakers such as George Lucas had begun using digital cameras for original photography and, with the release of *Star Wars Episode I: The Phantom Menace* in 1999, Lucas spearheaded the advent of digital projection in motion picture theatres" (Belton, 2013, p. 417).

As Mark Zoradi, former President of Walt Disney Studios Motion Pictures Group stated, "The key to a good film has always been story, story, story; but in today's environment, it's story, story, story and blow me away" (cited in Kolesnikov-Jessop 2009).

Background

Until recently "no matter how often the face of the cinema… changed, the underlying structure of the cinematic experience… remained more or less the same" (Belton, 2013, p. 6). Yet, even that which is most closely associated with cinema's identity—sitting with others in a darkened theater watching "larger-than-life" images projected on a big screen—was not always the norm.

From Novelty to Narrative

The origins of cinema as an independent medium lie in the development of mass communication technologies evolved for other purposes (Cook, 2004). Specifically, photography (1826-1839), roll film (1880), the Kodak camera (1888), George Eastman's motion picture film (1889), the motion picture camera (1891-1893), and the motion picture projector (1895-1896) each had to be invented in succession for cinema to be born.

Figure 10.1

Zoopraxiscope Disc by Eadweard Muybridge, 1893

Source: Wikimedia Commons

Early experiments in series photography for capturing motion were an important precursor to cinema's emergence. In 1878, Eadweard Muybridge set up a battery of cameras triggered by a horse moving through a set of trip wires. Adapting a Zoëtrope (a parlor novelty of the era) for projecting the photographs, Muybridge arranged his photograph plates around the perimeter of a disc that was manually rotated. Light from a "Magic Lantern" projector was shown through each slide as it passed in front of a lens. The image produced was then viewed on a large screen (Neale, 1985). If rotated rapidly enough, a phenomenon known as *persistence of vision* (an image appearing in front of the eye lingers a split second in the retina after removal of the image), allowed the viewer to experience smooth, realistic motion.

Perhaps the first movie projector, Muybridge called his apparatus the Zoopraxiscope, which was used to project photographic images in motion for the first time to the San Francisco Art Association in 1880 (Neale, 1985).

In 1882, French physiologist and specialist in animal locomotion, Étienne-Jules Marey, invented the Chronophotographic Gun in order to take series photographs of birds in flight (Cook, 2004). Shaped like a rifle, Marey's camera took 12 instantaneous photographs of movement per second, imprinting them on a rotating glass plate coated with a light-sensitive emulsion. A year later, Marey switched from glass plates to paper roll film. But like Muybridge, "Marey was not interested in cinematography... . In his view, he had invented a machine for dissection of motion similar to Muybridge's apparatus but more flexible, and he never intended to project his results" (Cook, 2004, p. 4).

In 1887, Hannibal Goodwin, an Episcopalian minister from New Jersey, first used celluloid roll film as a base for light-sensitive emulsions. George Eastman later appropriated Goodwin's idea and in 1889, began to mass-produce and market celluloid roll film on what would eventually become a global scale (Cook, 2004). Neither Goodwin nor Eastman were initially interested in motion pictures. However, it was the introduction of this durable and flexible celluloid film, coupled with the technical breakthroughs of Muybridge and Marey, that inspired Thomas Edison to attempt to produce recorded moving images to accompany the recorded sounds of his newly-invented phonograph (Neale, 1985). It is interesting to note that, according to Edison's own account (cited in Neale, 1985), the idea of making

motion pictures was never divorced from the idea of recording sound. "The movies were intended to talk from their inception so that, in some sense, the silent cinema represents a thirty-year aberration from the medium's natural tendency toward a total representation of reality" (Cook, 2004, p. 5).

Capitalizing on these innovations, W.K.L. Dickson, a research assistant at the Edison Laboratories, invented the first authentic motion picture camera, the Kinetograph—first constructed in 1890 with a patent granted in 1894. The basic technology of modern film cameras is still nearly identical to this early device. All film cameras, therefore, have the same five basic functions: a "light tight" body that holds the mechanism which advances the film and exposes it to light; a motor; a magazine containing the film; a lens that collects and focuses light on to the film; and a viewfinder that allows the cinematographer to properly see and frame what they are photographing (Freeman, 1998).

Thus, using Eastman's new roll film, the Kinetograph advanced each frame at a precise rate through the camera, thanks to sprocket holes that allowed metal teeth to grab the film, advance it, and hold the frame motionless in front of the camera's aperture at split-second intervals. A shutter opened, exposing the frame to light, then closed until the next frame was in place. The Kinetograph repeated this process 40 times per second. Throughout the silent era, other cameras operated at 16 frames per second; it wasn't until sound was introduced that 24 frames per second became standard, in order to improve the quality of voices and music (Freeman, 1998). When the processed film is projected at the same frame rate, realistic movement is presented to the viewer.

However, for reasons of profitability alone, Edison was initially opposed to projecting films to groups of people. He reasoned (correctly, as it turned out) that if he made and sold projectors, exhibitors would purchase only one machine from him—a projector—instead of several Kinetoscopes (Belton, 2013) that allowed individual viewers to look at the films through a magnifying eyepiece. By 1894, Kinetographs were producing commercially viable films. Initially the first motion pictures (which cost between $10 and $15 each to make) were viewed individually through Edison's

Kinetoscope "peep-shows" for a nickel apiece in arcades (called Nickelodeons) modeled on the phonographic parlors that had earlier proven so successful for Edison (Belton, 2013).

It was after viewing the Kinetoscope in Paris that the Lumière brothers, Auguste and Louis, began thinking about the possibilities of projecting films on to a screen for an audience of paying customers. In 1894, they began working on their own apparatus, the Cinématograph. This machine differed from Edison's machines by combining both photography and projection into one device at the much lower (and thus, more economical) film rate of 16 frames per second. It was also much lighter and more portable (Neale, 1985).

In 1895, the Lumière brothers demonstrated their Cinématograph to the Société d'Encouragement pour l'Industries Nationale (Society for the Encouragement of National Industries) in Paris. The first film screened was a short actuality film of workers leaving the Lumière factory in Lyons (Cook, 2004). The actual engineering contributions of the Lumière brothers were quite modest when compared to that of W.K.L. Dickson—they merely synchronized the shutter movement of the camera with the movement of the photographic film strip. Their real contribution is in the establishment of cinema as an industry (Neale, 1985).

The early years of cinema were ones of invention and exploration. The tools of visual storytelling, though crude by today's standards, were in hand, and the early films of Edison and the Lumière brothers were fascinating audiences with actuality scenes—either live or staged—of everyday life. However, an important pioneer in developing narrative fiction film was Alice Guy Blaché. Remarkable for her time, Guy Blaché was arguably the first director of either sex to bring a story-film to the screen with the 1896 release of her one-minute film, *La Fée aux Choux* (The Cabbage Fairy) that preceded the story-films of Georges Méliès by several months.

One could argue that Guy Blaché was cinema's first story "designer." From 1896 to 1920 she wrote and directed hundreds of short films including over 100 synchronized sound films and 22 feature films and produced hundreds more (McMahan, 2003). In the first half of her career, as head of film production for the Gaumont Company (where she was first employed as a secretary), Guy Blaché almost single-handedly developed the art of cinematic narrative (McMahan, 2003) with an emphasis on storytelling to create meaning.

Figure 10.2

First Publicly Projected Film: Sortie des Usines Lumière à Lyon, 46 seconds, 1895

Source: Screen capture courtesy Lumière Institute

Figure 10.3

Alice Guy Blaché Portrait, 1913

Source: Collection Solax, Public Domain

By the turn of the century, film producers were beginning to assume greater editorial control over the narrative, making multi-shot films and allowing for greater specificity in the story line (Cook, 2004). Such developments are most clearly apparent in the work of Georges

Méliès. A professional magician who owned and operated his own theater in Paris, Méliès was an important early filmmaker, developing cinematic narrative which demonstrated a created cause-and-effect reality. Méliès invented and employed a number of important narrative devices, such as the fade-in and fade-out, "lap" (overlapping) dissolve as well as impressive visual effects such as stop-motion photography (*Parce qu'on est des geeks!* 2013). Though he didn't employ much editing within individual scenes, the scenes were connected in a way that supported a linear, narrative reality.

By 1902, with the premiere of his one-reel film *Le Voyage Dans La Lune* (A Trip to the Moon), Méliès was fully committed to narrative filmmaking. Unfortunately, Méliès became embroiled in two lawsuits with Edison concerning issues of compensation over piracy and plagiarism of his 1902 film. Although he remained committed to his desire of "capturing dreams through cinema" (*Parce qu'on est des geeks!* 2013) until the end of his filmmaking career—and produced several other ground-breaking films (*Les Hallucinations Du Baron de Münchhausen*, 1911, and *A La Conquête Des Pôles*, 1912)—his legal battles left him embittered and by 1913, Méliès abandoned filmmaking and returned to performing magic.

Middle-class American audiences, who grew up with complicated plots and fascinating characters from such authors as Charles Dickens and Charlotte Brontë, began to demand more sophisticated film narratives. Directors like Edwin S. Porter and D.W. Griffith began crafting innovative films in order to provide their more discerning audiences with the kinds of stories to which theatre and literature had made them accustomed (Belton, 2013).

Influenced by Méliès, American filmmaker Edwin S. Porter is credited with developing the "invisible technique" of continuity editing. By cutting to different angles of a simultaneous event in successive shots, the illusion of continuous action was maintained. Porter's *Life of an American Fireman* and *The Great Train Robbery*, both released in 1903, are the foremost examples of this new style of storytelling through crosscutting (or, intercutting) multiple shots depicting parallel action (Cook, 2004).

Taking this a step further, D.W. Griffith, who was an actor in some of Porter's films, went on to become one of the most important filmmakers of all time, and truly the "father" of modern narrative form. Technologically and aesthetically, Griffith advanced the art form in ways heretofore unimagined. He altered camera angles, employed close-ups, and actively narrated events, thus shaping audience perceptions of them. Additionally, he employed "parallel editing"—cutting back and forth from two or more simultaneous events taking place in separate locations—to create suspense (Belton, 2013).

Figure 10.4

Méliès, Le Voyage Dans La Lune, 1902, 13 minutes

Source: Screen capture, M.R. Ogden.

Figure 10.5

Porter, The Great Train Robbery, 1903, 11 minutes

Source: Screen capture, M.R. Ogden.

Even though Edison's Kinetograph camera had produced more than 5,000 films (Freeman, 1998), by 1910, other camera manufacturers such as Bell and Howell, and Pathé (which acquired the Lumière patents in 1902) had invented simpler, lighter, more compact cameras that soon eclipsed the Kinetograph. In fact, "it has been estimated that, before 1918, 60% of all films were shot with a Pathé camera" (Cook, 2004, p. 42).

Nearly all of the cameras of the silent era were hand-cranked. Yet, camera operators were amazingly accurate in maintaining proper film speed (16 fps) and could easily change speeds to suit the story. Cinematographers could crank a little faster (over-crank) to produce slow, lyrical motion, or they could crank a little slower (under-crank) and when projected back at normal speed, they displayed the frenetic, sped-up motion apparent in the silent slapstick comedies of the Keystone Film Company (Cook, 2004).

Figure 10.6

Mitchell Standard Model A 35mm Camera, Circa 1920s

Source: mitchellcamera.com.

By the mid-1920s, the Mitchell Camera Corporation began manufacturing large, precision cameras that produced steadier images than previously possible. These cameras became the industry standard for almost 30 years until overtaken by Panavision cameras in the 1950s (Freeman, 1998).

In the United States, the early years of commercial cinema were tumultuous as Edison sued individuals and enterprises over patent disputes in an attempt to protect his monopoly and his profits (Neale, 1985). However, by 1908, the film industry was becoming more stabilized as the major film producers "banded

together to form the Motion Picture Patents Company (MPPC) which sought to control all aspects of motion picture production, distribution and exhibition" (Belton, 2013, p. 12) through its control of basic motion picture patents.

In an attempt to become more respectable, and to court middle-class customers, the MPPC began a campaign to improve the content of motion pictures by engaging in self-censorship to control potentially offensive content (Belton, 2013). The group also provided half-price matinees for women and children and improved the physical conditions of theaters. Distribution licenses were granted to 116 exchanges that could distribute films only to licensed exhibitors who paid a projection license of two dollars per week.

Unlicensed producers and exchanges continued to be a problem, so in 1910 the MPPC created the General Film Company to distribute their films. This development proved to be highly profitable and "was... the first stage in the organized film industry where production, distribution, and exhibition were all integrated, and in the hands of a few large companies" (Jowett, 1976, p. 34) presaging the emergence of the studio system 10 years later.

The Studio System

For the first two decades of cinema, nearly all films were photographed outdoors. Many production facilities were like that of George Méliès, who constructed a glass-enclosed studio on the grounds of his home in a Paris suburb (Cook, 2004). However, Edison's laboratory in West Orange, New Jersey, dubbed the "Black Maria," was probably the most famous film studio of its time. It is important to note that the "film technologies created in [such] laboratories and ateliers would come to offer a powerful system of world building, with the studio as their spatial locus" (Jacobson, 2011, p. 233). Eventually, the industry outgrew these small, improvised facilities and moved to California, where the weather was more conducive to outdoor productions. Large soundstages were also built in order to provide controlled staging and more control over lighting.

Figure 10.7
Edison's Black Maria, World's First Film Studio, circa 1890s

Source: Wikimedia Commons

By the second decade of the 20th Century, dozens of movie studios were operating in the U.S. and across the world. A highly secialized industry grew in southern California, honing sophisticated techniques of cinematography, lighting, and editing. The Hollywood studios divided these activities into preproduction, production, and post-production. During preproduction, a film was written and planned. The production phase was technology intensive, involving the choreography of actors, cameras and lighting equipment. Post-production consisted of editing the films into coherent narratives and adding titles—in fact, film editing is the only art that is unique to cinema.

The heart of American cinema was now beating in Hollywood, and the institutional machinery of filmmaking evolved into a three-phase business structure of production, distribution, and exhibition to get their films from studios to theater audiences. Although the MPPC was formally dissolved in 1918 as a result of an antitrust suit initiated in 1912 (Cook, 2004), powerful new film companies, flush with capital, were emerging. With them came the advent of vertical integration.

Through a series of mergers and acquisitions, formerly independent production, distribution, and exhibition companies congealed into five major studios; Paramount, Metro-Goldwin-Mayer (MGM), Warner Bros, RKO (Radio-Keith-Orpheum), and Fox Pictures. "All of the major studios owned theater chains; the minors—Universal, Columbia, and United Artists—

did not" (Belton, 2013, p. 68), but distributed their pictures by special arrangement to the theaters owned by the majors. The resulting economic system was quite efficient. "The major studios produced from 40 to 60 pictures a year… [but in 1945 only] owned 3,000 of the 18,000 theaters around the country… [yet] these theaters generated over 70% of all box-office receipts" (Belton, 2013, p. 69).

As films and their stars increased in popularity, and movies became more expensive to produce, studios began to consolidate their power, seeking to control each phase of a film's life. However, since the earliest days of the Nickelodeons, moralists and reformers had agitated against the corrupting nature of the movies and their effects on American youth (Cook, 2004). A series of scandals involving popular movie stars in the late 1910s and early 1920s resulted in ministers, priests, women's clubs, and reform groups across the nation encouraging their membership to boycott the movies.

In 1922, frightened Hollywood producers formed a self-regulatory trade organization—the Motion Picture Producers and Distributors of America (MPPDA). By 1930, the MPPDA adopted the rather draconian Hayes Production Code. This "voluntary" code, intended to suppress immorality in film, proved mandatory if the film was to be screened in America (Mondello, 2008). Although the code aimed to establish high standards of performance for motion-picture producers, it "merely provided whitewash for overly enthusiastic manifestations of the 'new morality' and helped producers subvert the careers of stars whose personal lives might make them too controversial" (Cook, 2004, p. 186).

Sound, Color and Spectacle

Since the advent of cinema, filmmakers hoped for the chance to bring both pictures and sound to the screen. Although the period until the mid-1920s is considered the silent era, few films in major theaters actually were screened completely silent. Pianists or organists—sometimes full orchestras—performed musical accompaniment to the projected images. At times, actors would speak the lines of the characters and machines and performers created sound effects. "What these performances lacked was fully synchronized sound contained within the soundtrack on the film itself" (Freeman, 1998, p. 408).

Figure 10.8
Cinema History Highlights

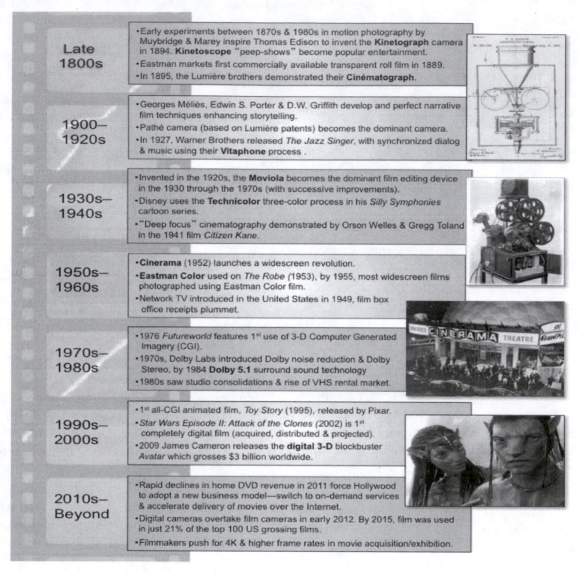

Late 1800s	• Early experiments between 1870s & 1980s in motion photography by Muybridge & Marey inspire Thomas Edison to invent the **Kinetograph** camera in 1894. **Kinetoscope** "peep-shows" become popular entertainment. • Eastman markets first commercially available transparent roll film in 1889. • In 1895, the Lumière brothers demonstrated their **Cinématograph.**
1900– 1920s	• Georges Méliés, Edwin S. Porter & D.W. Griffith develop and perfect narrative film techniques enhancing storytelling. • Pathé camera (based on Lumière patents) becomes the dominant camera. • In 1927, Warner Brothers released *The Jazz Singer*, with synchronized dialog & music using their **Vitaphone** process .
1930s– 1940s	• Invented in the 1920s, the **Moviola** becomes the dominant film editing device in the 1930 through the 1970s (with successive improvements). • Disney uses the **Technicolor** three-color process in his *Silly Symphonies* cartoon series. • "Deep focus" cinematography demonstrated by Orson Welles & Gregg Toland in the 1941 film *Citizen Kane*.
1950s– 1960s	• **Cinerama** (1952) launches a widescreen revolution. • **Eastman Color** used on *The Robe* (1953), by 1955, most widescreen films photographed using Eastman Color film. • Network TV introduced in the United States in 1949, film box office receipts plummet.
1970s– 1980s	• 1976 *Futureworld* features 1st use of 3-D Computer Generated Imagery (CGI). • 1970s, Dolby Labs introduced Dolby noise reduction & Dolby Stereo, by 1984 **Dolby 5.1** surround sound technology • 1980s saw studio consolidations & rise of VHS rental market.
1990s– 2000s	• 1st all-CGI animated film, *Toy Story* (1995), released by Pixar. • *Star Wars Episode II: Attack of the Clones* (2002) is 1st completely digital film (acquired, distributed & projected). • 2009 James Cameron releases the **digital 3-D** blockbuster *Avatar* which grosses $3 billion worldwide.
2010s– Beyond	• Rapid declines in home DVD revenue in 2011 force Hollywood to adopt a new business model—switch to on-demand services & accelerate delivery of movies over the Internet. • Digital cameras overtake film cameras in early 2012. By 2015, film was used in just 21% of the top 100 US grossing films. • Filmmakers push for 4K & higher frame rates in movie acquisition/exhibition.

Source: M.R. Ogden

By the late 1920s, experiments had demonstrated the viability of synchronizing sound with projected film. When Warner Bros Studios released The *Jazz Singer* in 1927, featuring synchronized dialog and music using their Vitaphone process, the first "talkie" was born. Vitaphone was a sound-on-disc process that issued the audio on a separate 16-inch phonographic disc. While the film was projected, the disc played on a turntable indirectly coupled to the projector motor (Bradley, 2005). Other systems were also under development during this time, and Warner Bros Vitaphone process had competition from Movietone, DeForest Phonofilm, and RCA's Photophone.

Though audiences were excited by this new novelty, from an aesthetic standpoint, the advent of sound actually took the visual production value of films backward. Film cameras were loud and had to be housed in refrigerator-sized vaults to minimize the noise; as a result, the mobility of the camera was suddenly limited. Microphones had to be placed very near the actors, resulting in restricted blocking and the odd phenomenon of actors leaning close to a bouquet of flowers as they spoke their lines; the flowers, of course, hid the microphone. No question about it, though, sound was here to stay.

Figure 10.9
Vitaphone Projection Setup, 1926 Demonstration

Source: Wikimedia Commons.

Once sound made its appearance, the established major film companies acted cautiously, signing an agreement to only act together. After sound had proved a commercial success, the signatories adopted Movietone as the standard system—a sound-on-film method that recorded sound as a variable-density optical track on the same strip of film that recorded the pictures (Neale, 1985). The advent of the "talkies" launched another round of mergers and expansions in the studio system. By the end of the 1920s, more than 40% of theaters were equipped for sound (Kindersley, 2006), and by 1931, "…virtually all films produced in the United States contained synchronized sound-tracks" (Freeman, 1998, p. 408).

Movies were gradually moving closer to depicting "real life." But life isn't black and white, and experiments with color filmmaking had been conducted since the dawn of the art form. Finally, however, in 1932, Technicolor introduced a practical three-color dye-transfer process that slowly revolutionized moviemaking and dominated color film production in Hollywood until the 1950s (Higgins, 2000). Though aesthetically beautiful the Technicolor process was extremely expensive, requiring three times the film stock, complicated lab processes, and strict oversight by the Technicolor Company, who insisted on strict secrecy during every phase of production. As a result, most movies were still produced in black and white well into the 1950s; that is, until single-strip Eastman Color Negative Safety Film (5247) and Color Print Film (5281) were released, "revolutionizing the movie industry and forcing Technicolor strictly into the laboratory business" (Bankston, 2005, p. 5).

By the late 1940s, the rising popularity of television and its competition with the movie industry helped drive more changes. The early impetus for widescreen technology was that films were losing money at the box office because of television. In the early years of film, the 4:3 aspect ratio set by Edison (4 units wide by 3 units high, also represented as 1.33:1) was assumed to be more aesthetically pleasing than a square box; it was the most common aspect ratio for most films until the 1960s (Freeman, 1998) and that adopted for broadcast by television until the switch to HDTV in 1998. However, the studios' response was characteristically cautious, initially choosing to release fewer but more expensive films (still in the standard Academy aspect ratio) hoping to lure audiences back to theaters with quality product (Belton, 2013). However, "[it] was not so much the Hollywood establishment… as the independent producers who engineered a technological revolution that would draw audiences back" (Belton, 2013, p. 327) to the movie theaters.

It was through the efforts of independent filmmakers during the 1950s and early 1960s, that the most pervasive technological innovations in Hollywood since the introduction of sound were realized. "A series of processes changed the size of the screen, the shape of the image, the dimensions of the films, and the recording and reproduction of sound" (Bordwell, Staiger & Thompson, 1985, p. 358).

Cinerama (1952) launched a widescreen revolution that would permanently alter the shape of the motion picture screen. Cinerama was a widescreen process that required filming with a three-lens camera and projecting with synchronized projectors onto a deeply curved screen extending the full width of most movie theaters. This viewing (yielding a 146° by 55° angle of view) was meant to approximate that of human vision (160° by

60°) and fill a viewer's entire peripheral vision. Mostly used in travelogue-adventures, such as *This is Cinerama* (1952) and *Seven Wonders of the World* (1956), the first two Cinerama fiction films—*The Wonderful World of the Brothers Grimm* and *How the West Was Won*—were released in 1962, to much fanfare and critical acclaim. However, three-camera productions and three-projector system theaters like Cinerama and CineMiracle (1957) were extremely expensive technologies and quickly fell into disuse.

Figure 10.10
Cinerama's 3-Camera Projection

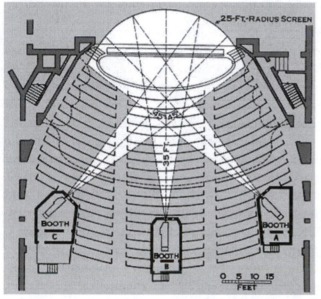

Source: Wikimedia Commons.

Anamorphic processes used special lenses to shoot or print squeezed images onto the film as a wide field of view. In projection, the images were unsqueezed using the same lenses, to produce an aspect ratio of 2.55:1—almost twice as wide as the Academy Standard aspect ratio (Freeman, 1998). When Twentieth Century-Fox released *The Robe* in 1953 using the CinemaScope anamorphic system, it was a spectacular success and just the boost Hollywood needed. Soon, other companies began producing widescreen films using similar anamorphic processes such as Panascope and Superscope. Nearly all these widescreen systems—including CinemaScope—incorporated stereophonic sound reproduction.

Figure 10.11
Film Aspect Ratios

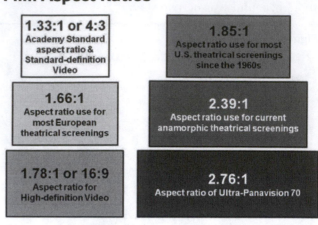

Source: M.R. Ogden

If widescreen films were meant to engulf audiences, pulling them into the action, "3D assaulted audiences—hurling spears, shooting arrows, firing guns, and throwing knives at spectators sitting peacefully in their theatre seats" (Belton, 2013, p. 328).

The technology of 3D is rooted in the basic principles of binocular vision. Early attempts at reproducing monochromatic 3D used an anaglyphic system: two strips of film, one tinted red, the other cyan, were projected simultaneously for an audience wearing glasses with one red and one cyan filtered lens (Cook, 2004). When presented with slightly different angles for each eye, the brain processed the two images as a single 3D image. The earliest 3D film using the anaglyphic process was *The Power of Love* in 1922.

In the late 1930s, MGM released a series of anaglyphic shorts, but the development of polarized filters and lenses around the same time permitted the production of full-color 3D images. Experiments in anaglyphic films ceased in favor of the new method. In 1953, Milton Gunzberg released *Bwana Devil*, a "dreadful" film shot using a polarized 3D process called Natural Vision. It drew in audiences and surprisingly broke box office records, grossing over $5 million by the end of its run (Jowett, 1976). Natural Vision employed two interlocked cameras whose lenses were positioned to approximate the distance between the human eyes and record the scene on two separate negatives. In the theater, when projected simultaneously onto the screen, spectators wearing disposable glasses with polarized lenses perceived a single three-dimensional image (Cook, 2004). Warner Bros released the second Natural Vision feature, *House of Wax* (1953), which featured six-track stereophonic sound and was a critical and popular success, returning $5.5 million on an investment of $680,000 (Cook, 2004). "Within a matter of months after the initial release of *Bwana Devil*, more than 4,900 theaters were converted to 3D" (Belton, 2013, p. 329).

Although Hollywood produced 69 features in 3D between 1953 and 1954, most were cheaply made exploitation films. By late 1953, the stereoscopic 3D craze had peaked. Two large budget features shot in Natural Vision, MGM's *Kiss Me Kate* (1953) and Alfred Hitchcock's *Dial M for Murder* (1954) were released "flat" because the popularity of 3D had fallen dramatically. Although 3D movies were still made decades later for special short films at Disney theme parks, 3D was no longer part of the feature-film production process (Freeman, 1998). One reason for 3D's demise was that producers found it difficult to make serious narrative films in such a gimmicky process (Cook, 2004). Another problem was the fact that audiences disliked wearing the polarized glasses; many also complained of eyestrain, headaches and nausea.

But, perhaps, the biggest single factor in 3D's rapid fall from grace was cinematographers' and directors' alternative use of deep-focus widescreen photography—especially anamorphic processes that exploited depth through peripheral vision—and compositional techniques that contributed to the feeling of depth without relying on costly, artificial means. Attempts to revive 3D, until most recently, met with varying degrees of success, seeing short runs of popularity in the 1980s with films like *Friday the 13th, Part III* and *Jaws 3D* (both 1983).

In 1995, with the release of the IMAX 3D film, *Wings of Courage*—and later, *Space Station 3D* in 2002—the use of active display LCD glasses synchronized with the shutters of dual-filmstrip projectors using infrared signals presaged the eventual rise of digital 3D films 10 years later.

Hollywood Becomes Independent

"The dismantling of the studio system began just before World War II when the U.S. Department of Justice's Antitrust Division filed suit against the [five] major [and three minor] studios, accusing them of monopolistic practices in their use of block booking, blind bidding, and runs, zones, and clearances" (Belton, 2013, p. 82).

In 1948, in the case of *U.S. vs. Paramount*, the Supreme Court ruled against the block booking system and recommended the breakup of the studio–theater monopolies. The major studios were forced to divorce their operations from one another, separate production and distribution from exhibition, and divest themselves of their theater chains (Belton, 2013). "RKO and other studios sold their film libraries to television stations to offset the losses from the *Paramount* case. The studios also released actors from contracts who became the new stars of the television world" (Bomboy, 2015). Other factors also contributed to the demise of the studio system, most notably changes in leisure-time entertainment, the aforementioned competition with television, and the rise of independent production (Cook, 2004). Combined with the extreme form of censorship Hollywood imposed upon itself through the Hayes Production Codes—and "after World War II, with competition from TV on the family front, and from foreign films with nudity on the racy front" (Mondello, 2008)—movie studios were unable (or unwilling) to rein in independent filmmakers who chafed under the antiquated Code.

In another landmark ruling, the U.S. Supreme Court decided in 1952 (also known as the "Miracle decision") that films constitute "a significant medium of communication of ideas" and were therefore protected by both the First and Fourteenth Amendments (Cook, 2004, p. 428). By the early 1960s, supported by subsequent court rulings, films were "guaranteed full freedom of expression" (Cook, 2004, p. 428). The influence of the Hayes Production Code had all but disappeared by the end of the 1960s, replaced by the MPAA ratings system (MPAA, 2011) instituted in 1968, revised in 1972, and now in its latest incarnation since 1984.

Though the 1960s still featured big-budget, lavish movie spectacles, a parallel movement reflected the younger, more rebellious aesthetic of the "baby boomers." Actors and directors went into business for themselves, forming their own production companies, and taking as payment lump-sum percentages of the film's profits (Belton, 2013). The rise and success of independent filmmakers like Arthur Penn (*Bonnie and Clyde*, 1967), Stanley Kubrick (*2001: A Space Odyssey*, 1968), Sam Peckinpah (*The Wild Bunch*, 1969), Dennis Hopper (*Easy Rider*, 1969), and John Schlesinger (*Midnight Cowboy*, 1969), demonstrated that filmmakers outside the studio system were freer to experiment with style and content. The writing was on the wall, the major studios would no longer dominate popular filmmaking as they had in the past.

In the early 1960s, an architectural innovation changed the way most people see movies—the move from single-screen theaters (and drive-ins) to multi-screen cineplexes. Although the first multi-screen house with two theaters was built in the 1930s, it was not until the late 1960s that film venues were built with four to six theaters. What these theaters lacked was the atmosphere of the early "movie palaces." While some owners put effort into the appearance of the lobby and concessions area, in most cases the "actual theater was merely functional" (Haines, 2003, p. 91). The number of screens in one location continued to grow; 1984 marked the opening of the first 18-plex theater in Toronto (Haines, 2003).

The next step in this evolution was the addition of stadium seating—offering moviegoers a better experience by affording more comfortable seating with unobstructed views of the screen (EPS Geofoam, n.d.). Although the number of screens in a location seems to have reached the point of diminishing returns, many theaters are now working on improving the atmosphere they create for their patrons. From bars and restaurants to luxury theaters with a reserved $29 movie ticket, many theater owners are once again working to make the movie-going experience something different from what you can get at home (Gelt & Verrier, 2009).

Recent Developments

Since the films of Georges Méliès thrilled audiences with inventive cinematic "trickery," much has changed in the past century of moviemaking. According to Box Office Mojo, of the top ten highest worldwide grossing films of all time (not adjusted for inflation), all of them featured heavy use of visual effects with James Cameron's *Avatar* (2009) and *Titanic* (1997) occupying the top two positions (respectively) and 2015's *Star Wars: The Force Awakens* appearing on the list at third place while popular comic book franchise films filled out most of the next five places with Disney's *Frozen* (2013) coming in ninth and *Star Wars: The Last Jedi* (2017) rounding out the top ten (Box Office Mojo, 2017a). When examining the US Domestic top ten highest grossing films—and adjusting for inflation (Box Office Mojo, 2017b)—the list is topped by *Gone with the Wind* (1939), a film that featured innovative matte shots as well as other "trickery." The second film on the list is George Lucas' *Star Wars* (1977), arguably the most iconic visual effects movie ever made (Bredow, 2014).

Such "trickery," special effects, or more commonly referred to now as "visual effects" (VFX), are divided into mechanical, optical, and computer-generated imagery (CGI). "Mechanical effects include those devices used to make rain, wind, cobwebs, fog, snow, and explosions. Optical effects allow images to be combined… through creation of traveling mattes run through an optical printer" (Freeman, 1998, p. 409).

In the early sound era, miniatures and rear projection became popular along with traveling mattes (Martin, 2014), like those employed in the landmark VFX film of the 1930s, *King Kong* (1933).

Four years in the making, the 1968 Stanley Kubrick film, *2001: A Space Odyssey*, created a new standard for VFX credibility (Martin, 2014). Kubrick used sophisticated traveling mattes combined with "hero" miniatures of spacecraft (ranging from four to 60 feet in length) and live-action to stunning effect (Cook, 2004). The film's "star gate" sequence dazzled audiences with controlled streak photography, macrophotography of liquids, and deliberate misuse of color-records. Also, throughout the opening "Dawn of Man" sequence, audiences witnessed the first major application of the front projection technique (Martin, 2014).

Arguably the first movie ever to use computers to create a visual effect—a two-dimensional rotating structure on one level of the underground lab—was *The Andromeda Strain* in 1971. This work was considered extremely advanced for its time.

In 1976, American International Pictures released *Futureworld*, which featured the first use of 3D CGI—a brief view of a computer-generated face and hand. In 1994, this groundbreaking effect was awarded a Scientific and Engineering Academy Award. Since then, CGI technology has progressed rapidly.

"In the history of VFX, there is a before-and-after point demarcated by the year 1977—when *Star Wars* revolutionized the industry" (Martin, 2014, p. 71-72). VFX supervisor John Dykstra invented an electronic motion-controlled camera capable of repeating its movements (later called the "Dykstraflex") and developed methodologies for zero-gravity explosions. Likewise, George Lucas' visual effects company, Industrial Light & Magic (ILM), took a big step forward for CGI with the rendering of a 3D wire-frame view of the Death Star trench depicted as a training aid for rebel pilots in *Star Wars* (1977).

Star Trek: The Wrath of Kahn (1982) incorporated a one-minute sequence created by Pixar (a LucasFilm spin-off), that simulated the "Genesis Effect" (the birth and greening of a planet) and is cinema's first totally computer-generated VFX shot. It also introduced a fractal-generated landscape and a particle-rendering system to achieve a fiery effect (Dirks, 2009).

Tron (1982) was the first live-action movie to use CGI for a noteworthy length of time (approximately 20 minutes) in the most innovative sequence of its 3D graphics world inside a video game and "showed studios that digitally created images were a viable option for motion pictures" (Bankston, 2005, p. 1).

In *Young Sherlock Holmes* (1985), LucasFilm/Pixar created perhaps the first fully photorealistic CGI character in a full-length feature film with the sword-wielding medieval "stained-glass" knight who came to life when jumping out of a window frame.

Visual impresario James Cameron has always relied on VFX in his storytelling dating back to the impressive low-budget miniature work on *The Terminator* (1984), later expanded on for *Aliens* (1986). Cameron's blockbuster action film, *Terminator 2: Judgment Day* (1991) received a Best Visual Effects Oscar thanks to its

depiction of Hollywood's first CGI main character, the villainous liquid metal T-1000 cyborg (Martin, 2014).

Toy Story (1995), was the first successful animated feature film from Pixar, and was also the first all-CGI animated feature film (Vreeswijk, 2012).

In *The Lord of the Rings* trilogy (2001, 2002 and 2003), a combination of motion-capture performance and key-frame techniques brought to life the main digital character Gollum (Dirks, 2009) by using a motion-capture suit (with reflective sensors) and recording the movements of actor Andy Serkis.

CGI use has grown exponentially and hand-in-hand with the increasing size of the film's budget it occupies. *Sky Captain and the World of Tomorrow* (2004) was the first big-budget feature to use only "virtual" CGI back lot sets. Actors Jude Law, Gwyneth Paltrow, and Angelina Jolie were filmed in front of blue screens; everything else was added in post-production (Dirks, 2009).

More an "event" than a movie, James Cameron's *Avatar* (2009) ushered in a new era of CGI. Many believe that *Avatar*, a largely computer-generated, 3D film—and the top-grossing movie in film history, earning nearly $3 billion worldwide—changed the movie-going experience (Muñoz, 2010). New technologies used in the film included CGI Performance Capture techniques for facial expressions, the Fusion Camera System for 3D shooting, and the Simul-Cam for blending real-time shoots with CGI characters and environments (Jones, 2012).

One of the greatest obstacles to CGI has been effectively capturing facial expressions. In order to overcome this hurdle, Cameron built a technology he dreamed up in 1995, a tiny camera on the front of a helmet that was able to "track every facial movement, from darting eyes and twitching noses to furrowing eyebrows and the tricky interaction of jaw, lips, teeth and tongue" (Thompson, 2010).

The 2018 Oscar nominations for Best Visual Effects (*Blade Runner 2049, Guardians of the Galaxy 2, Kong: Skull Island, Star Wars: The Last Jedi,* and *War for the Planet of the Apes*) illustrate that today's VFX-heavy films point the way toward a future where actors could perform on more advanced virtual sets where the director and the cinematographer are in complete control of the story-telling environment. For example, *Guardians of the Galaxy 2* (2017)—the first feature film shot in 8K resolution on RED Epic cameras to accommodate the heavy VFX elements of the story (Pennington, 2018)—had only 60 non-VFX shots and over 2,300 VFX shots (Frei, 2017).

Advances in motion-capture and the ability to provide realistic facial expressions to CGI characters captured from the actor's performance were recently demonstrated to great emotional effect by actor Andy Serkis who did the motion-capture performances for Cesar in the *Planet of the Apes* trilogy (2011, 2014 and 2017). In the last film of the trilogy, *War for the Planet of the Apes* (2017), VFX supervisor Dan Lemmon from Weta Digital, stated that "[every] time we reviewed shots with Caesar, we had Andy Serkis' performance side-by-side with Caesar to get the emotion right" (Verhoeven, 2017). For *Blade Runner 2049* (2017), the stand out among many stunning VFX scenes is the digital recreation of the Rachael replicant played by Sean Young in the original movie and body-doubled by actress Loren Peta (in costume, makeup and dotted face!) in the 2-minute scene with Harrison Ford and Jared Leto (Desowitz, 2017). Likewise, VFX artists are now able to overlay "digital makeup" on actors, as was done for Dan Stevens' portrayal of Beast in Disney's 2017 live-action remake of *Beauty and the Beast*, (Failes, 2017), or the ability to apply digital "de-aging" techniques on older actors playing younger versions of themselves in a film's flashback scenes (Ward, 2016).

Of course, the unstated goal of the VFX industry has always been to make a photorealistic, CGI character so realistic that the audience can't tell the difference between the CGI character and a real one (Media Insider, 2018). This was done in *Star Wars: Rogue One* (2016) to bring back an iconic character from the original *Star Wars* trilogy, the Grand Moff Tarkin, originally played by Peter Cushing who died in 1994 (Ward, 2016).

Multichannel Sound

Sound plays a crucial role in making any movie experience memorable. Perhaps none more so than the 2018 Oscar nominated *Dunkirk* (2017), in which sound editor Richard King combines the abstract and experimental score of Hans Zimmer with real-world soundscapes to create a rich, dense, and immersive ex-perience (Andersen, 2017). As impressive as it is, *Dunkirk* "doesn't break new ground, but the sound design is still impressive, taking full advantage of the powerful subwoofers and speakers in IMAX cinemas" (Hardawar, 2017).

There's no mistaking the chest-rumbling crescendo associated with THX sound in theaters. It's nearly as recognizable as its patron's famous *Star Wars* theme. Developed by Lucasfilm and named after *THX1138* (George Lucas' 1971 debut feature film), THX is not a cinema sound format, but rather a standardization system that strives to "reproduce the acoustics and ambience of the movie studio, allowing audiences to enjoy a movie's sound effects, score, dialogue, and visual presentation with the clarity and detail of the final mastering session" (THX, 2016).

At the time of THX's initial development in the early 1980s, most of the cinemas in the U.S. had not been updated since World War II. Projected images looked shoddy, and the sound was crackly and flat. "All the work and money that Hollywood poured into making movies look and sound amazing was being lost in these dilapidated theaters" (Denison, 2013).

Even if movies were not being screened in THX-certified theaters, the technical standards set by THX illustrated just how good the movie-going experience could be and drove up the quality of projected images and sound in all movie theaters. THX was introduced in 1983 with *Star Wars Episode VI: Return of the Jedi* and quickly spread across the industry. To be a THX Certified Cinema, movie theaters must meet the standards of best practices for architectural design, acoustics, sound isolation and audio-visual equipment performance (THX, 2016). As of mid-2016, self-reported company data states that there are about 4,000 THX certified theaters worldwide (THX, 2016).

While THX set the standards, Dolby Digital 5.1 Surround Sound is one of the leading audio delivery technologies in the cinema industry. In the 1970s, Dolby Laboratories introduced Dolby noise reduction (removing hiss from magnetic and optical tracks) and Dolby Stereo—a highly practical 35mm stereo optical release print format that fit the new multichannel soundtrack into the same space on the print occupied by the traditional mono track (Hull, 1999).

Dolby's unique quadraphonic matrixed audio technique allows for the encoding of four channels of information (left, center, right and surround) on just two physical tracks on movie prints (Karagosian & Lochen, 2003). The Dolby stereo optical format proved so practical that today there are tens of thousands of cinemas worldwide equipped with Dolby processors, and more than 25,000 movies have been released using the Dolby Stereo format (Hull, 1999).

By the late 1980s, Dolby 5.1 was introduced as the cinematic audio configuration documented by various film industry groups as best satisfying the requirements for theatrical film presentation (Hull, 1999). Dolby 5.1 uses "five discrete full-range channels—left, center, right, left surround, and right surround—plus a… low-frequency [effects] channel" (Dolby, 2010a). Because this low-frequency effects (LFE) channel is felt more than heard, and because it needs only one-tenth the bandwidth of the other five channels, it is refered to as a ".1" channel (Hull, 1999).

Dolby also offers Dolby Digital Surround EX, a technology developed in partnership with Lucasfilm's THX that places a speaker behind the audience to allow for a "fuller, more realistic sound for increased dramatic effect in the theatre" (Dolby, 2010b).

Dolby Surround 7.1 is the newest cinema audio format developed to provide more depth and realism to the cinema experience. By resurrecting the full range left extra, right extra speakers of the earlier Todd-AO 70mm magnetic format, but now calling them left center and right center, Dolby Surround 7.1 improves the spatial dimension of soundtracks and enhances audio definition thereby providing full-featured audio that better matches the visual impact on the screen.

Film's Slow Fade to Digital

With the rise of CGI-intensive storylines and a desire to cut costs, celluloid film is quickly becoming an endangered medium for making movies as more filmmakers use digital cinema cameras capable of creating high-quality images without the time, expense, and chemicals required to shoot and process on film. "While the debate has raged over whether or not film is dead, ARRI, Panavision, and Aaton quietly ceased production of film cameras in 2011 to focus exclusively on design and manufacture of digital cameras" (Kaufman, 2011).

But, film is not dead—at least, not yet. Separate from its unlikely continued viability as an exhibition format, 35mm film was used in shooting nearly 100 movies (in whole or in part) in 2015. Steve Bellamy, the president of Motion Picture and Entertainment for Kodak, wrote in an open letter, "There have been so many people instrumental in keeping film healthy and vibrant. Steven Spielberg, JJ Abrams and … [generations] of filmmakers will owe a debt of gratitude to Christopher Nolan and Quentin Tarantino for what they did with the studios. Again, just a banner year for film!" (cited in Fleming Jr., 2016). 35mm is making a comeback as a prestige choice among A-list directors who embrace the medium's warm natural aesthetic or wish to emulate a period look. With recent efforts by the likes of J.J. Abrams, Christopher Nolan, Quentin Tarantino, and Judd Aptow, who lead a group of other passionate film supporters to urge Hollywood to keep film going (Giardiana, 2014), it looks like—for now, anyway—35mm is "still the stock of choice for really huge productions that want to look good and have enough complications (and a big enough budget) that the cost of digital color correction vs. the expense of shooting film no longer becomes a factor" (Rizov, 2015).

"Of course, facts are funny things… Digital cinema has a very short history—*Star Wars Episode II: Attack of the Clones* (2002) was the first full-on 24p [high definition digital] release… and… [10 years later], more than one-third of the films [up for] the industry's highest honors were shot digitally" (Frazer, 2012). In 2013, only four of the nine films nominated for Best Picture at the Academy Awards were shot on film, and *The Wolf of Wall Street* was the first to be distributed exclusively digitally (*Stray Angel Films*, 2014). And the trend toward increasing digital acquisition seems to be continuing. At the 2018 Oscars, of the ten nominations for Best Picture, seven were shot digitally (ARRI Alexa XT and ARRI Mini), including *The Shape of Water* which won the coveted award. However, the three other nominated films were shot on film. The 2018 Oscar nominations in the technical categories, however, are very good news for ARRI and the Alexa cameras; four of the five films nominated for Best Cinematography were shot using the ARRI Alexa. Only *Dunkurk* (2017), directed by Christopher Nolan with Dutch cinematographer Hoyte van Hoytema, was the sole nomination shot using film; 65mm IMAX to be exact (Pennington, 2018). Contrast this with, *Blade Runner 2049* (2017), directed by Denis Villeneuve and shot by British cinematographer Roger Deakins, using the ARRI Alexa XT with Zeiss Prime lenses and a large 2.40:1 aspect ratio for the film's release in IMAX (Pennington, 2018) and you can see that digital cameras can hold their own against the "gold standard" of traditional film.

Although visual aesthetics are important in cinematic storytelling, so is budget. Rachel Morrison, the cinematographer for director Dee Rees' 1940s period film *Mudbound* (2017)—and the first female to be nominated in the cinematography category—originally wanted to shoot on celluloid, but budget constraints forced a rethink. Eventually, Morrison used the ARRI Alexa Mini and 50-year old anamorphic lenses to create the vintage look (Pennington, 2018). Digital presents a significant savings for low-budget and independent filmmakers. Production costs using digital cameras and non-linear editing are a fraction of the costs of film production; sometimes as little as a few thousand dollars, and lower negative costs mean a faster track to profitability.

At Sundance 2018, all award-wining fiction films were shot digitally with the dominant camera being the ARRI Alexa or the Alexa Mini followed by the Panasonic VariCam and RED Weapon Helium 8K (O'Falt, 2018a). For Sundance documentaries, given their unique shooting situations, digital was the camera of choice and ranged from iPhones and GoPro action cameras to DSLRs (mostly Canon 5D Mark III), Canon C300, mirrorless Sony A7s, Sony FS7, and the ARRI Amira (O'Falt, 2018b). However, perhaps one of the most innovative films to come out of Sundance, was Sean Baker's 2015 feature film, *Tangerine*. The film was full of surprises. As a social realist film, the treatment of its subject matter (transgender prostitutes) raised a few eyebrows. But, perhaps most surprising of all was the fact that the entire film was shot in anamorphic widescreen aspect ratio (2.35:1) using the iPhone 5s (3, in fact), a then $8 app (Filmic Pro), a hand-held Steadicam Smoothee, and a set of prototype lenses from Moondog Labs (Newton, 2015). Although the decision to use the iPhone 5s was initially made as a matter of

budget necessity, Baker, and his co-cinematographer Radium Cheung, observed that the mobile phones added considerably to their goal of realism (Thomson, 2015). Sometimes, the best technology for making a film is what you have in your hand!

It is obvious that digital acquisition "offers many economic, environmental, and practical benefits" (Maltz, 2014). For the most part, this transition has been a boon for filmmakers, but as "born digital" productions proliferate, a huge new headache emerges: preservation (Maltz, 2014). This is why the Academy of Motion Picture Arts and Sciences sounded a clarion call in 2007 over the issue of digital motion picture data longevity in the major Hollywood studios. In their report, titled *The Digital Dilemma*, the Academy concluded that, although digital technologies provide tremendous benefits, they do not guarantee long-term access to digital data compared to traditional filmmaking using motion picture film stock. Digital technologies make it easier to create motion pictures, but the resulting digital data is much harder to preserve (Science & Technology Council, 2007).

The Digital Dilemma 2, the Academy's 2012 update to this initial examination of digital media archiving, focused on the new challenge of maintaining long-term archives of digitally originated features created by the burgeoning numbers of independent and documentary filmmakers (Science & Technology Council, 2012). Digital preservation issues notwithstanding, film preservation has always been of concern. According to the Library of Congress Film Preservation Study, less than half of the feature films made in the United States before 1950 and less than 20 percent from the 1920s are still around. Even films made after 1950 face danger from threats such as color-fading, vinegar syndrome, shrinkage, and soundtrack deterioration (Melville & Simmon, 1993). However, the "ephemeral" nature of digital material means that "digital decay" is as much a threat to films that were "born digital" as time and the elements are to nitrate and celluloid films of the past. The good news is that there is a lot of work being done to raise awareness of the risk of digital decay generally and to try to reduce its occurrence (Maltz, 2014). Among the things discovered from exploring issues surrounding digital preservation is "the crucial role of metadata for maintaining long-term access to digital materials... [The] information that gets stored needs to include a description of its contents, its format, what hardware and software were used to create it, how it was encoded, and other detailed 'data about the data.' And digital data needs to be managed forever" (Maltz, 2014).

Along with digital acquisition technology, new digital distribution platforms have emerged making it easier for independent filmmakers to connect their films with target audiences and revenue streams (through video-on-demand, pay-per-view, and online distribution). However, these platforms have not yet proven themselves when it comes to archiving and preservation (Science & Technology Council, 2012). "Unless an independent film is picked up by a major studio's distribution arm, its path to an audiovisual archive is uncertain. If a filmmaker's digital work doesn't make it to such a preservation environment, its lifespan will be limited—as will its revenue-generating potential and its ability to enjoy the full measure of U.S. copyright protection" (Science & Technology Council, 2012, p. 6). For now, at least, the "digital dilemma" seems far from over.

As physical film acquisition yields to digital recording, so too has film editing made the digital shift.

With digital cinema cameras like the ARRI Alexa and the RED cameras coming into more common use, and the advanced, computer-based non-linear editing systems having the capability of working with the digital footage at full resolution; it is now possible to shoot, edit, and project a movie without ever having to leave the digital environment.

Digital 3D

The first digital 3D film released was Disney's *Chicken Little*, shown on Disney Digital 3D (PR Newswire, 2005). Dolby Laboratories outfitted about 100 theaters in the 25 top markets with Dolby Digital Cinema systems in order to screen the film. The idea of actually shooting live-action movies in digital 3D did not become a reality until the creation of the Fusion Camera, a collaborative invention by director James Cameron and Vince Pace (*Hollywood Reporter*, 2005). The camera "fuses" two Sony HDC-F950 HD cameras "2½ inches apart to mimic the stereoscopic separation of human eyes" (Thompson, 2010). The

camera was used to film 2008's *Journey to the Center of the Earth* and 2010's *Tron Legacy*.

Cameron used a modified version of the Fusion Camera to shoot 2009's blockbuster *Avatar*. The altered Fusion allows the "director to view actors within a computer-generated virtual environment, even as they are working on a 'performance-capture' set that may have little apparent relationship to what appears on the screen" (Cieply, 2010).

Figure 10.12

Vince Pace and James Cameron with Fusion 3D Digital Cinema Camera

Source: CAMERON|PACE Group
Photo credit: Marissa Roth

Another breakthrough technology born from *Avatar* is the swing camera. For a largely animated world such as the one portrayed in the film, the actors must perform through a process called motion capture which records 360 degrees of a performance, but with the added disadvantage that the actors do not know where the camera will be (Thompson, 2010). Likewise, in the past, the director had to choose the shots desired once the filming was completed. Cameron tasked virtual-production supervisor Glenn Derry with creating the swing camera, which "has no lens at all, only an LCD screen and markers that record its position and orientation within the volume [the physical set space] relative to the actors" (Thompson, 2010). An effects switcher feeds back low-resolution CG images of the virtual world to the swing camera allowing the director to move around shooting the actors photographically or even capturing other camera angles on the empty stage as the footage plays back (Thompson, 2010).

When Hollywood studio executives saw the $2.8 billion worldwide gross receipts for *Avatar* (2009), they took notice of the film's ground-breaking use of 3D technology; and when Tim Burton's 2010 *Alice in Wonderland*—originally shot in 2D—had box office receipts reach $1 billion thanks to 3D conversion and audiences willing to pay an extra $3.50 "premium" up-charge for a 3D experience, the film industry decided that 3D was going to be their savior. Fox, Paramount, Disney, and Universal collectively shelled out $700 million to help equip theaters with new projectors, and the number of 3D releases jumped from 20 in 2009 to 45 in 2011. However, there was a slight drop to 41 in 2012, that fell further to only 35 in 2013 (Movie Insider, 2014a & 2014b respectively), and 3D releases for 2014 topped out at only 39 films while 2015 saw a sharp decline with only 28 3D releases that year (Box Office Mojo, 2017c). 2016 and 2017 saw a slight up-tick in 3D releases (35 and 34, respectively), and projected 3D film releases for 2018-2019 presently total 27, (Box Office Mojo, 2017c).

What accounts for this up and down—but, mostly down—cycle in 3D film releases? Does the success of landmark 3D films like *Avatar* (2009), *Hugo* (2011), *Gravity* (2013), *The Martian* (2015), and Disney's 3D release of *Star Wars: The Last Jedi* (2017), represent true innovations in cinematic storytelling, or has 3D been subsumed by moneymaking gimmickry? Perhaps, the answer draws from both perspectives. "Some movies use the technology to create more wonderfully immersive scenes, but others do little with the technology or even make a movie worse." (Lubin, 2014). In part, it could also be that the novelty has worn off with film audiences tired of having their expectations go unfulfilled. Citing the analysis of 1,000 random moviegoers' Tweets, Phillip Agnew, an analyst with Brandwatch (a social media monitoring and analytics company), observed that, "Before taking their seats,

the majority of cinema goers [600 out of 1,000] were positive about 3D films" (Agnew, 2015). However, comments made after screening the 3D movie showed that viewers were mostly disappointed. "Positive mentions decreased by 50%, whilst the amount of negative mentions more than doubled. Consumers expecting to see the 'future of cinema' were instead paying more for a blurrier viewing experience, and in many cases, headaches" (Agnew, 2015). Likewise, moviegoers may have tired of paying the hefty surcharge for 3D films that failed to meet their expectations. The latter could be the result of studios rushing post-production conversions of 2D films into 3D resulting in inferior products.

Digital Theater Conversions

In 1999, *Star Wars I: The Phantom Menace* was the first feature film projected digitally in a commercial cinema—although there were very few theaters capable of digital projection at the time.

January 2012, marked the film industry's crossover point when digital theater projection surpassed 35mm. Nearly 90,000 movie theaters were converted to digital by the end of 2012 (Hancock, 2014).

By 2014, Denmark, Hong Kong, Luxembourg, Norway, the Netherlands, Canada, and South Korea had achieved nearly full digital conversion. Meanwhile, Belgium, Finland, France, Indonesia, Switzerland, Taiwan, the U.K., and the U.S. are all well above 90 percent penetration for digital screens, while China accounted for over half of the 32,642 digital theater screens in the Asia-Pacific region (Hancock, 2014).

Australia, Singapore and Malaysia reported having achieved near 100 percent digitization in 2015, leaving South America as the only region lagging behind the rest of the world (Sohail, 2015).

The Motion Picture Association of America's *Theatrical Market Statistics 2016* report, observed that, in 2016, the number of digital screens in the United States now account for 98 percent of all US screens and the number of 3D screens increased two percent from 2015 (MPAA, 2017). Interestingly, analog screens were also up by 18 percent between 2015-2016 (MPAA, 2017), perhaps reflecting a resurgence in niche "art house" screenings of "classic" films. Continued rapid growth in global digital cinema conversion is still due in large measure to double digit growth in the Asia Pacific region (over 18 percent) with China accounting for 10 percent of the growth (MPAA, 2017). It appears that this slightly slower rate of digital screen proliferation than past years could be reaching its limit due to market penetration (MPAA, 2017).

Roger Frazee, Regal Theater's Vice President of Technical Services, having observed the changes, stated, "Films are now 200-gigabyte hard drives, and projectors are those big electronic machines in the corridor capable of working at multiple frame rates, transmitting closed-captioned subtitles and being monitored remotely. The benefit of digital is, you don't have damaged film. You don't have scratched prints, and it looks as good in week six as it does on day one" (cited in Leopold, 2013).

But, the global digital theater conversion is approaches it's endgame, as the MPAA reports, "Today, 95 percent of the world's cinema screens are digital, up [nearly] two percentage points from 2015" (2017, P. 8). So now exhibitors and technology manufacturers have turned their attention to the developing other technologies that can enhance the audience experience, drive innovation, and—hopefully—new revenues… the most visible (and audible) technologies being explored include laser projection and immersive audio (Hancock, 2015a).

Current Status

Global box office revenues in 2016 reached $38.6 billion, only a one percent rise from 2015, while domestic U.S. and Canadian box office revenues grew two percent over 2015 receipts, topping out at $11.4 billion on ticket sales of 1.32 billion ((MPAA, 2017). In 2017, gross domestic (U.S.) revenues were $11 billion, down 2.7 percent, while ticket sales were down 6.2 percent at just over 1.2 billion (Box Office Mojo, 2017d). Non-U.S. and Canadian box office revenues held steady in 2016 compared to 2015 ($27.2 billion), despite the increased strength of the U.S. dollar and slowed growth in China (MPAA, 2017). Still, global box office revenues are expected to continue their steady growth and are expected to reach $49.3 billion by the end of 2020 (Statista, 2018a).

In 2016, the average price of a movie ticket in the U.S. was $8.65 with just over 1.3 billion tickets sold for a total box office gross of $11.25 billion (The Numbers, 2018a). For 2017, U.S. movie ticket prices were up 3.6 percent to $8.97 while ticket sales were down to just over 1.2 billion reflecting total box office gross revenues at just under $11 billion (The Numbers, 2018b). The top five grossing films for the U.S. market were again dominated by Disney with three films, *Star Wars: The Last Jedi* ($605 million), *Beauty and the Beast* ($504 million), and *Guardians of the Galaxy 2* ($390 million), while Warner Brothers broke in at number three with *Wonder Woman* ($413 million), and Sony Pictures *Spiderman: Homecoming* ($334 million) came in fifth (The Numbers, 2018b).

The MPAA's Classification and Rating Administration (CARA) issue ratings (G, PG, PG-13, R, NC-17) on the basis of graphic sex or violence, drug use and dark or adult themes prior to a movie's release in order to provide parents with guidelines for deciding what films they would allow their children to watch (CARA, 2013). Nine of the 2016 MPAA rated films in the top ten received a PG or PG-13 rating, while only one, *Deadpool, was* rated R (The Numbers, 2018a). Continuing the trends of past years, there were 140 PG-13 rated films in 2016 worth $4.3 billion in gross domestic box office receipts (The Numbers, 2018a), and 2017 saw 129 PG-13 rated films in 2016 with gross domestic box office revenues of nearly $5.5 billion (The Numbers, 2018b). Interestingly, in 2016 there were more R rated films released than in past years (223), but they only accounted for $2.6 billion in gross box office revenue (The Numbers, 2018a). 2017 saw 178 R rated films with gross box office receipts totaling almost $2.7 billion (The Numbers, 2018b). There were only a combined 17 G rated films released over 2016-2017. All of the top ten films of 2016 were Action/Adventure films (The Numbers, 2018a), while the number one genre for 2017 was Action/Adventure films, the number two film in terms of gross revenues was a Musical (*Beauty and the Beast*) and the R rated Horror film *It* came in sixth (The Numbers, 2018b). Writing for *The Wrap*, Todd Cunningham perhaps stated it best, "It's not a coincidence that most films are rated PG-13. Not just because they're the most lucrative, but because life is unrated… Try to find a G-rated war movie, for example (Cunningham, 2015).

As part of the second phase of the digital cinema technology rollout, cinema operators have turned their attention to the electronic distribution of movies. In 2013, the Digital Cinema Distribution Coalition (DCDC) was formed by "AMC Theatres, Cinemark Theatres, Regal Entertainment Group, Universal Pictures and Warner Bros. to provide the industry with theatrical digital delivery services across North America through a specially created network comprised of next-generation satellite and terrestrial distribution technologies. It is capable of supporting feature, pro-motional, pre-show and live content distribution into theaters" (DCDC, 2016). Randy Blotky, CEO of DCDC, described the movie distribution network as a "…'smart pipe' made up of sophisticated electronics, software and hardware, including satellites, high-speed terrestrial links, with hard drives used as backup" (Blotky, 2014). "DCDC pays for all of the equipment that goes in the theatres [sic], we maintain all of that equipment, and we install all of that equip-ment at no cost to exhibition "(Blotky, cited it Fuchs, 2016). To recover their costs, DCDC charges fees for de-livery to both the theater and content providers that is "way less expensive than for delivering hard drives and physical media to the theatres [sic]. And also, we priced it below the normal return freight for exhibitors" (Blotky, cited it Fuchs, 2016). According to the DCDC, roughly three years after it began offering movies and other content to theaters via satellite distribution, it ended 2016 serving about three of every four U.S. screens—up from 62 percent in 2015 (Lieberman, 2017). By 2015, the DCDC had reached 23,579 screens, and in 2016, the coalition was under contract with 179 exhibi-tors with more than 2,650 theaters and 30,000-plus screens (Lieberman, 2017). Electronic distribution is not only taking hold in the U.S., from Canada, to Europe, Asia, Australia, and South America, all are examining options and solutions. In India, "the old, celluloid prints that where physically ferried to cinema halls have disappeared from India's 12,000 screens… . In-stead, the digital file of the film is downloaded by sat-ellite or other means to those cinema theatres [sic] that have paid for it… . Also, [electronic] distribution of d-cinema is simpler, faster, cheaper, and piracy can be better controlled" (BW Online Bureau, 2015).

Growth in digital 3D screens in 2016 continued for most regions and increased by 17 percent compared to 2015 which had a slightly more modest 15 percent growth rate (MPAA, 2017). The global portion of 3D digital screens increased three percentage points to 56 percent of all digital screens in 2016 (MPAA, 2017). As in past years, the Asia Pacific regions had the highest 3D digital proportion of total digital screens at 78 per-cent (MPAA, 2017). The number of digital 3D screens in the U.S. and Canada increased by only 2 percent (1415,783,318) in 2014 2016 compared to 2013 2015 (1415,483,077) even as 3D US box office revenues in 2016 were down 8% compared to 2015 (MPAA, 20142017).

So, why does Hollywood continue to make 3D movies if domestic revenues and audience attendance are down? The answer is quite simple, the total global box office receipts for 3D films have grown 44 percent since 2012 largely due to growth in the Asian markets; China, Japan and India specifically, but mostly China (MPAA, 2017). In 2015, China's 3D market was worth over $1 billion more than all of North America's. When *Jurassic World* was released in 2015, 3D screen-ings were responsible for a modest 48% of the film's U.S. domestic box office gross revenues. Yet, in China, 95 percent of the film's gross revenues came from 3D screenings. Hollywood studios have now started re-leasing 3D versions of some big films exclusively in China, knowing there's an audience that will "eat them up" (Epstein, 2017).

It is not all doom and gloom for U.S. domestic 3D cinema, however. Although it seems to have followed the *Gartner Hype Cycle* of rapidly reaching the "Peak of Inflated Expectation" after the two-fold "Technol-ogy Trigger" of the rising sophistication in CGI and motion capture animation combining with the in-creasing ubiquity of digital projection, only to plum-met into the "Trough of Disillusionment" in 2010 (Gyetvan, *n.d.*). However, it appears that some filmmakers now see 3D as a serious storytelling tool and not a visual gimmick. Thus, it won't be too long before the industry is ready to start up the "Slope of Enlightenment" they hope will eventually leads to the "Plateau of Productivity" and sustainable profits. Writing for *Wired*, Jennifer Wood observed that, after wringing their hands and lamenting whether 3D was even viable any longer in the domestic movie market-place, a film comes along to silence the naysayers

(Wood, 2013). "In 2009 it was James Cameron's *Avatar*. With 2011 came Martin Scorsese's *Hugo*... The take away should be that 3D must assume the role of supporting element and enabler of an otherwise outstanding story. If this is so, audiences will come and awards will follow—witness the 7 out of 10 Academy Awards given to *Gravity*—including Best Cinematography—at the 2014 Oscars as validation.

If domestic box office revenues are any indication of 3D's slow climb up the "Slope of Enlightenment," then the 3D release of *Star Wars: The Force Awakens* (2015) set a new benchmark earning a total of $936.66 million (lifetime gross) in North America, eclipsing the previous record holder, *Avatar* (2009) which slipped to second place (Statista, 2018b). The next three domestic top grossing 3D releases for the U.S. & Canada, as of January 2018, include: *Jurassic World* (2015) at $652.27 million, *The Avengers* (2012) with $623.36 million, and 2017's *Star Wars: The Last Jedi* coming in with $574.48 million (Statista, 2018b). If the *Gartner Hype Cycle* is correct, domestic 3D films should reach the "Plateau of Productivity" in another eight years.

Factors to Watch

Of all of the current trends in digital cinema camera designs, perhaps one of the most interesting is the shift toward larger sensors. Full-frame sensors (36mm x 24mm) have been used to shoot video in DSLR cameras since the Canon 5D Mark II was introduced in September 2008. Large format sensors, however, are a completely different matter. As Rich Lackey, writing for *Digital Cinema Demystified* observed, "It is clear that more is more when it comes to the latest high-end cinema camera technology. More pixels, more dynamic range, larger sensors, and more data. Enough will never be enough... however there are points at which the technology changes so significantly that a very definite generational evolution takes place" (Lackey, 2015b). Since the introduction of the full-frame sensor, the quest for higher resolution, higher dynamic range and better low light performance in digital cinema cameras inevitably resulted in a "devil's choice" in sensor design trade-offs. "High dynamic range and sensitivity require larger photosites, and simple mathematics dictate that this places limits on overall resolution for a given size sensor... unless, of course, you increase the size of the sensor" (Lackey, 2015a). Answering the call for larger sensors that can provide more pixels as well as greater dynamic range, the Phantom 65 was introduced with a 4K (4096 x 2440) sensor (52.1mm x 30.5mm), while ARRI Alexa 65, used to shoot 2015's *The Revenant* had a 54.12mm x 25.59mm sensor yielding 6.5K (6560 x 3100), and 2017's *Guardians of the Galaxy Vol. 2* was the first feature film shot using the 8K (8192 x 4320) RED Weapon using a 40.96mm x 21.6mm sensor. Of course, large frame sensors require optics that can cover the sensor and for this, the lens manufacturers have had to come to the table and join the discussion.

If 4K and 8K-plus image resolutions seems inevitable, then apparently, so are high frame rates (HFR) for digital acquisition and screening. Since the introduction of synchronous sound in movies, the standard frame rate has been 24 frames per second (fps). However, this was primarily a financial decision. In the shift from the variable, and more forgiving, hand-cranked 14- to 24-fps of the silent days, the technical demands of the "talkies" required a constant playback speed to keep the audio synchronized with the visuals. Using more frames meant more costs for film and processing, and studio executives found that 24-fps was the cheapest, minimally acceptable frame rate they could use for showing sound-synchronous movies with relatively smooth motion. Nearly a century later, it is still the standard. However, since the late 1920s, projectors have been using shutter systems that show the same frame two or three times to boost the overall frame rate and reduce the "flicker" an audience would otherwise experience. But even this is not enough to keep up with the fast motion of action movies and sweeping camera movements or fast panning. The visual artifacts and motion-blur that's become part of conventional filmmaking is visually accentuated in 3D, because our eyes are working particularly hard to focus on moving objects.

For most of the digital 3D movies already running in theaters, the current generation of digital projectors can operate at higher frame rates with just a software update, but theaters are still showing 3D movies at 24-fps. To compensate for any distractions that can result from barely perceivable flashing due to the progression of frames (causing eyestrain for some viewers), each frame image is shown three times per eye. Called "triple flashing," the actual frame rate is tripled, resulting in viewers receive 72 frames per second per eye for a total frame rate of 144-fps. Audiences watching a 3D film produced at 48-fps would see the same frame flashed twice per second (called, "double flashing"). This frame rate results in each eye seeing 96-fps and 192-fps overall, nearly eliminating all flicker and increasing image clarity. Any 3D films produced and screened at 60-fps, and double-flashed for each eye, would result in movie-goers seeing a 3D film at an ultra-smooth 240-fps.

As a leading proponent of HFR, director Peter Jackson justified the release of his *The Hobbit* film trilogy (2012, 2013, and 2014) in a 2014 interview, stating, "science tells us that the human eye stops seeing individual pictures at about 55 fps. Therefore, shooting at 48 fps gives you much more of an illusion of real life. The reduced motion blur on each frame increases sharpness and gives the movie the look of having been shot in 65mm or IMAX. One of the biggest advantages is the fact that your eye is seeing twice the number of images each second, giving the movie a wonderful immersive quality. It makes the 3D experience much more gentle and hugely reduces eyestrain. Much of what makes 3D viewing uncomfortable for some people is the fact that each eye is processing a lot of strobing, blur and flicker. This all but disappears in HFR 3D" (Jackson, 2014).

However, HFR film has its detractors who opine that 24-fps films deliver a depth, grain and tone that is unique to the aesthetic experience and not possible to recreate with digital video—this lack of "graininess" is often jarring and uncomfortable to first-time viewers of HFR 3D visual images. Such adjectives as "blurry," or "Hyper-real," "plastic-y," "weirdly sped-up," or the more derisive "Soap Opera Effect" are thrown around a lot by film critics and purists who perceive the 3D digital images presented at HFRs as "cool" and "sterile." When confronted with such criticisms leveled at *The Hobbit: An Unexpected Journey* (2012), Peter Jackson continued to use the 48-fps 3D digital format for his two other *Hobbit* films, but added lens filters and post-production work to relax and blur the imagery, "hoping it wouldn't look so painfully precise" (Engber, 2016).

Cinematic futurist like Peter Jackson, James Cameron (who plans to use HFR 3D for his four *Avatar* sequels), and Ang Lee—whose *Billy Lynn's Long Halftime Walk* (2016) broke new ground shooting in 3D with 4K resolution at 120-fps—the rejection of HFR digital technology seems bizarre. As James Cameron himself said, "I think [HFR] is a tool, not a format… I think it's something you want to weave in and out and use when it soothes the eyes, especially in 3D during panning, movements that [create] artifacts that I find very bothersome. I want to get rid of that stuff, and you can do it through high frame rates" (cited in Giardina, 2016). Perhaps then, Ang Lee's *Billy Lynn's Long Halftime Walk* breaks as much ground in the storytelling front as in the technical. As Daniel Engber, a columnist for *Slate Magazine* noted in his extensive review of Lee's film, "Most exciting is the way Lee modulates the frame rate from one scene to the next. At certain points he revs the footage up to 120 fps, while at [other times] the movie slides toward more familiar speeds. the most effective scenes in *Billy Lynn*… the ones that look the best in… HFR/4K/3D, are those that seem the least movie-like. First, a glimpse of action on the football field, just enough to advertise the format for showing live sports events. (Trust me, high-frame-rate 3-D sports will be extraordinary.) Second, Billy tussling, *Call of Duty*–style, with a soldier in a broken sewage pipe, hinting at a brighter future for virtual reality…. [Yet] Lee saves the most intense, hyperreal effects for Billy's wartime flashbacks, where the format works to sharpen the emotion. These are Billy's post-traumatic visions, the result of too much information stored inside his head, and the movie shows them as they seem to him—overly graphic, drenched in violent clarity, bleeding out minutiae. If there's any future for HFR in Hollywood, this must be it—not as a hardware upgrade on the endless path to total cinema, but as a tool that can be torqued to fill a need" (Engber, 2016).

A relatively recent innovation that could become a potential game-changer is that of light-field technology—also referred to as computational or plenoptic imaging and better understood as volumetric capture—and is expected to be a "disruptive" technology for cinematic virtual reality (VR) filmmakers. Whereas, ordinary cameras are capable of receiving 3D light and focusing this on an image sensor to create a 2D image, a plenoptic camera samples the 4D light field on its sensor by inserting a microlens array between the sensor and main lens (Ng, et al., 2005). Not only does this effectively yield 3D images with a single lens, but the creative opportunities of light-field technology will enable such typical production tasks as refocusing, virtual view rendering, shifting focal plains, and dolly zoom effects all capable of being accomplished during post-production. Companies including Raytrix and Lytro are currently at the forefront of light field photography and videography. In late 2015, "Lytro announced… the *Lytro Immerge*… a futuristic-looking sphere with five rings of light field cameras and sensors to capture the entire light field volume of a scene. The resulting video [was intended to] be compatible with major virtual reality platforms and headsets such as *Oculus*, and allow viewers to look around anywhere from the *Immerge's* fixed position, providing an immersive, 360 degree live-action experience" (Light-Field Forum, 2015). However, by February 2017, having raised an additional $60 million to continue developing the light-field technology—and with feedback from early productions using the spherical approach—the company decided to switch to a flat (planar) capture design (Lang, 2017). With this approach, capturing a 360-degree view requires the camera to be rotated to individually shoot each side of an eventual pentagonal capture volume and then stich the image together in post-production. The advantage of this is that the director and the crew can remain behind the camera and out of the shot throughout the production process.

In 2016, the cinematic world was introduced to the prototype Lytro Cinema Camera at the National Association of Broadcasters Convention in Las Vegas, NV. The Lytro Cinema Camera is large "…and unwieldy enough to remind DPs of the days when cameras and their operators were encased in refrigerator-sized sound blimps. But proponents insist the Lytro has the potential to change cinematography as we know it… It produces vast amounts of data, allowing the generation of thousands of synthetic points of view. With the resulting information, filmmakers can manipulate a range of image characteristics, including frame rate, aperture, focal length, and focus—simplifying what can be a lengthy, laborious process." (Heuring, 2016). The Lytro Cinema Camera has 755 RAW megapixels of resolution per frame utilizing a high resolution active scanning system, up to 16 stops of dynamic range and can shoot at up to 300-fps and generates a data stream of 300Gb per second while processing takes place in the cloud where Google spools up thousands of CPUs to compute each thing you do, while you work with real-time proxies (Sanyal, 2016). But, independent filmmakers will probably not be using a Lytro Cinema Camera any time soon. Rental packages for the Lytro Cinema Camera are reported to start at $125K. Thus, for now, the camera will probably only be attractive to high-end VFX companies willing to experiment with this technology (Ward, 2016).

Only a few years ago terms such as "immersive", "experiential", "volumetric" and "virtual reality" (VR) were on the outer fringe of filmmaking. Today, however, predictions are that VR will change the face of cinema in the next decade—but only if content keeps up with the advances in technology. It is projected that, by 2021, augmented reality (AR) and VR spending will increase from 2017's $11.4 billion to $215 billion worldwide (International Data Corporation, 2017). VR is already being heavily promoted by the tech giants, with Facebook and Microsoft launching new headsets they hope will ensure the format goes mainstream. As if to emphasize this point, in early 2017, IMAX opened its first VR cinema in Los Angeles (Borruto, 2017), while the leading film festivals—including Sundance, Cannes, Venice, and Tribeca—now have sections dedicated to recognizing ground-breaking work in VR. According to James George, co-founder of Scatter and technical director of *Blackout*, which the studio showed at Tribeca's 2017 International Film Festival, calls VR an "experience" rather than a film. "This whole desire to move into immersive is a generational shift. A new generation is demanding participation in their media—which you can call interactivity" (Dowd, 2017). Patric Palm, the CEO and co-founder of *Favro*, a collaboration application used by VR and AR studios, weighed in on how he sees technology and VR pushing the boundaries of filmmaking, stating that "[when] you're making a film, as a filmmaker, you are in the driver's seat of what's going to happen. You control everything. With a 360 movie, you have a little bit less of that because where the person's going to look is going to be a little different. More interactive experiences, where the consumer is in the driver's seat, are the single most interesting thing. Right now, the conversation is all about technology. What will happen when brilliant creative minds start to get interested in this space and they're trying to explore what kind of story we can do in this medium? There's a lot of money stepping into the VR and AR space, but a lot of creative talent is moving in this direction, too" (Volpe, 2017). Some believe that AR and VR are about to crest the "Peak of Inflated Expectation" of the *Gartner Hype Cycle* (if not already) and is destined to plunge into the "Trough of Disillusionment" in the coming year, but for now, it is a virtual cornucopia of possibilities. For more on VR and AR, see Chapter 15

In September 2016, 20th Century Fox announced that it used the artificial intelligence (AI) capabilities of IBM's Watson to make a trailer for its 2016 sci-fi thriller *Morgan*. Not surprisingly, everyone was eager to see how well Watson could complete such a creative and specialized task (Bludov, 2017). IBM research scientists "taught" Watson about horror movie trailers by feeding it 100 horror trailers, cut into scenes. Watson analyzed the data and "learned" what makes a horror trailer scary. The scientists then uploaded the entire 90-minute film and Watson "instantly zeroed in on 10 scenes totaling six minutes of footage" (Kaufman, 2017). To everyone's surprise, Watson did remarkably well. However, it's important to point out that a real person did the actual trailer editing using the scenes that were selected by Watson, so AI did not actually "cut" the trailer—a human was still needed to do that.

Figure 10.13

Morgan (2016), First Movie Trailer Created by IBM's Watson AI System

Source: 20th Century FOX, Official Trailer

For the 2016 Sci-Fi London 48 Hour Film Challenge, director Oscar Sharp submitted, *Sunspring*, a science fiction short written entirely by an AI system that named itself "Benjamin." Sharp and his longtime collaborator, Ross Goodwin, an AI researcher at New York University, wanted to see if they could get a computer to generate an original script. Benjamin is a Long Short-Term Memory recurrent neural network, a type of AI used for text recognition. "To train Benjamin in screenplay writing, Goodwin fed the AI dozens of sci-fi screenplays he found online… Benjamin learned to imitate the structure of a screenplay, producing stage directions and well-formatted character lines" (Newitz, 2016). Unfortunately, the script had an incoherent plot, quirky and enigmatic dialogue and almost impossible stage directions like "He is standing in the stars and sitting on the floor" (Newitz, 2016). The actors did the best they could and when Sharp and Goodwin submitted *Sunspring*, they were surprised the film placed in the top ten. "One judge, award-winning sci-fi author Pat Cadigan, said, 'I'll give them top marks if they promise never to do this again'" (cited in Newitz, 2016).

Jack Zhang, through his company Greenlight Essentials, has taken a different approach to AI script generation. Using what Zhang calls "augmented intelligence software" to analyze audience response data to help writers craft plot points and twists that connect with viewer demand (Busch, 2016). The patent-pending predictive analytics software helped write, *Impossible Things*, a feature-length screenplay Zhang describes as "the scariest and creepiest horror film out there" (Nolfi, 2016). Additionally, the AI software suggested a specific type of trailer which included key scenes from the script to increase the likelihood that the target audience would like it. Greenlight Essentials produced the trailer to use in a Kickstarter campaign to fund the film.

Presently, there are more "misses" then there are "hits" with the use of AI or "machine learning." Machine learning was coined in 1959 by computer game pioneer Author Samuels to describe a computer capable of being "trained" to do any repetitive task. Machine learning already plays an increasingly significant role in VFX and post-production, performing such repetitive tasks as 3D camera-match moves and rotoscoping. "Adobe, for example, is working with the Beckman Institute for Advanced Science and Technology to use a kind of machine learning to teach a software algorithm how to distinguish and eliminate backgrounds…[with] the goal of automating compositing"

(Kaufman, 2017). Philip Hodgetts, founder of Intelligent Assistance and Lumberjack System, two integrated machine learning companies, argues that "there's a big leap from doing a task really well to a generalized intelligence that can do multiple self-directed tasks" (cited in Kaufman, 2017). So, it appears that machine learning is here, and AI is likely to arrive, inevitably, to the entertainment industry as well. Nonetheless, the very human talent of creativity—a specialty in the entertainment industry—appears to be safe for the foreseeable future. (Kaufman, 2017).

There is little doubt that emerging technologies will continue to have a profound impact on cinema's future. This will also force the industry to look at an entirely new kind of numbers game that has nothing to do with weekend grosses, HFR, 3D, VR, or AI.

As Steven Poster, ASC and President of the International Cinematographers Guild stated, "Frankly I'm getting a little tired of saying we're in transition. I think we've done the transition and we're arriving at the place we're going to want to be for a while. We're finding out that software, hardware and computing power have gotten to the point where it's no longer necessary to do the things we've always traditionally done… [and] as the tools get better, faster and less expensive… [what] it allows for is the image intent of the director and director of photography to be preserved in a way that we've never been able to control before" (Kaufman, 2014b). When discussing *Avatar*, James Cameron stated that his goal is to render the technological presence of cinema invisible. "[Ideally], the technology is advanced enough to make itself go away. That's how it should work. All of the technology should wave its own wand and make itself disappear" (cited in Isaacs, 2013, P. 246).

Indeed, moviegoers of the future might look back on today's finest films as quaint, just as silent movies produced a century ago seem laughably imperfect to moviegoers today (Hart, 2012). Cinematographer David Stump, noting the positive changes brought about by the transition from analog to digital, states that "[the] really good thing that I didn't expect to see… is that the industry has learned how to learn again…. We had the same workflow, the same conditions and the same parameters for making images for 100 years. Then we started getting all these digital cameras and

workflows and… [now] we have accepted that learning new cameras and new ways of working are going to be a daily occurrence" (Kaufman, 2014a).

Getting a Job

University film programs can teach you a lot about the art and craft of visual storytelling. But, there is a big difference between an "student" film/video set and a "professional" set. "Until you are on a professional set and experience the daily 12-hour plus grind of making movies… knowing the craft and practicing the craft can seem like the difference between knowing how a camera works and building one from parts in a box" (Ogden, 2014).

Few individuals walk out of a university film program and on to a professional production set—most get their start through an internship. Hollywood production companies like Disney/ABC or LucasFilms, and many others, have active (and competitive!) paid internship programs looking to place eager creative minds in to such areas as technology (*e.g.,* IT operations, engineering, DIT, computer visual effects, etc.), to post-production and editorial, to camera, sound design, art and animation, to production management, as well as research and development or public relations and marketing and even accounting. Film and media interns can find themselves working in all areas of the entertainment industry from films, to video games, television to augmented and virtual reality programming, some even find themselves working on the design of immersive theme park ride experiences.

Once you obtain an internship, you must demonstrate your willingness and ability to work and perform every reasonable task, however menial the task may seem at the beginning of your internship. "…[Being] a production intern is a 'test' with only one question; are you willing to become the best intern you can even though you know that you do not want to be an intern for long? If you have chosen your internship well and continue to display enthusiasm for your work, you may be given a bit more responsibility and an opportunity to gain experience with a greater variety of work areas, tasks, and duties as your internship progresses" (Musburger & Ogden, 2014, p. 242). The logic behind this is that those individuals who are above you paid their dues and proved their passion, and they expect you to do the same.

It isn't easy to "break in" to the business—even with a successful internship. But the great thing about the film industry is that once you have a good start in it, and have established yourself and built a reliable network, the possibilities for high earnings and success are well within your grasp.

Projecting the Future

It's 2031. You and your companion settle into your seats in the movie theater for the latest offering from Hollywood's dream factory. You're there for the experience, to sit in the dark with a group of intimate strangers and lose yourself in the story. Though the movie is simultaneously available at home, on your mobile device, or projected on your retina by the latest incarnation of Google Glass—it's not the same. Here, the 4D motion seating engulfs you in comfort and responds according to the projected image (you've opted to forgo the X, or extra sensory add-ons like smell, however); the large, curved screen encompasses your full field of view (Hancock, 2015a). The lights dim, the murmur of conversation dies down. Everyone is anticipating the immersive, experiential spectacle that is the "new cinema." You reflect on how far cinema technology has come. Although film still exists, it's only a niche market in art-house screening rooms. Today's cinema experience is fully digital—from acquisition to intermediary, distribution to projection. Funny, what actually "saved film" isn't its "organic" aesthetic or even hipster nostalgia; it was its longevity as an archive medium.

The first two decades of the 21st Century saw digital cinema camera manufactures push the envelope of image resolution beyond 8K (Frazer, 2014; Lackey, 2015a) while the dynamic range expanded, contrast ratios improved, and color depth increased; all allowing for the creation of images that "popped" off the screen with blacker blacks and richer colors than the human eye could even detect (Koll, 2013).

The early laser-illuminated projectors developed by Barco, Christie and IMAX have also improved substantially since 2015 (Hancock, 2015a). Today's laser projectors yield images as sharp as sunlight and with more "organic" colors that result in the screen itself seeming to simply vanish (Collin, 2014). Likewise, laser projectors and faster frame-rates (up to 120fps) saved 3D. Film director Christopher Noland commented that, "Until we get rid of the glasses or until we really massively improve the process... I'm a little weary of [3D]" (De Semlyn, 2015). His words mirrored what you also thought at the time. However, today the uncomfortable glasses and post-screening headaches are gone and new generations of technology and cinematic storytellers have really advanced the art of 3D cinema.

The rumble of the immersive audio system snaps you from your reverie. Amazing how much audio influences your cinematic experience. Beginning with the 2011 Barco Auro-3D system and the roll-out of Dolby Atmos in 2015, as well as the DST:X systems soon after; immersive audio spread rapidly as audiences became enveloped in a spatial environment of three-dimensional sound (Boylan, 2015). Real advancements in "3D audio" actually came from virtual reality and gaming industries (now, basically the same) thanks to Facebook's acquisition of Oculus in 2014 and a desire to expand the technology beyond VR gaming (Basnicki, 2015). You recall a quote you read from Varun Nair, founder of Two Big Ears, "All of a sudden immersive audio stopped being a technology that existed for [its own sake, and became something]... ultimately very crucial: you've got great 3D visuals creating a sense of realism, and the audio needs to match up to it" (cited it Basnicki, 2015).

It's clear that technology has become a fundamental part of both cinematic storytelling and exhibition in a way that was at times fiercely resisted in the early part of the 21st Century. However, innovation in cinema technology has proven necessary and inexorable. For storytelling to maintain its rightful place at the head of pop-culture, it had to embrace the future of "new cinema" and use it to tell stories in new and unique ways.

Bibliography

Adorno, T. (1975). Culture industry reconsidered. *New German Critique,* (6), Fall. Retrieved from http://libcom.org/library/culture-industry-reconsidered-theodor-adorno.

Agnew, P. (2015, May 19). Research: The rapid rise and even more rapid fall of 3D movies. *Brandwatch Blog.* Retrieved from https://www.brandwatch.com/2015/05/4-reasons-why-3d-films-have-failed/.

Andersen, A. (2017, August 2). Behind the Spectacular Sound of *Dunkirk* — With Richard King. *A Sound Effect*. Retrieved from https://www.asoundeffect.com/dunkirk-sound/.

Andrae, T. (1979). Adorno on film and mass culture: The culture industry reconsidered. *Jump Cut: A Review of Contemporary Media*, (20), 34-37. Retrieved from http://www.ejumpcut.org/archive/onlinessays/JC20folder/AdornoMassCult.html.

ARRI Group. (2016). *Digital Camera Credits*. Retrieved from http://www.arri.com/camera/digital_cameras/credits/.

Bankston, D. 2005, April). The color space conundrum part two: Digital workflow. *American Cinematographer*, 86, (4). Retrieved from https://www.theasc.com/magazine/april05/conundrum2/index.html.

Belton, J. (2013). *American cinema/American culture* (4th Edition). New York: McGraw Hill.

Blotky, R. (2014 March 24). Special delivery: DCDC network promises to revolutionize cinema content distribution. Film Journal International. Retrieved from http://www.filmjournal.com/content/special-delivery-dcdc-network-promises-revolutionize-cinema-content-distribution.

Bludov, S. (2017, August 16). Is Artificial Intelligence Poised to Revolutionize Hollywood? *Medium*. Retrieved from, https://medium.com/dataart-media/is-artificial-intelligence-poised-to-revolutionize-hollywood-e088257705a3.

Bomboy, S. (2015, May 4). The day the Supreme Court killed Hollywood's studio system. *Constitution Daily* [Blog]. Retrieved from http://blog.constitutioncenter.org/2015/05/the-day-the-supreme-court-killed-hollywoods-studio-system/

Bordwell, D., Staiger, J. & Thompson, K. (1985). *The classical Hollywood cinema: Film style & mode of production to 1960*. New York: Columbia University Press.

Borruto, A. (2017, February 15). IMAX'S First Virtual Reality Theater is Transforming the Film Industry. *Resource*. Retrieved from, http://resourcemagonline.com/2017/02/imaxs-first-virtual-reality-theater-is-transforming-the-film-industry/76083/.

Box Office Mojo. (2017a). *All Time Box Office Worldwide Grosses*. Retrieved http://www.boxofficemojo.com/alltime/world/.

Box Office Mojo. (2017b). *All Time Box Office Domestic Grosses: Adjusted for Ticket Price Inflation*. Retrieved from http://www.boxofficemojo.com/alltime/adjusted.htm.

Box Office Mojo. (2017c). Genres: 3D, 1980-Present. Retrieved from http://www.boxofficemojo.com/genres/chart/?id=3d.htm.

Box Office Mojo. (2017d). *Yearly Box Office*. Retrieved from http://www.boxofficemojo.com/yearly/.

Boylan, C. (2015, January 24). What's up with 3D immersive Sound: Dolby Atmos, DTS:X and AURO-3D? *Big Picture, Big Sound*. Retrieved from http://www.bigpicturebigsound.com/What-s-Up-with-3D-Immersive-Sound.shtml.

Bradley, E.M. (2005). *The first Hollywood sound shorts, 1926-1931*. Jefferson, NC: McFarland & Company.

Bredow, R. (2014, March). Refraction: Sleight of hand. *International Cinematographers Guild Magazine*, 85(03), p. 24 & 26.

Busch, A. (2016, August 25). Wave Of The Future? How A Smart Computer Is Helping To Craft A Horror Film. *Deadline Hollywood*. Retrieved from, http://deadline.com/2016/08/concourse-productivity-media-horror-film-impossible-things-smart-computer-1201808227/.

BW Online Bureau (2015, November 17). Battle of the box office. *BW Businessworld*. Retrieved from http://businessworld.in/article/Battle-Of-The-Box-Office/17-11-2015-82087/.

Cieply, M. (2010, January 13). For all its success, will "Avatar" change the industry? *The New York Times*, C1.

Classification and Rating Administration (CARA). (2013). *How: Tips to be "screenwise."* Retrieved from http://filmratings.com/how.html.

Cook, D.A. (2004). *A history of narrative film* (4th ed.). New York, NY: W.W. Norton & Company.

Cunningham, T. (2015, March 4). PG-13 vs. R movies: How each rating stacks up at the box office. *The Wrap*. Retrieved from http://www.thewrap.com/pg-13-vs-r-movies-how-each-rating-stacks-up-at-the-box-office/.

De La Merced, M. (2012, January 19). Eastman Kodak files for bankruptcy. *The New York Times*. Retrieved from http://dealbook.nytimes.com/2012/01/19/eastman-kodak-files-for-bankruptcy/.

Denison, C. (2013, March 1). THX wants to help tune your home theater, not just slap stickers on it. *Digital Trends*. Retrieved from http://www.digitaltrends.com/home-theater/thx-wants-to-help-tune-your-home-theater-not-just-slap-stickers-on-it/.

Desowitz, B. (2017 October 17). *Blade Runner 2049*: How VFX Masters Replicated Sean Young as Rachael. *IndieWire*. Retrieved from http://www.indiewire.com/2017/10/blade-runner-2049-vfx-replicant-sean-young-rachael-1201889072/.

Dirks, T. (2009, May 29). Movie history—CGI's evolution From *Westworld* to *The Matrix* to *Sky Captain and the World of Tomorrow*. *AMC Film Critic*. Retrieved from http://www.filmcritic.com/features/2009/05/cgi-movie-milestones/.

Digital Cinema Distribution Coalition (DCDC). (2016). *About us*. Retrieved from http://www.dcdcdistribution.com/about-us/.

Digital Cinema Initiative (DCI). (2016). *About DCI*. Retrieved from http://www.dcimovies.com.

Dolby. (2010a). *Dolby Digital Details*. Retrieved from http://www.dolby.com/consumer/understand/playback/dolby-digital-details.html.

Dolby. (2010b). 5.1 Surround sound for home theaters, TV broadcasts, and cinemas. Retrieved from http://www.dolby.com/consumer/understand/playback/dolby-digital-details.html.

Dowd, V. (2017, April 19). Is 2017 the year of virtual reality film-making? *BBC News*. Retrieved from, http://www.bbc.com/news/entertainment-arts-39623148.

Engber, D. (2016, October 20). It Looked Great. It Was Unwatchable. *Slate*. Retrieved from http://www.slate.com/articles/arts/movies/2016/10/billy_lynn_s_long_halftime_walk_looks_fantastic_it_s_also_unwatchable.html

EPS Geofoam raises Stockton theater experience to new heights. (n.d.). Retrieved from http://www.falcongeofoam.com/Documents/Case_Study_Nontransportation.pdf.

Epstein, A. (2017, March 24). Americans are over 3D movies, but Hollywood hasn't got the memo. *Quartz Media*. Retrieved from https://qz.com/940399/americans-are-over-3d-movies-but-hollywood-hasnt-got-the-memo/.

Failes, I. (2017, June 6). What it Took to Make Emma Watson's *Beauty and the Beast* Costar Look Like A Beast. *Thrillist Entertainment*. Retrieved from, https://www.thrillist.com/entertainment/nation/beauty-and-the-best-movie-dan-stevens-special-effects.

Fleming Jr., M. (2016, January 7). Picture this: Hateful 8 caps strong year for movies for Kodak. *Deadline Hollywood*. Retrieved from http://deadline.com/2016/01/kodak-the-hateful-eight-star-wars-the-force-awakens-bridge-of-spies-1201678064/.

Frazer, B. (2012, February 24). Oscars favor film acquisition, but digital looms large. *Studio Daily*. Retrieved from http://www.studiodaily.com/2012/02/oscars-favor-film-acquisition-but-digital-looms-large/.

Freeman, J.P. (1998). Motion picture technology. In M.A. Blanchard (Ed.), *History of the mass media in the United States: An encyclopedia*, (pp. 405-410), Chicago, IL: Fitzroy Dearborn.

Frei, V. (2017, May 19). Interviews: *Guardians of the Galaxy 2*, Christopher Townsend, Production VFX Supervisor, Marvel Studios. *Art of VFX*. Retrieved from, http://www.artofvfx.com/guardians-of-the-galaxy-vol-2-christopher-townsend-production-vfx-supervisor-marvel-studios/.

Fuchs, A. (2016, January 21). Delivering on the Promise: DCDC connects content and cinemas large and small. *Film Journal International*. Retrieved from http://www.filmjournal.com/features/delivering-promise-dcdc-connects-content-and-cinemas-large-and-small.

Fulford, R. (1999). *The triumph of narrative: Storytelling in the age of mass culture*. Toronto, ON: House of Anansi Press.

Gelt, J. and Verrier, R. (2009, December 28) "Luxurious views: Theater chain provides upscale movie-going experience." *The Missoulian*. Retrieved from http://www.missoulian.com/business/article_934c08a8-f3c3-11de-9629-001cc4c03286.html.

Giardina, C. (2014, July 30). Christopher Nolan, J.J. Abrams Win Studio Bailout Plan to Save Kodak Film. *The Hollywood Reporter*. Retrieved from https://www.hollywoodreporter.com/behind-screen/christopher-nolan-jj-abrams-win-722363.

Giardina, C. (2016, October 29). James Cameron Promises Innovation in *Avatar* Sequels as He's Feted by Engineers. *The Hollywood Reporter*. Retrieved from https://www.hollywoodreporter.com/behind-screen/james-cameron-promises-innovation-avatar-sequels-as-hes-feted-by-engineers-942305.

Gyetvan, A. (n.d.). The Dog Days of 3D. *CreativeCOW*. Retrieved from https://library.creativecow.net/article.php?author_folder=gyetvan_angela&article_folder=magazine_29_Dog-Days-of-3D&page=1.

Haines, R. W. (2003). *The Moviegoing Experience, 1968-2001*. North Carolina: McFarland & Company, Inc.

Hancock, D. (2014, March 13). Digital Cinema approaches end game 15 years after launch. *IHS Technology*. Retrieved from https://technology.ihs.com/494707/digital-cinema-approaches-end-game-15-years-after-launch.

Hancock, D. (2015a, April 17). Advancing the cinema: Theatres reinforce premium status to ensure their future. *Film Journal International*. Retrieved from http://www.filmjournal.com/features/advancing-cinema-theatres-reinforce-premium-status-ensure-their-future.

Hancock, D. (2015b, March 27). The final push to global digital cinema conversion. *IHS Technology*. Retrieved from https://technology.ihs.com/527960/the-final-push-to-global-digital-cinema-conversion.

Hardawar, D. (2017, July 24). Dunkirk demands to be experienced in a theater. Engadget. Retrieved from https://www.engadget.com/2017/07/24/dunkirk-movie-theater-experience/.

Hart, H. (2012, April 25). Fast-frame Hobbit dangles prospect of superior cinema, but sill theaters bite? *Wired*. Retrieved from http://www.wired.com/underwire/2012/04/fast-frame-rate-movies/all/1.

Heuring, D. (2016, June 15). Experimental Light-Field Camera Could Drastically Change the Way Cinematographers Work. *Variety*. Retrieved from, http://variety.com/2016/artisans/production/experimental-light-field-camera-1201795465/.

Higgins, S. (2000). Demonstrating three-colour Technicolor: Early three-colour aesthetics and design. *Film History*, 12, (3), Pp. 358-383.

Hollywood Reporter. (2005, September 15). *Future of Entertainment.* Retrieved from http://www.hollywoodreporter.com/hr/search/article_display.jsp?vnu_content_id=1001096307.

Horkheimer, M. & Adorno, T. (1969). *Dialectic of enlightenment.* New York: Herder & Herder.

Hull, J. (1999). *Surround sound: Past, present, and future.* Dolby Laboratories Inc. Retrieved from http://www.dolby.com/uploadedFiles/zz-_Shared_Assets/English_PDFs/Professional/2_Surround_Past.Present.pdf.

Indiewire. (2014, January 27). *How'd They Shoot That? Here's the Cameras Used By the 2014 Sundance Filmmakers.* Retrieved from http://www.indiewire.com/article/how-they-shot-that-heres-what-this-years-sundance-filmmakers-shot-on.

International Data Corporation (2017, August 7). *Worldwide Spending on Augmented and Virtual Reality Expected to Double or More Every Year Through 2021, According to IDC.* Retrieved from, https://www.idc.com/getdoc.jsp?containerId=prUS42959717.

Isaacs, B. (2013). *The Orientation of Future Cinema: Technology, Aesthetics, Spectacle.* New York, London: Bloomsbury.

Jackson, P. (2014). *See It in HFR 3D: Peter Jackson HFR Q&A.* Retrieved from http://www.thehobbit.com/hfr3d/qa.html.

Jacobson, B. (2011, June). The Black Maria: Film studio, film technology (cinema and the history of technology). *History and Technology*, 27, (2), 233-241.

Jones, B. (2012, May 30). New technology in AVATAR—Performance capture, fusion camera system, and simul-cam. AVATAR. Retrieved from http://avatarblog.typepad.com/avatar-blog/2010/05/new-technology-in-avatar-performance-capture-fusion-camera-system-and-simulcam.html.

Jowett, G. (1976). *Film: The Democratic Art.* United States: Little, Brown & Company.

Karagosian, M. & Lochen, E. (2003). *Multichannel film sound.* MKPE Consulting, LLC. Retrieved from http://mkpe.com/publications/d-cinema/misc/multichannel.php.

Kaufman, D. (2011). Film fading to black. *Creative Cow Magazine.* Retrieved from http://magazine.creativecow.net/article/film-fading-to-black.

Kaufman, D. (2014b). Technology 2014 | Production, Post & Beyond: Part TWO. *Creative Cow Magazine.* Retrieved from http://library.creativecow.net/kaufman_debra/4K_future-of-cinematography-2/1.

Kaufman, D. (2014a). Technology 2014 | Production, Post & Beyond: Part ONE. *Creative Cow Magazine.* Retrieved from http://library.creativecow.net/kaufman_debra/4K_future-of-cinematography/1.

Kaufman, D. (2017, April 18). Artificial Intelligence Comes to Hollywood. *Studio Daily.* Retrieved from, http://www.studiodaily.com/2017/04/artificial-intelligence-comes-hollywood/.

Kindersley, D. (2006). *Cinema Year by Year 1894-2006.* DK Publishing.

Kolesnikov-Jessop, S. (2009, January 9). Another dimension. *The Daily Beast.* Retrieved from http://www.thedailybeast.com/newsweek/2009/01/09/another-dimension.html.

Lackey, R. (2015a, November 24). Full frame and beyond—Large sensor digital cinema. *Cinema5D.* Retrieved from https://www.cinema5d.com/full-frame-and-beyond-large-sensor-digital-cinema/.

Lackey, R. (2015b, September 9). More resolution, higher dynamic range, larger sensors. *Digital Cinema Demystified.* Retrieved from http://www.dcinema.me/2015/09/more-resolution-higher-dynamic-range-larger-sensors/.

Lang, B. (2017, April 11). Lytro's Latest VR Light-field Camera is Huge, and Hugely Improved. *Road To VR.* Retrieved from https://www.roadtovr.com/lytro-immerge-latest-light-field-camera-shows-major-gains-in-capture-quality/.

Leiberman, D. (2017, March 23). About 75% Of Screens Receive Movies Via Satellite, Digital Cinema Group Says. *Deadline Hollywood.* Retrieved from http://deadline.com/2017/03/about-75-screens-receive-movies-via-satellite-digital-cinema-dcdc-1202048869/.

Leopold, T. (2013, June 3). Film to digital: Seeing movies in a new light. *CNN Tech.* Retrieved from http://www.cnn.com/2013/05/31/tech/innovation/digital-film-projection/.

Light-Field Forum (2015, November 7). *Lytro Immerge: Company focuses on cinematic virtual reality creation.* Retrieved from http://lightfield-forum.com/2015/11/lytro-immerge-company-focuses-on-cinematic-virtual-reality-creation/.

Lubin, G. (2014, July 3). Here's when 3D movies work, when they don't, and what the future holds. *Business Insider.* Retrieved from http://www.businessinsider.com/are-3d-movies-worth-it-2014-7.

Lytro (2018). *Lytro Cinema: The Ultimate Cinema Tool.* Retrieved from, https://www.lytro.com/cinema.

Maltz, A. (2014, February 21). Will today's digital movies exist in 100 years? *IEEE Spectrum.* Retrieved from http://spectrum.ieee.org/consumer-electronics/standards/will-todays-digital-movies-exist-in-100-years.

Martin, K. (2014, March). Kong to Lift-Off: A history of VFX cinematography before the digital era. *International Cinematographers Guild Magazine*, 85(03), p. 68-73.

McMahan, A. (2003). *Alice Guy Blaché: Lost Visionary of the Cinema*. New York: Continuum 2003.

Media Insider (2018). The Future Trends in VFX Industry. *Insights Success*. Retrieved from, http://www.insightssuccess.com/the-future-trends-in-vfx-industry/.

Melville, A. & Simmon, S. (1993). *Film preservation 1993: A study of the current state of American film preservation*. Report of the Librarian of Congress. Retrieved from https://www.loc.gov/programs/national-film-preservation-board/preservation-research/film-preservation-study/current-state-of-american-film-preservation-study/.

Mondello, B. (2008, August 12). Remembering Hollywood's Hays Code, 40 years on. *NPR*. Retrieved from http://www.npr.org/templates/story/story.php?storyId=93301189.

Motion Picture Association of America (MPAA). (2011). *What each rating means*. Retrieved from http://www.mpaa.org/ratings/what-each-rating-means.

Movie Insider. (2014a*). 3-D Movies 2012*. Retrieved from http://www.movieinsider.com/movies/3-D/2012/.

Movie Insider. (2014b). *3-D Movies 2013*. Retrieved from http://www.movieinsider.com/movies/3-D/2013/.

MPAA (Motion Picture Association of America) (2017, March). Theatrical market statistics 2016. Retrieved from https://www.mpaa.org/wp-content/uploads/2017/03/MPAA-Theatrical-Market-Statistics-2016_Final-1.pdf.

Muñoz, L. (2010, August). James Cameron on the future of cinema. *Smithsonian Magazine*. Retrieved from http://www.smithsonianmag.com/specialsections/40th-anniversary/James-Cameron-on-the-Future-of-Cinema.html.

Musburger, R. & Ogden, M. (2014). *Single-camera video production*, (6th Ed.). Boston, MA & Oxford, UK: Focal Press.

Nagy, E. (1999). The triumph of narrative: Storytelling in the age of mass culture (Parts 1-5), *Ideas with Paul Kennedy: Robert Fulford's 1999 CBC Massey lectures* [Radio lecture series]. Retrieved from http://www.cbc.ca/radio/ideas/the-1999-cbc-massey-lectures-the-triumph-of-narrative-storytelling-in-an-age-of-mass-culture-1.2946862

Neale, S. (1985). *Cinema and technology: Image, sound, colour*. Bloomington, IN: Indiana University Press.

New Jersey Hall of Fame (2018). *2013 Inductees*. Retrieved from https://njhalloffame.org/hall-of-famers/2013-inductees/.

Newitz, A. (2016, June 9). Movie written by algorithm turns out to be hilarious and intense. *Ars Technica*. Retrieved from, https://arstechnica.com/gaming/2016/06/an-ai-wrote-this-movie-and-its-strangely-moving/.

Newton, C. (2015, January 28). How one of the best films at Sundance was shot using an iPhone 5s. *The Verge*. Retrieved from http://www.theverge.com/2015/1/28/7925023/sundance-film-festival-2015-tangerine-iphone-5s.

Ng, R., Levoy, M., Brüdif, M., Duval, G., Horowitz, M. & Hanrahan, P. (2005, April). *Light Field Photography with a Hand-Held Plenoptic Camera*. Retrieved from http://graphics.stanford.edu/papers/lfcamera/.

Nolfi, J. (2016, July 26). Artificial intelligence writes 'perfect' horror script, seeks crowdfunding. *Entertainment Weekly*. Retrieved from, http://www.ew.com/article/2016/07/26/artificial-intelligence-writes-perfect-movie-script/.

O'Falt, C. (2018a, January 17). Sundance 2018: Here Are the Cameras Used to Shoot This Year's Narrative Films. *IndieWire*. Retrieved from http://www.indiewire.com/2018/01/sundance-2018-cameras-movies-1201918085/.

O'Falt, C. (2018b, January 22). Sundance 2018: Here Are the Cameras Used to Shoot This Year's Documentaries. *IndieWire*. Retrieved from http://www.indiewire.com/2018/01/sundance-documentary-camera-films-non-fiction-1201918709/.

Ogden, M. (2014). *Website: Single-camera video production*, (6th Edition). Focal Press. Retrieved from http://routledgetextbooks.com/textbooks/9780415822589/future.php.

Parce qu'on est des geeks! [Because we are geeks!]. (July 2, 2013). *Pleins feux sur—Georges Méliès, le cinémagicien visionnaire* [Spotlight—Georges Méliès, the visionary cinema magician]. Retrieved March 16, 2014 from http://parce-qu-on-est-des-geeks.com/pleins-feux-sur-georges-melies-le-cinemagicien-visionnaire/.

Pennington, A. (2018, January 24). 2018 Oscar nominations: Inside the technical categories. International Broadcasting Convention. Retrieved from https://www.ibc.org/production/2018-oscar-nominations-inside-the-technical-categories/2640.article.

Poster, S. (2012, March). President's letter. *ICG: International Cinematographers Guild Magazine*, 83(03), p. 6.

PR Newswire. (2005, June 27). The Walt Disney Studios and Dolby Bring Disney Digital 3-D(TM) to Selected Theaters Nationwide With CHICKEN LITTLE. Retrieved from http://www.prnewswire.co.uk/cgi/news/release?id=149089.

RED Digital Cinema. (2016). Shot on RED. Retrieved from http://www.red.com/shot-on-red?genre=All&sort=release_date_us:desc.

Renée, V. (2017, January 24). Which cameras were used on the Oscar-nominated films of 2017? *No Film School*. Retrieved from https://nofilmschool.com/2017/01/which-cameras-were-used-oscar-nominated-films-2017.

Rizov, V. (2015, January 15). 39 movies released in 2014 shot on 35mm. *Filmmaker Magazine*. Retrieved from http://filmmakermagazine.com/88971-39-movies-released-in-2014-shot-on-35mm/#.Vvu2-2P0gdc.

Sanyal, R. (2016, April 20). Lytro poised to forever change filmmaking: debuts Cinema prototype and short film at NAB. *Digital Photography Review*. Retrieved from, https://www.dpreview.com/news/6720444400/lytro-cinema-impresses-large-crowd-with-prototype-755mp-light-field-video-camera-at-nab.

Science & Technology Council. (2007). *The Digital Dilemma*. Hollywood, CA: Academy of Motion Picture Arts & Sciences. Retrieved from http://www.oscars.org/science-technology/council/projects/digitaldilemma/index.html.

Science & Technology Council. (2012). *The Digital Dilemma 2*. Hollywood, CA: Academy of Motion Picture Arts & Sciences. Retrieved from http://www.oscars.org/science-technology/council/projects/digitaldilemma2/.

Schank, R. (1995). *Tell me a story: Narrative and intelligence*. Evanston, IL: Northwest University Press.

Sohail, H. (2015, March 27). The state of cinema. *Qube: Events & News*. Retrieved from http://www.qubecinema.com/events/news/2015/state-cinema.

Statista (2018a). Leading film markets worldwide in 2016, by gross box office revenue (in billions U.S. dollars). *Statista: The Statistics Portal*. Retrieved from https://www.statista.com/statistics/252730/leading-film-markets-worldwide--gross-box-office-revenue/.

Statista (2018b). Highest grossing 3D movies in North America as of January 2018 (in million U.S. dollars). *Statista: The Statistics Portal*. Retrieved from https://www.statista.com/statistics/348870/highest-grossing-3d-movies/.

Stray Angel Films (2014, January 21). What cameras were the 2013 best picture Oscar nominees shot on? *Cinematography*. Retrieved from http://www.strayangel.com/blog/2014/01/21/what-cameras-were-the-2013-best-picture-oscar-nominees/.

The Numbers. (2018a). *Top 2016 Movies at the Domestic Box Office*. Retrieved from https://www.the-numbers.com/box-office-records/domestic/all-movies/cumulative/released-in-2016.

The Numbers. (2018b). *Top 2017 Movies at the Domestic Box Office*. Retrieved from https://www.the-numbers.com/box-office-records/domestic/all-movies/cumulative/released-in-2017.

Thompson, A. (2010, January). How James Cameron's innovative new 3-D tech created Avatar. *Popular Mechanics* Retrieved from http://www.popularmechanics.com/technology/digital/visual-effects/4339455.

Thomson, P. (2015, February). Sundance 2015: Inspiring indies—Tangerine. *American Cinematographer*, 96, (2). Retrieved from http://www.theasc.com/ac_magazine/February2015/Sundance2015/page5.php#

THX. (2016). *THX Certified Cinemas*. Retrieved from http://www.thx.com/professional/cinema-certification/thx-certified-cinemas/.

Verhoeven, B. (2017, July 17) *War for the Planet of the Apes*: How the VFX Team Created the Most Realistic Apes Yet. *The Wrap*. Retrieved from https://www.thewrap.com/war-for-the-planet-of-the-apes-vfx-team-more-real/.

Vreeswijk, S. (2012). A history of CGI in movies. *Stikkymedia.com*. Retrieved from http://www.stikkymedia.com/articles/a-history-of-cgi-in-movies.

Volpe, A. (2017, September 17). How VR Is Changing the Film Industry. *Backstage*. Retrieved from, https://www.backstage.com/advice-for-actors/inside-job/how-vr-changing-film-industry/.

Ward, P. (2016, December 16). Visual Effects Bring Back Actors and Their Youth In *Rogue One* and *Westworld*. *The Culture Trip*. Retrieved from, https://theculturetrip.com/north-america/usa/articles/visual-effects-bring-back-actors-and-their-youth-in-rogue-one-and-westworld/.

Wood, J. (2013, October 22). What *Gravity's* Box Office Triumph Means for the Future of 3-D Film. *Wired*. Retrieved from http://www.wired.com/underwire/2013/10/gravity-future-3d-movies/.

Computers & Consumer Electronics

11

Computers

Glenda Alvarado, Ph.D.[*]

Overview

Compared to other communication technologies, computers have gone through extremely rapid changes. From the 30-ton ENIAC to the 3-pound laptop, information processing in a couple of hours to instant responses after a verbal command, personal computing devices have gone from "business only" to "everyone has one" in a matter of 50 years. Modern societies can't imagine life without some type of computer, with all access, all the time.

Introduction

What's a computer? That's the opening phrase in several editions of the textbook *Introducing computers: concepts, systems, and application* (Blissmer, 1991) from the early 1990s. It's asked again decades later in a commercial for Apple's iPad Pro (Miller, 2017). It is a valid question given the current state of personal computers. Ackerman (2018) reported from CES 2018 that the "long-predicted PC-phone convergence is happening, but rather than phones becoming more like computers, computers are becoming more like phones." Perhaps we should be asking, "What's NOT a computer?"

Background

Blissmer (1991) traces the roots of computers back to the 1600s with the invention of logarithms that led to the development of slide rules and early calculators. Blaise Pascal and Gottfried von Leibniz led the computing pack with "mechanical calculators" but the industrial revolution was the driving force toward development of a system that could deal with large amounts of information (Campbell-Kelly, Aspray, & Wilkes, 1996). Increasingly elaborate desk calculators and then punch-card tabulating machines enabled the U.S. government to process the census data of 1890 in months instead of years. The punch-card system gained wide commercial application during the first few decades of the twentieth century and was the foundation of International Business Machines (IBM) (Campbell-Kelly et al., 1996).

Herman Hollerith won the contract for analyzing the 1890 census data and created the, Tabulating Machine Company (TMC) in 1896 (Swedin & Ferro, 2005). TMC leased or sold its tabulating machines to other countries for census taking, as well as making inroads to the private sector in the U.S. These computational machines were then put to use to compile statistics for railroad freight and agriculture (Swedin & Ferro, 2005). Through a series of mergers and a shift to focus on big leasing contracts over small office equipment, IBM was established by Thomas Watson in 1924 (Campbell-Kelly et al., 1996). Early computing machines were built with specific purposes in mind —

---

[*] Visiting Assistant Professor of Marketing, School of Business, Claflin University (Orangeburg, South Carolina)

counting people or products, calculating coordinates, breaking codes, or predicting weather.

Toward the end of World War II, general purpose computers began to emerge (Swedin & Ferro, 2005). The Electronic Numerical Integrator and Calculator (ENIAC) became operational in 1946 and the Electronic Discrete Variable Automatic Computer (EDVAC) concept was published the same year (Blissmer, 1991). The first commercially available computer was the UNIVAC (UNIVersal Automatic Calculator). UNIVAC gained fame for accurately predicting the outcome of the 1952 presidential election (Swedin & Ferro, 2005). IBM and UNIVAC battled for commercial dominance throughout the 1950s. Smaller companies were able to participate in the computer business in specialized areas of science and engineering, but few were able to profitably complete with IBM in the general business market.

Until the 1970s, the computer market was dominated by large mainframe computer systems. The Altair 8800 emerged in 1974. It graced the cover of *Popular Electronics* in 1975, touted as the "World's First Minicomputer Kit to Rival Commercial Models" (Roberts & Yates, 1975). The Altair had to be assembled by the user and was programmed by entering binary code using hand switches (Campbell-Kelly et al., 1996). This do-it-yourself model was the forerunner to the machines most people associate with computers today. Paul Allen and Bill Gates owe the success of Microsoft to the origins of the Altair 8800 (Swedin & Ferro, 2005). Steve Wozniak and Steve Jobs founded Apple on April Fool's Day in 1976, and their Apple II microcomputer was introduced in 1977. IBM did not join the microcomputer market until 1981 with the IBM PC (Swedin & Ferro, 2005).

Fast-forward to the twenty-first century and personal computing devices are commonplace in business and the private sector. In 1996, a home computer was considered by most a luxury item. By 2006, 51% of the adult public considered a home computer a necessity (Taylor, Funk, & Clark, 2006). In 2017, Blumberg and Luke reported that 80% of U.S. households contain either a desktop or laptop computer, and 23% have three or more computers. When considering all connected technology (smartphone, desktop/laptop, tablet, or streaming media device), 90% of households contain some form of personal computing device. The typical household contains five such devices (Blumberg & Luke, 2017), many of them the mobile variety. (The author confesses to eight devices in a three-person household.)

Mobile computing gained ground in the early 1990s with a hand-held device introduced by Hewlett-Packard (MacKenzie & Soukoreff, 2002). The first-generation models were primary operated with a stylus on a touchscreen or slate and had limited applications. Apple's Newton (aka Apple MessagePad) emerged in 1993, but one of the most significant entrants to the mobile computing market was the release of the Palm™Pilot in 1996 (Swedin & Ferro, 2005). Microsoft launched the Pocket PC in 2000, with Bill Gates using the term "Tablet PC" during its trade show introduction (Atkinson, 2008). Apple's iPad came to the consumer market in 2010 and changed the landscape of personal computing devices (Gilbert, 2011). For many school children in the United States, iPads are issued instead of textbooks, and homework is done online rather than with pen and paper.

Computer ownership levels have remained roughly the same for more than a decade; however all-purpose devices are taking the place of specialized components (music-players, game systems, electronic readers) (Anderson, 2015). The newest trends in portable computing combine features of devices that were previously sold as separate items. These hybrids, convertibles and 2-in-1s are emerging as crossover devices, which share traits of both tablets and laptops. Combined features include a touchscreen display designed to allow users to work in a tablet mode as well as a keyboard that can be removed or folded out of the way.

How a computer works

The technology ecosystem model presented in Chapter 1 starts with breaking down the hardware of a computer to its basic components. In simplest terms, a computer has four fundamental components—input, output, processing, and storage. Hardware elements or the physical components of which a computer is comprised have become smaller and more compact, but are essentially the same across platforms. These main pieces are the central processing unit (CPU), power circuits, memory, storage, input, and output. Software or application components determine how information or content is used and manipulated.

The CPU is the brain of the computer—a series of switches that are essentially either on or off—a binary system or string of ones and zeroes. Early processors were constructed using vacuum tubes that were hot, fragile, bulky and expensive. This meant that it took and entire room (or rooms) to accommodate a single computer. The 1940s-era ENIAC weighed 30 tons and contained 18,000 vacuum tubes with the tubes being run at a lower than designed capacity in an attempt to lessen their rate of failure (Swedin & Ferro, 2005). The EDVAC used a great deal less tubes (5,000) but the heat they generated caused technicians and engineers to work in their underwear during the summer months.

Computer development followed a path very similar to that of radio discussed in Chapter 8. As can be seen there, science and technological developments are often tied to war—military battling to best their foes. In 1947, Bell Telephone scientists invented "point-contact" transistors made out of a semiconductor material—much smaller, lighter and more durable than tubes (Swedin & Ferro, 2005). Military funding sped up this development of transistors that were key components in computers built throughout the 1950s. By the 1960s "integrated circuits" contained multiple tiny transistors on a single silicon "chip." Continued miniaturization has enabled complex processing systems to be reduced from the size of a room to a single chip, often smaller than a pinkie fingernail.

Hardware

Modern computers are built with a motherboard that contains all the basic electrical functions. A motherboard is the physical foundation for the computer. Additional "boards" can be tied into the mother to provide supplementary components—extra memory, high-resolution graphics, improved video, and audio systems. Memory of a computer determines how much information it can store, how fast it can process that information and how complex the manipulations can be.

For a computer to operate, the CPU needs two types of memory, random access memory (RAM) and storage memory. Most forms of memory are intended for temporary storage and easy access. RAM space is always at a premium—all data and programs must be transferred to random access memory before any processing can occur. Frequently run program data and information are stored in temporary storage known as cache memory; users typically access the same information or instructions over and over during a session. Running multiple programs at the same time creates a "RAM cram" (Long & Long, 1996) situation where there is not enough memory to run programs. That's the reason many troubleshooting instructions include clearing "cache" memory before re-attempting an operation.

As of early 2018, new computers are being manufactured with approximately 4GB or 8GB RAM ("How much RAM do you need? It's probably less than you think," 2018). Budget tablets or PCs have around 2GB; 8GB is recommended for casual gamers and general-purpose users. Computer professionals and hard-core gamers will be looking for machines with 16GB that provide optimal performance. High-end, specific purpose-built workstations can be obtained with 32GB RAM. Typically, the more complex an application or program, the more RAM that is used. However, more RAM does not necessarily mean better performance. RAM should be considered for immediate access, but not long-term storage. For large amounts of data, a hard drive or external storage device is more appropriate.

Most computers have built-in internal hard drives that contain its operating system and key software programs. External storage systems typically have high capacity and are used to back up information or serve as an archive. These systems connect to a computer using USB, thunderbolt, or firewire connection. Increasingly, additional memory is part of a "cloud," or virtual network system that is digitally stored in a remote location.

Small-scale storage devices called flash drives or thumb drives allow users to share or transport information between computers or systems, especially when network connections are unreliable or not available. Other memory or storage devices include SD cards, CDs, DVDs, and Blu-ray discs. SD cards are most often associated with digital cameras or other small recording devices. Information is easily deleted or can be transferred to a permanent location before the contents are erased and written over. CDs and DVDs are considered "optical discs" and are commonly associated with movies and music files—although some versions have read/write capacity, most are used for a single purpose or one time only. Information in this format is accessed using an optical disc drive that uses laser light to read and write to or from the discs. For many years, software programs were produced as a CD-ROM (Read Only Memory) version.

Manufacturers are building machines without optic drives to save space and lessen battery drain. Headphone jacks, USB sockets, and other input/output ports are being eliminated as laptops are being constructed less than 15mm in thickness. (Murray & Santo Domingo, 2017). USB-C ports are emerging as the industry standard connector for power and data. These ports are 2.6mm tall versus the 7.5mm of a standard USB. Murray and Santo Domingo (2017) tout USB-C as the emerging industry standard with no up/down or in/out orientation offering greater bandwidth for data transfer and reduced power consumption.

Input and output devices are the last pieces of the computer anatomy. Input devices include keyboards, mice, microphones, scanners, and touchpads. Sophisticated computer operators may also have joysticks, hand held controllers, and sensory input devices. Output options include printers, speakers, monitor, and increasingly, virtual reality headsets.

Most current computers or personal computing devices are also equipped with Bluetooth, a form of technology that uses short-range wireless frequencies to connect and sync systems. Users looking for a wireless experience enjoy the "uncabled" functionality of such systems. Wireless options are also available for input devices (mice and keyboards), often using radio frequency (RF) technology whereby the input item contains a transmitter that sends messages to a receiver that plugs into a USB port.

Software

For all the sophistication and highly technical components of a computer, it is essentially useless without software. These are the programs that enable the computer to perform its operations. At the core of each computer is the operating system—the Windows versus Mac debate continues to rage, but user preference remains a personal choice. These systems have more similarities than differences at this point in time, however Windows is the most used operating system worldwide (Ghosh, 2017).

Operating systems control the look and feel of your devices—and to a certain extend how you connect your multiple-device households. One reason for the dominance of Windows is the ability to connect devices from a variety of manufacturers. Mac/Apple products tend to only connect with each other—although there are "compatible" technologies that adapt across platforms. Unfortunately for users, frequent updates or system changes can render software obsolete. MacOS allows users free and continual updates to its operating systems. Windows users lose the ability to get systems support or updates after operating systems are phased out. *Business Insider* suggests that there were at least 140 million computers still running the popular Windows XP (introduced in 2001) when support was ended in 2014 (Ghosh, 2017). Without support, users were forced to upgrade the operating system or purchase a new computer.

There are hundreds of thousands of individual programs and applications for use on either system. Users can write, calculate, hear, watch, learn, play, read, edit, control, and produce…virtually anything. Outside of a "computer" the most popular operating system for personal computing devices is Google's Android platform (Simpson, 2017).

Recent Developments

Popular commentary (Barel, 2017; Pratap, 2017) suggests that the personal computer industry has been in a decline and smartphones are replacing traditional devices. Laptops and desktops continue to be the top choices for gamers, but large screen phones and tablets are the first choice for people looking to use a computer for entertainment purposes (Pratap, 2017). A desktop or laptop is still the device of choice for work-related tasks, but the rise of cloud-based computing has lessened the reliance on bulky equipment.

Figure 11.1
Number of Smartphone Users in the United States from 2010–2022

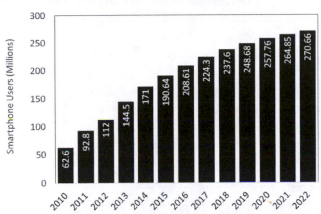

For 2017, the number of smartphone users in the U.S. is estimated to reach 224.3 million, with the number of users worldwide forecast to exceed 2 billion users by that time.

Source: Statista, 2018

Hardware

Users can purchase a basic computer for less than $100—for many users in the U.S., this is their first interaction with computing technology as a toddler. Most tablets more closely resemble a cell phone than a computer. High-end tablets with more versatile and sophisticated operations are available for $900–1200, but the most popular models are $250–400. Entry-level laptops—perfect for younger families to use for homework, web surfing, and movie watching—can often be found in the same price range.

PC Magazine rated the best value ultraportable laptops for 2018 the Acer brand's Swift models at $350-$400. For general purpose usage, Dell and HP earned the best reviews with a price tag ranging from $800-1200 (Santo Domingo, 2018). These new generation lightweight computers weigh three pounds or less. Budget-friendly Chromebooks offer a personal computing device option for students and casual users that spend most of their time in a web-browser. According to Murray (2018), this low-cost Windows alternative doesn't have an impressive hardware system, but functionality beyond basic web-centric operations is being introduced. Touch and large screen displays and more Windows-like functionality will soon be options in high-end Chromebook versions (Murray, 2018).

Software

Laptops and desktops have been trapped in a Windows or MacOS battle since the 80s. Even students who might otherwise be permanently attached to their phones are still likely to use a MacBook, Windows laptop, or Chromebook some of the time (Mossberg, 2017). Ackerman (2018) reported from the *2018 Consumer Electronics Show* that tablets have been "subsumed" by large screen phones, and laptops are becoming more like phones with always-on LTE connectivity. Cloud computing—the delivery of on-demand computing resources via connection to the Internet—has eliminated the need for many people to purchase hardware and individual software programs. Storage and access are available for free or on a pay-as-you-go basis.

Common cloud computing examples include Microsoft's Office 365 and OneDrive that marry local and online services; Google Drive—docs, sheets and slides that are purely online; and Apple's iCloud—primarily storage, but offering cloud-based versions of popular programs (Griffith, 2016). Hybrid services like Box, DropBox, and SugarSync are considered cloud-based because they store synchronized information. Synchronization is the key to cloud computing, regardless of where or how information is accessed.

In short, with an online connection, cloud computing can be done anywhere, at any time with any number of devices. At least that's how it impacts typical consumers—people at home or school, and small-to-medium office environments. For big business, a subscription (Software-as-a-service, SaaS) or custom application (Platform-as-a-service, PaaS) over the Internet might be used (Griffith, 2016). Tech industry giants Amazon, Microsoft, and Google also operate an Infrastructure-as-a-Service (IaaS) that can be leased by other companies. As can be expected, cloud-computing services are moneymaking ventures. Griffith (2016) reported that $100 billion was generated in 2012 and the market is expected to reach $500 billion by 2020.

Current Status

Worldwide shipments of computing devices (computers, tablets, and phones) totaled 2.28 billion units in 2017 and are expected to increase 2.1% to 2.32 billion units in 2018 (van der Meulen & Bamiduro, 2018). Overall growth is expected in the mobile market and premium ultramobile market where thin, lightweight devices are in high demand. Researchers predict that Apple smartphone sales will increase by more than the market average due to release of new models. Shipments of traditional personal computers are expected to decline by 5–6% (van der Meulen & Bamiduro, 2018). A forecast from International Data Corporation (IDC) shows a decline of more than 2% for 2017 in global shipments of personal computers (Chou, et al., 2017). Additionally, the prediction of a 4% decline year-over-year for 2018 continues the five-year downward trend. Conversely, detachable tablets are expected to achieve double-digit growth over the five-year period through 2021.

The Apple HomePod went on sale in February 2018, aiming to compete with Amazon market-leading Echo (Richter, 2018). The smart-speaker market is a growing category among home internet-users and replaces phones for search and general entertainment purposes. Smart-speaker penetration among U.S. households rose from 8% in June to 11% in October 2017. It is expected to reach 15% during the first quarter of 2018 (Richter, 2017). Personal assistants are discussed in more detail in Chapter 12.

Technology stocks fell dramatically in early 2018 in what many investors viewed as an overdue market correction (Rabe, 2018). Alphabet (Google), Microsoft, and Apple all suffered losses in value after having posted steady growth since 2015. Kim (2018) reported that the losses totaled more than $90 billion, with Alphabet suffering the greatest drop at 5.1% or $39.5 billion. The report goes on to state that the tech stocks were still up for the year-to-date.

Education has outpaced other businesses in adoption of digital technology and computers. According to Gurr (2016), educational buildings hold nearly twice as many computers per square foot as other commercial buildings. The number of computers per square foot increased 71% between 1999 and 2012. Data collected by the U.S. Energy Information Administration (EIA) indicates that the report did not include information about smartphones, tablets or personally owned computers that were brought into buildings for periodic use (Gurr, 2016). The most recently completed survey included additional questions about computers and the EIA continues to monitor building energy usage patterns. New computers are generally more energy-efficient than older ones, but the increasing total number of devices plays a large role in energy consumption patterns.

Factors to Watch

Technological advances are targeting seniors. One such example is the grandPad—a tablet that utilizes a closed network system to help older people stay in touch (Kaye, 2017). Approximately two-thirds of seniors (those aged 65+) use the Internet at home and 42% now report owning smartphones (Anderson & Perrin, 2017). Baby boomers are the first wave of retirees that are comfortable with personal computing devices as a part of their daily lives. Tech savvy boomers are taking advantage of fitness trackers and home healthcare diagnostics. They expect technology to improve medical treatments, track outcomes, and reduce overall medical costs (Jain, 2018).

Income and education remain the major factors that predict device ownership. In 2016, 32% of seniors reported owning a tablet computer—a double digit increase since 2013 (Anderson & Perrin, 2017). For many low-income families, making less than $30K per year, a smartphone is the only Internet access in the household. This number has risen from 12% in 2013 to 20% in 2016 (Rainie & Perrin, 2017).

Artificial intelligence, immersive experiences, and adaptive security systems are the categories of expansion that should be watched. As noted in the opening paragraph, nearly every aspect of modern society is touched by computer technology. We are constantly connected and increasingly reliant on personal computing devices.

Getting a Job

The majority (85%) of Americans believe that understanding computer technology is one of the most important traits for workplace success (Rainie, 2016). The same report indicated that knowing a computer or programming language was more important that knowing a foreign language (64% v. 36%, respectively). "New collar" jobs (coined by IBM CEO Rometty) are emerging with apprenticeships and skills-centered training being favored over traditional four-year college degrees (CompTIA, 2018).

Bibliography

Ackerman, D. (2018). PC-phone convergence is happening, but not how you think. Retrieved from https://www.cnet.com/news/pc-phone-convergence-is-happening-not-how-you-think-ces-2018/.

Anderson, M. (2015). *Technology Device Ownership: 2015*. Retrieved from Pewresearch.org: http://assets.pewresearch.org/wp-content/uploads/sites/14/2015/10/PI_2015-10-29_device-ownership_FINAL.pdf.

Anderson, M., & Perrin, A. (2017). *Tech Adoption Climbs Among Older Adults*. Retrieved from pewresearch.org: http://assets.pewresearch.org/wp-content/uploads/sites/14/2017/05/16170850/PI_2017.05.17_Older-Americans-Tech_FINAL.pdf.

Atkinson, P. (2008). A bitter pill to swallow: the rise and fall of the tablet computer. *Design issues, 24*(4), 3-25.

Barel, A. (2017). Your Smartphone vs. Your PC. Retrieved from https://medium.com/adventures-in-consumer-technology/smartphones-will-replace-pcs-soon-541b5c8a4f48.

Blissmer, R. H. (1991). *Introducing computers: concepts, systems, and* applications (1991–92 ed.). Austin, TX: John Wiley & Sons, Inc.

Blumberg, S., & Luke, J. (2017). *Wireless substitution: Early release of estimates from the National Health Interview Survey, July–December 2016*. Retrieved from http://www.cdc.gov/nchs/nhis.htm.

Campbell-Kelly, M., Aspray, W., & Wilkes, M. V. (1996). Computer A History of the information Machine. *Nature, 383*(6599), 405-405.

Chou, J., Reith, R., Loverde, L., & Shirer, M. (2017). Despite Pockets of Growth the Personal Computing Device Market Is Expected to Decline at a -2% CAGR through 2021, According to IDC [Press release]. Retrieved from https://www.idc.com/getdoc.jsp?containerId=prUS43251817.

CompTIA. (2018). *IT Industry Outlook 2018*. Retrieved from CompTIA.org: https://www.comptia.org/resources/it-industry-trends-analysis.

Ghosh, S. (2017, May 15, 2017). Windows XP is still the third most popular operating system in the world. Retrieved from http://www.businessinsider.com/windows-xp-third-most-popular-operating-system-in-the-world-2017-5.

Gilbert, J. (2011, December 6, 2017). HP TouchPad Bites The Dust: Can Any Tablet Dethrone The IPad? Retrieved from https://www.huffingtonpost.com/2011/08/19/hp-touchpad-ipad-tablet_n_931593.html.

Griffith, E. (2016). What is Cloud Computing. https://www.pcmag.com/article2/0,2817,2372163,00.asp.

Gurr, T. (2016). Computer and Technology Use in Education Buildings Continues to Increase. Retrieved from http://www.pacifictechnology.org/blog/in-the-news-computer-and-technology-use-in-education-buildings-continues-to-increase/.

How much RAM do you need? It's probably less than you think. (2018). Retrieved from https://www.digitaltrends.com/computing/how-much-ram-do-you-need/.

Jain, A. (2018). Forget the Alexa-powered toasters at CES, these innovations will really shape 2018. Retrieved from https://techcrunch.com/2018/01/12/forget-the-alexa-powered-toasters-at-ces-these-innovations-will-really-shape-2018/.

Kaye, G. (2017). GrandPad Keeps the Senior Set Social. Retrieved from https://www.tech50plus.com/grandpad-keeps-the-senior-set-social/.

Kim, T. (2018). More than $90 billion in value wiped out from the popular 'FANG' tech stocks. https://www.cnbc.com/2018/02/05/more-than-90-billion-in-value-wiped-out-from-the-popular-fang-tech-stocks.html

Long, L. E., & Long, N. (1996). *Introduction to Computers and Information Systems*. Upper Saddle River, NY: Prentice Hall PTR.

MacKenzie, I. S., & Soukoreff, R. W. (2002). Text entry for mobile computing: Models and methods, theory and practice. *Human–Computer Interaction, 17*(2-3), 147-198.

Miller, C. (2017). Apple challenges the definition of a PC with new 'What's a computer' iPad Pro ad [Video]. Retrieved from https://9to5mac.com/2017/11/16/ipad-pro-whats-a-computer-ad/.

Mossberg, W. (2017). The PC is being redefined. Retrieved from https://www.theverge.com/2017/3/1/14771328/walt-mossberg-pc-definition-smartphone-tablet-desktop-computers.

Murray, M. (2018). The Best Chromebooks of 2018. *PC Magazine*. Retrieved from https://www.pcmag.com/article2/0,2817,2413071,00.asp.

Murray, M., & Santo Domingo, J. (2017, March 2, 2017). What is USB-C? *PC Magazine*.

Pratap, A. (2017). Personal Computing Industry PESTEL Analysis. Retrieved from https://www.cheshnotes.com/2017/02/personal-computing-industry-pestel-analysis/.

Rabe, L. (2018). "Flash Crash" Wreaks Havoc on Major Tech Stocks https://www.statista.com/chart/12815/tech-company-market-cap-losses/.

Rainie, L. (2016). *The State of American Jobs*. Retrieved from Pewresearch.org: http://www.pewinternet.org/2017/10/10/10-facts-about-jobs-in-the-future/.

Rainie, L., & Perrin, A. (2017). 10 facts about smartphones as the iPhone turns 10. Retrieved from http://www.pewresearch.org/fact-tank/2017/06/28/10-facts-about-smartphones/.

Richter, F. (2017). Are Smart Speakers the Key to the Smart Home? Retrieved from https://www.statista.com/chart/12261/smart-speakers-and-other-smart-home-products/.

Richter, F. (2018). The U.S. Smart Speaker Landscape Prior to Apple's Entrance . Retrieved from https://www.statista.com/chart/12858/smart-speaker-ownership/.

Roberts, H. E., & Yates, W. (1975). Altair 8800 minicomputer. *Popular Electronics, 7*(1), 33-38.

Santo Domingo, J. (2018). The Best Laptops of 2018. *PC Magazine*. Retrieved from https://www.pcmag.com/article2/0,2817,2369981,00.asp.

Simpson, R. (2017). Android overtakes Windows for first time. Retrieved from http://gs.statcounter.com/press/android-overtakes-windows-for-first-time.

Swedin, E. G., & Ferro, D. L. (2005). Computers: the life story of a technology, Westport, CT: Greenwood Publishing Group.

Taylor, P., Funk, C., & Clark, A. (2006). Luxury or Necessity? Things We Can't Live Without: The List Has Grown in the Past Decade [Press release]. Retrieved from http://assets.pewresearch.org/wp-content/uploads/sites/3/2010/10/Luxury.pdf.

van der Meulen, R., & Bamiduro, W. (2018). Gartner Says Worldwide Device Shipments Will Increase 2.1 Percent in 2018. Retrieved from https://www.gartner.com/newsroom/id/3849063.

12

Internet of Things (IoT)

Jeffrey S. Wilkinson, PhD*

Overview

The Internet of Things (IoT) refers to the growing global trend of connecting sensors—objects and devices—to computer-based networks. Active and passive chipsets are becoming embedded into virtually all aspects of life across every imaginable sector. Some of the most notable devices today are the Home Assistant (Echo, HomePod, and others) and Nest thermostats. Other applications include inventory management, supply chain, logistics, and security. Global migration to 5G and narrow-band IoT will result in more advanced developments in the coming year. Globally, many cities are testing large-scale IoT applications such as city water metering using smart technologies.

Introduction

The Internet of Things (IoT) refers to the global trend of connecting as many devices as possible to computer-based networks. According to Gartner (2017), there were an estimated 8 billion connected devices in 2017 which will grow to 20.4 billion by 2020. These connected items range from refrigerators to jet engines, all operating together every minute of every day. The intention is to make life more efficient. But, as always, there is a very real danger of abuse.

Although the Internet has been in wide use since the 1990s, according to Evans (2011) IoT was officially "born" around 2009 when the ratio of connected objects-to-people became equal. Experts say eventually there will be trillions of trackable objects—thousands for each individual—to help us navigate our world. But one of many hurdles is interoperability. There is no coherent set of standards, and some nations are determined to establish their own views of what IoT should be for their citizens.

The result is a world of chaos that some view through a lens of blue sky predictions of global standardization. We continue to slowly move toward a time when personal and family medical records, home appliances, vehicles, and media entertainment devices may seamlessly meld with government, medical, and business institutions. In practical terms we have to change our views and expectations of privacy, because privacy has to be balanced with our equally important and changing view of security. Our habits, behaviors, purchases, and even communication are increasingly being tracked, linked, mined, and archived to the extent allowed by privacy laws, which vary significantly around the globe.

The boundless benefits of IoT are balanced with predictions of equally dire consequences. This chapter offers an overview for understanding the potential, the benefits, and the dangers this connectivity brings. IoT may be the defining technology issue of the 21st century.

---

* Associate Dean, Academic Affairs, Sino-U.S. College, Beijing Institute of Technology, Zhuhai, China

Background

Since the beginning of commerce, companies have dreamed of being the provider of everything a consumer would want. Manufacturers and service providers are competing to be part of the system that provides IoT products, from familiar giants like Samsung and Sony to relatively unknown newcomers like Red Hat and Zebra Technologies.

The concept of the Internet of Things was anticipated long before the Internet was invented. As early as 1932, Jay B. Nash anticipated smart machines in *Spectatoritis*:

> These mechanical slaves jump to our aid. As we step into a room, at the touch of a button a dozen light our way. Another slave sits twenty-four hours a day at our thermostat, regulating the heat of our home. Another sits night and day at our automatic refrigerator. They start our car; run our motors; shine our shoes, and curl our hair. They practically eliminate time and space by their very fleetness (p.265).

According to Press (2014), the thinking behind IoT evolved throughout the 20th century, starting with things like Dick Tracy's 2-way wrist radio in the 1940s to wearable computers (1955), head mounted displays (1960), and familiar tech acronyms ARPANET (1969), RFID (1973), and UPC (1974). As these converged with existing technologies, new applications were conceived which begat newer advances.

How IoT works

In all the configurations, the internet of things is essentially made up of four key elements; mass creation and diffusion of smart objects, hyper-fast machine-to-machine communication, widespread RF technologies, and an ever-increasing capacity to store information at a central hub.

The first step in having so many objects connected to the Internet is to have a unique identifier for each object—specifically a unique IP (Internet Protocol) address. Each object must have a unique IP address. In moving from IPv4 (32-bit numbering system) to IPv6 (128-bit) (see Chapter 23), the capacity for unique addresses went from 4.3 billion to over 340 undecillion addresses (340 billion billion billion billion…).

The next factor is whether the smart object is to be active or passive. A passive connection uses RFID (radio frequency identification) tags which contain a small amount of information needed simply to identify the object. For tracking inventory like clothing, toys, toiletries, or office supplies, knowing where the object is and when it was bought is the primary function of the RFID tag.

An active connection is more robust and makes our devices "smart." Using embedded sensors and computing power allows the object to gather and process information from its own environment and/or through the Internet. These objects include cars, appliances, medical devices (anything involving insurance!), and pretty much anything needing the capability of initiating the sending and receiving of information.

Thus, when you put billions and billions of passive and active objects together with ever-more-powerful computers with almost limitless storage capacity, you have the Internet of Everything. Any individual, business, institution, or government has the capability to monitor anything that can be tagged. Whether it is food, clothing, shelter, or services, these can be analyzed individually or in conjunction with each other. The most common application used today is the relatively innocuous function of inventory management. Future applications are only limited by our imagination (concerns about privacy are discussed later in this chapter).

The Technology (Alphabet Soup)

To be an IoT object or device, it must be able to connect and transmit information using Internet transfer protocols. Efforts are underway to test narrow-band IoT using cellular networks in several cities around the world. IoT platforms serve as a bridge between the devices' sensors and data networks (Meola, 2016).

There is aggressive competition among companies offering interoperability standards to connect everything together in a seamless tapestry of services. In 2016, it looked like there were two dominant initiatives, the AllSeen Alliance and the Open Interconnect Consortium. The AllSeen Alliance was started by Qualcomm and included LG, Sharp, Panasonic, HTC,

Sony, Electrolux, Cisco, and Microsoft. Qualcomm came out with the AllJoyn protocol in 2011 and gave it to the Linux Foundation in 2013 which became the basis for the AllSeen Alliance (Bradbury, 2017). Nature seems to love a good struggle, so a competing group, the Open Interconnect Consortium, was formed by a group that included Intel, Samsung, and GE. After some wrangling and negotiations, the two groups resolved their differences and merged to create the Open Connectivity Foundation (OCF) in 2016.

Still, the protocols from the two groups (AllJoyn from AllSeen, and IoTivity from OIC) remain part of the Linux Foundation effort to manage IoT development and some software bridges have been built to bring them together (Bradbury, 2017). In Fall 2017, Linux announced the creation of an effort to pull vendors together, calling it the EdgeX Foundry (Bradbury, 2017). Since Linux champions open-source software, their hope is to move vendors to common frameworks that interact with each other.

The world of standards-seekers seems to be dividing up into industrial and consumer-oriented sectors. Industrial vendors don't want to control so much as integrate with the wider field. On this side, one notable effort to harmonize connected components is the Industrial Internet Consortium (IIC) founded by AT&T, Cisco, GE, Intel, and IBM.

The consumer side is, as one might expect, more wide-open. Vendors release their own framework and then try to get others to jump on. Examples are Google's Works with Nest and Apple's Homekit. Google chose Nest, the smart thermostat, as the central hub of a smart home because the Nest Learning Thermostat can handle many core home maintenance tasks (Burns, 2014). The thermostat has a motion detector and the ability to learn user behavior. Because Nest is integrated with wireless protocols it doesn't need a control panel which would in turn require constant user interaction. Nest operates "serendipitously" which means it is activated when it needs to do something; it doesn't need a human to constantly tell it what to do (Burns, 2014).

According to Meola (2016), the top IoT platforms include Amazon Web Services, Microsoft Azure, IoT Platform, ThingWorx IBM's Watson, Cisco IoT Cloud Connect, Oracle Integrated Cloud, and GE Predix.

Recent Developments

IoT is so big and moves so fast we can only provide the briefest of snapshots. Tata Consultancy Services (TCS) suggested in 2015 that the four core areas businesses should focus on regarding IoT are (1) Premises monitoring, (2) Product monitoring, (3) Supply chain monitoring, and (4) Customer monitoring. Premises monitoring tracks customers' experiences at physical locations like stores, offices, kiosks, and hotels. Product monitoring tracks products or services after the customer purchases them. Supply chain monitoring tracks the production and distribution of operations. Customer monitoring tracks how customers are using products and services as well as what they say about them. These four areas probably cover the bulk of the current work being done to implement IoT.

Figure 12.1
Image of Smartphone Purchase with QR Code

Source: J. Wilkinson

The wall between the industry side and the consumer side is wafer thin, exemplified by the push to develop so-called "smart homes." Smart homes involve several types of overlapping industries, goods, and services (automobile/transportation, electric/water/utilities, police/fire/security/safety, media/entertainment/communication, to name a few).

So other than automobiles (see Chapter 13), the family home is the best starting point to see the impact of IoT technology. There are two main approaches, the fully-connected 'smart home' and the piecemeal 'connected home.' In practical terms, given the rate of change in technology and devices, most people live in a connected home that strives to becoming 'smarter.'

A smart home is a closed system that can respond—after proper programming—to those who live and manage the home itself with each device "geared for central control" (Sorrel, 2014). The connected home takes those closed systems and integrates them with vendors and service providers so they can share information. This enables the businesses to optimize services and revenue opportunities. Today's fully-connected "smart home" should have off-the-shelf innovations that are fully integrated with the building. Probably the leaders in this area today are General Electric, and LG with its SmartThinQ line of appliances. At CES 2018, LG demonstrated how the fully integrated showers, kitchen, laundry, and more can be managed with either Google's Home or Amazon's Echo (Gibbs, 2018).

Most people are probably living in a 'connected home,' adopting IoT technology as a one-at-a-time exercise of experimentation, trial, and error. Whether you live in a smart home or a connected home, efficiencies are in the eyes of the beholder. The idea of dozens of independent objects working off and on around your home is science fiction myth. In practical terms, we mostly want one device to rule them all. Enter the Home Assistant.

Home Assistants

Since the 2017 Consumer Electronics Show when Amazon's Echo made headlines, several companies have since entered the market. According to Patrick Moorhead, principal analyst at Moor Insights & Strategy, a main theme of CES 2018 was connecting appliances, "TVs, refrigerators and vacuums controllable by voice and to work together. Samsung, Amazon and Google are all competing for this space." (Newcomb, January 12, 2018).

Figure 12.2
Smart Home

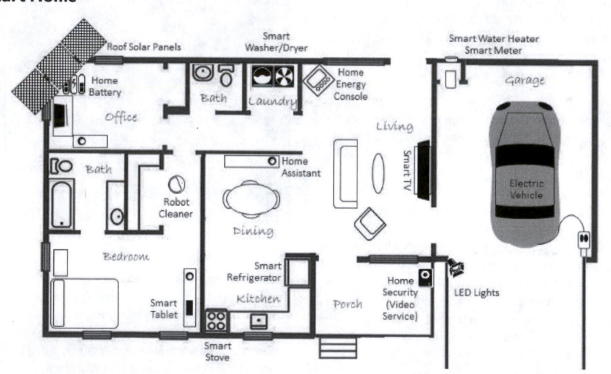

Source: J. Wilkinson & Technology Futures, Inc.

"Computer, OBEY"
Levels of Interaction for Personal Assistants?
August E. Grant

It is important to separate science fiction from science fact. We are all on a journey, and IoT personal assistants are rapidly improving in their capability to serve us. There are at least six stages of evolution for these automated helpers before they attain human status and personhood. But don't ask them to prepare your personal jetpack for work just yet.

The Six Stages of Personal Assistants:

1. **Simple Commands:** One clearly defined question or task.
   Example: "What time is it?" [time given]. "Play Mandy by Barry Manilow." [song plays].

2. **Multiple Commands:** A short series of clearly defined tasks.
   Example: "Play three songs from U2, then play two songs from Foo Fighters." [tasks done].

3. **Conditional Commands:** A clearly defined task with new parameters.
   Example: "Read the latest news, but don't include any sports stories." [task done].

4. **Serial Commands:** Ability to execute multiple commands at the same time, resolving conflicts between the commands.
   Example: "Play that great new song from Taylor Swift about her last boyfriend, then read the latest news...Oh, and I want to see this morning's phone interview segment with the president from 'Fox and Friends'." [song played, segue to news, then cut to full interview segment].

5. **Anticipatory Commands:** Use of AI to anticipate commands, so the PA automatically changes the thermostat, starts playing music, or executes another command after predicting that the user might issue such a command
   Example: "It's cold in here and it's been a rough day." [cue thermostat adjustment, then play soft jazz, then refrigerator pops ice cube into prepared glass of favorite beverage. Advanced feature might include a confidential text being sent to best friends recommending a quick casual call.]

6. *__Full Autonomy:__ The ability to execute all of the above at the same time, prioritizing commands. Extra feature: ability to predict a warp core breach or initiate photon torpedo launch with confirmation voice command from another bridge officer.*
   Example: "Siri, beam me up NOW!" [trouble eliminated, refrigerator pops ice cube into glass...]

Figure 12.3
Amazon Echo

Source: J. Meadows

As of January 2018, Amazon Echo seems to be the current frontrunner, with its Bluetooth speaker powered by Alexa. Alexa works with many devices directly and IFTTT (If This Then That) to control others via 'recipes' you can create yourself (Griffith and Colon, 2017). The Echo (with Alexa) is voice-activated and voice-controlled, but touchscreens can also be added to help in environments where sound might be an issue.

Figure 12.4
Home Control

An integrated IoT operating system will allow you to run smart house appliances with a finger and any tablet or smartphone.

Source: A.Grant

The industry is watching closely how the home assistant market grows. Analysts say if Amazon reaches more than 100 million installed Echo devices, it then becomes a "platform that can generate its own revenue" (Kharpal, 2017). A smart home assistant has built in abilities to integrate with apps like Uber or Spotify. Based upon ever increasingly sophisticated voice command technology, the Echo can evolve into a self-contained marketplace. At that stage, the consumer can ask Alexa to buy goods or services from vendors, and Amazon will be able to charge those vendors to appear prominently in its app store. Investment Bank RBC also predicts there will likely be paid skills on the Echo, for which Amazon could collect revenue share payments (Kharpal, 2017).

Other home automation hubs include the Wink Hub 2 and Apple's Homekit. Wink Hub 2 retails for around $99 and is sold at chains like Home Depot or Target. Apple is trying to leverage its suite of internal interoperable devices with Homekit, which integrates home automation with tablets and smartphones. As of this writing the framework is reportedly still a big "buggy," but with improvements is expected at some point to take significant market share (Sargent and Caldwell, 2017).

Apple's connected-home platform is called HomeKit, and its home-speaker digital assistant is called HomePod. HomeKit was launched in 2014 and designed to reassure users of a secure home space by avoiding the cloud (unlike Google's Nest or Amazon's Alexa). HomePod was launched in 2017 as Apple's answer to Google's Home (using Siri, of course). Apple said research shows users typically use these assistants for music, so "Personal DJ" is but one popular feature. But like all other such devices, HomePod also handles lighting, thermostat adjustment, and other smart home features (Swanner, 2017). Apple is trying to expand the product lines for the connected home, but a barrier is the corporate "MFi" (made for iphone/ipod/ipad) policy which limits what products can be a part of Apple's empire.

Google Home is the home assistant speaker device competing against Amazon's Echo, Apple's HomePod, and others. But Google Home has additional skills that you might expect given the parent— anything relating to company business, searches,

directions, and list generation (Ludlow, 2018). Amazon's Echo is the basic unit; a smaller version is called Dot, and a version with a screen is called Echo Show. Google still has the basic Home device and the Home Mini which can be placed in different rooms as needed.

Other Smart Consumer Devices

IoT developments routinely appear almost daily in the popular press. A report from *USA Today* stated it costs around $2,000 to outfit your home with basic IoT features like thermostat control, smart lighting, and home security (Graham, 2017). There are several companies that offer home security and lighting. Vivint, for example, charges a monthly subscription beginning at $39.99 per month (around $500 per year). There are also several products that can be bought off the shelf at places like Best Buy, Walmart, or Target. For around $200 there is the "Ring" "video doorbell which allows you to see who is at the door and talk to them, too. For a smart lock which can be opened with your phone or via Alexa, prices range between $200-$300. So-called "smart lights" by Philips turn off and on by voice command and cost about $150. As mentioned earlier, one of the most popular products is the smart thermostat called Nest. It can be controlled by your phone from anywhere and can learn to adjust to individual needs. The Nest retails for around $250.

Security Concerns

As the home assistant market grows, so also is the concern about hacking. A team of security researchers in China found that hackers could easily exploit voice recognition systems like Siri and Alexa by using ultrasonic frequencies (Khandelwal, 2017). Humans can't hear sounds above 20kHz (like dog whistles), but the Home Assistants can. Labelled "DolphinAttack," the team at Zhejiang University took human voice commands and translated them into ultrasonic frequencies and played them back over a smartphone equipped with an amplifier, transducer and battery which cost less than $3. By this method an attacker could send inaudible voice commands to instruct a device to visit a malicious website. Or allow them to spy on your premises by activating outgoing video and sound. Or place fake information to emails, online posts, or calendars. Or take your phone offline by automatically setting it to "airplane mode." Or worst of all, hide such attacks by dimming the screen and lowering the volume on the device.

Those tests worked on Siri, Google Assistant, Samsung S Voice, Huawei, HiVoice, Cortana, and Alexa. The tests worked on smartphones, iPads, MacBooks, Amazon Echo and even an Audi Q3.

The team noted that the best antidote is to program your device to ignore commands at 20kHz or any other voice command at inaudible frequencies (Khandelwal, 2017). Otherwise, end users should look for an official patch for the device. But still, users should be on the lookout for this type of activity.

Home Assistants in China

IoT has captured the attention of tech industries worldwide. In China, almost every major tech company is working on its own Chinese-language voice assistant (Joffe, 2017). Chinese search giant Baidu unveiled its model in November 2017 which offered something western models didn't—a device that was visually attractive (Horwitz, 2017). Named the Raven H, it is a colorful stack of eight squares that can be tilted upward to become a forward-facing LED display, or used as a portable microphone for karaoke and more. Early models listed the Raven H at RMB1699 (US$255.95) compared to the American models which go for far less ($100-$180). In January 2018 the Raven H was listed online for even more—RMB2000 (US$310). Other Chinese home assistants include Echo clones like Alibaba's Tmall Genie and JD's Dingdong. Another factor is availability. As of this writing, the Raven H is only available from a prominent technology store in Beijing. The Raven H (or any other home assistant) is not available in cities like Zhuhai, with over a million people, down south in Guangdong province.

China's Migration to IoT and Narrowband IoT

Gartner (2017) and others note that China is showing leadership in developing IoT. According to China Daily, telecom companies are actively working with manufacturers and developers to seamlessly integrate life in the People's Republic of China. Drivers in Shanghai can use an app to find a space in a crowded parking lot, get there, and pay the bill with their smartphone. "Moreover, the app can calculate the probability of a space becoming empty using information from the parking ticket machine" (Ma, He, & Cheng, 2017). An important piece of this is the development of narrow-band IoT, a radio technology which connects billions of devices in a smarter way than Wi-Fi or Bluetooth (Ma, He, & Cheng, 2017).

China's Ministry of Industry and Information Technology unveiled a five-year plan to enable 1.7 billion connections by 2020. Several cities have been selected as testbeds for IoT applications and services. The 'smart city' projects are building databases and sensor networks to collect, store, and analyze information related to transportation, electricity, public safety, and the environment (Ma, He, Cheng, 2017). For example, the city of Wuxi is China's first high-standard all-optical network and a designated test site for IoT technologies.

China has particularly jumped into narrow-band IoT testing. According to Chinese officials they are conducting several test projects in a number of cities. Narrow-band IoT connects hidden things such as water and gas meters, even underground pipes. Shenzhen started its Smart Water metering project in March, 2017. The smart water meters intelligently read and upload data to Shenzhen's online platform. This helps residents avoid any service charge caused by false or missing meter readings or water loss from pipeline leakage. This project will enable the city to analyze water usage patterns of different consumer groups to build up or reconstruct the network to maximize efficiency and service.

European Narrowband IoT

Just as the "last mile" was the challenge for early internet connectivity, Earls (2017) notes the new challenge in IoT is the "last few feet." Bluetooth and Wi-Fi can be insecure and, because they use a crowded range of the radio spectrum, are subject to interference. Wireless or not, power limitations mean many IoT applications only become practical when a device can sleep most of the time, waking up only when needed. So whether it's for the home, transportation systems, agriculture, factory or distribution, developers have sought to find a safe efficient way to make IoT work in small spaces.

One solution has been found in Narrowband IoT, or NB-IoT, which enables connectivity via cellular network. Europe and China have embraced NB-IoT and are rolling out projects wherever possible. European carrier Vodafone launched NB-IoT in four European markets—Germany, Ireland, the Netherlands and Spain—and committed to rolling out the network standard across all of the countries in which it operates by 2020. Similarly, 3GPP, which involves seven different telecommunications standard development organizations, has come up with agreements to allow telecom carriers to compete in the space and move ahead with NB-IoT. Narrowband is a standardized technology that reuses most of a mobile operator's existing network. For example, Vodafone has stated that for around 85% of its sites, the move to NB-IoT is "simply a software upgrade, which should help to keep upgrade costs low."

An important benefit of Narrowband IoT is that it has a wider base of industry support than competing options. "The NB-IoT forum has telecom operators that represent more than 2 billion subscribers," (Earls, 2017) Furthermore, NB-IoT has the support of Huawei, Ericsson, Nokia, Intel, Qualcomm, and others, and so has a broader base of established vendors than the other technologies. Special NB-IoT device chipsets are also coming to the market. NB-IoT offers a lower cost of around $5 per device and $1 per year for connectivity. Initial interest comes from utility companies and metering, particularly water metering. Other tracking applications include child and pet trackers, consumer electronics, retail, and agriculture.

According to 3GPP, other benefits of Narrowband IoT include long battery life (a 5 watt/hour battery could last up to 10 years), the ability to support large numbers of devices; and uplink and downlink rates of around 200 kbps using only 200 kHz of available bandwidth; this is important because it means carriers won't have problems accommodating the service (Earls, 2017).

Current Status

Gartner (2017) forecasted total spending on endpoints and services reached almost $2 trillion by the end of 2017. The number of connected things worldwide increased over 30% during 2017 to an estimated 8.4 billion connected things in use.

Gartner (2017) estimates that two-thirds (67%) of the installed base of connected things is led by three regions, Greater China, North America, and Western Europe. Furthermore, the consumer segment is the largest user of connected things. Aside from automotive systems (see chapter 13), the most used applications or devices are smart TVs and digital set-top boxes. The same report noted businesses have moved most quickly to adopt smart electric meters and commercial security cameras.

According to Weinreich (2017), total global spending on IoT devices and appliances across all environments (work and home) was an estimated $737 billion in 2016 and is projected to reach up to $1.4 trillion by 2021. According to a McKinsey Global Institute report (cited by Weinreich), IoT is projected to have an economic impact of somewhere between $4 to $11 trillion on the global economy by 2025, when factoring in its impact in sectors like manufacturing, health, retail, and the smart home.

In March, 2017, the Amazon Echo was declared 'a megahit' capable of generating $10 billion in revenues (Kharpal, 2017). That same report noted Investment bank RBC predicted there will be 60 million Alexa-enabled devices sold in 2020 which would bring the total install base to around 128 million.

Factors to watch

According to Meola (2016), there are three major areas that will use IoT ecosystems: consumers, governments, and businesses. Manufacturing, transportation, defense, healthcare, food services, and dozens of others will gradually adopt IoT in ways to improve efficiency, economies of scale, and security.

The "four trends in smart home technology 2018" are big data analytics, smart security systems, more décor-friendly environments, and 5G connectivity (Luke, 2017). The floodgates of database creation will be wide open with IoT. Big data is generated by billions of devices providing information, sometimes moment-by-moment. Big data enables consumers to observe their behavior and be more efficient. Big data also allows service providers to look over your shoulder and do the same for you, probably without your knowledge.

Security cameras may become as ubiquitous in the home as they are in the public square. The ability to perform facial recognition will improve until any and every visitor to your door will be recorded for posterity no matter the purpose of the visit. In a similar way, smart devices are already changing from the drab circles and cubes to become more interesting, artistic, and still functional. Designers and manufacturers know the best way to increase market share is to make their smart object friendly, cool, and almost unnoticed. Finally, as the glut of smart objects into your home reaches capacity, you will need even more network bandwidth to keep it all going strong. A device that has to buffer is a device that will be replaced. Forecasters say 5G cellular networks (see Chapter 22) will transmit data 10 times faster than 4G.

According to Soergel (2018), moving from 4G to 5G (the fifth generation of wireless connectivity) will allow consumers to download things significantly faster on compatible devices. When 4G rolled out in 2009, it allowed consumers to download at a speed of about 100 megabits per second. At that speed, a two-hour-long movie downloaded in about six minutes. But 5G is currently being tested with full adoption likely by 2020 and would download that same movie in 3.6 seconds.

Getting a job

Because IoT is so vast and encompasses so many industries, working in it may demand a particular mindset more than a set of skills. According to Bridgewater (2017), there are seven characteristics, skillsets, and personality traits that one should have to be successful in an IoT-related business:

- **A mobile mindset**, as the things in IoT are generally small and mobile. The fluidity and pace of change with mobile devices and objects is almost limitless.

- **A nose for data analytics**, being comfortable with large data sets and able to accurately assess trends and outcomes for sub-groups and even individuals.

- **A key to security.** IT security is a growth area that has to look end-to-end rather than just guarding passwords.

- **A head in the clouds**. The ability to work with the "digital twin" data in the cloud and sync it back with the actual device is an important skill. In the Industrial IoT (IIoT) world, a digital twin represents a simulation of some real-world object or machine.

- **A gauge for scaling early**. Be comfortable with distributed data strategies, because it allows you to start small and simply add nodes over time. Don't build a tall tower; build a single story you can add one floor at a time.

- **A view into business process optimization**. With so many small moving parts, it's important to be able isolate and identify places where one small change positively affects the whole process.

- **A collaborative interpersonal touch**. Interpersonal skills are always in demand. Are you a diva? Do you think your opinion is the most important? See how others see you.

Conclusion:

One of the final thoughts about Internet of Things developments recently is the advances in robotics. We are coming ever closer to the world envisioned by Heinlein and others where humanoid robots become part of our family. Tests are currently underway with robot pets and helpers who can read to our children, help with the groceries, and even keep company with shut-ins. Whether this heralds utopia or digresses into Terminator territory is anyone's guess.

Bibliography:

Bradbury, D. (2017, August 2). Grab a fork! Unravelling the Internet of Things' standards spaghetti. The register.com. Retrieved from https://www.theregister.co.uk/2017/08/02/iot_standards_spaghetti/

Bridgewater, A. (2017, December 13). The 7 Personality types needed for the Internet of things. *Forbes*. Retrieved from https://www.forbes.com/sites/adrianbridgewater/2017/12/13/the-7-personality-types-needed-for-the-internet-of-things/.

Burns, M. (2014, June 23). Google Makes Its Nest At The Center Of The Smart Home Techcrunch.com. Retrieved from https://techcrunch.com/2014/06/23/google-makes-its-nest-at-the-center-of-the-smart-home/

Earls, A. (2017, February 27). Narrowband IoT may be the answer to cellular IoT woes. Techtarget.com. Retrieved from http://internetofthingsagenda.techtarget.com/feature/Narrowband-IoT-may-be-the-answer-to-cellular-IoT-woes

Evans, D. (April, 2011). The Internet of Things: How the next evolution of the Internet is changing everything. Cisco Internet Business Solutions Group (IBSG) White paper. Retrieved from www.cisco.com/c/dam/en.../IoT_IBSG_0411 FINAL.pdf

Gartner, Newsroom (2017, February 7). Gartner says 8.4 connected "Things" will be in use in 2017, up 31 percent from 2016 [Press release]. Retrieved from https://www.gartner.com/newsroom/id/3598917.

Gibbs, S. (2018, January 12). CES 2018: voice-controlled showers, non-compliant robots and smart toilets. The Guardian.com. Retrieved from https://www.theguardian.com/technology/2018/jan/12/ces-2018-voice-controlled-showers-robots-smart-toilets-ai

Graham, J. (2017, November 12). The real cost of setting up a smart home. *USA Today*. Retrieved from https://www.usatoday.com/story/tech/talkingtech/2017/11/12/real-cost-setting-up-smart-home/844977001/

Griffith, E., and Colon, A. (2017, December 1). The Best Smart Home Devices of 2018. Pcmag.com. Retrieved from https://www.pcmag.com/article2/0,2817,2410889,00.asp

Horwitz, J. (2017, November 16). The newest smart home assistant stands out for one reason—Quartz. Qz.com. Retrieved from https://qz.com/1131072/baidu-raven-h-the-newest-smart-home-assistant-stands-out-for-one-reason/

Joffe, B. (2017, July 17). Amazon may lose the war for voice assistants in China before it begins. Forbes.com. Retrieved from https://www.forbes.com/sites/benjaminjoffe/2017/07/17/amazon-may-lose-the-war-for-voice-assistants-in-china-before-it-begins/#30d7199a227a

Kharpal, A. (2017, March 10). Amazon's voice assistant Alexa could be a $10 billion 'mega-hit' by 2020: Research. CNBC.com. Retrieved from https://www.cnbc.com/2017/03/10/amazon-alexa-voice-assistan-could-be-a-10-billion-mega-hit-by-2020-research.html

Khandelwal, S. (2017, September 6). Hackers can silently control Siri, Alexa & Other Voice Assistants Using Ultrasound. The hacker news. Retrieved from https://thehackernews.com/2017/09/ai-digital-voice-assistants.html

Ludlow, D. (March 8, 2018). Google Home vs Amazon Echo: Which is the best smart speaker? Trusted Reviews.com. Retrieved from http://www.trustedreviews.com/news/google-home-vs-amazon-echo-2945424#BE6eyr8290Uo0VGF.99.

Luke (2017, December 22). 10 smart home devices should be on your 2018 wishlist. 4 trends in smart home technology 2018. Annke.com. Retrieved from http://www.annke.com/blog/2017/12/22/smart-home-devices-2018/

Ma, S., He, W., & Cheng, Y. (2017, September 11). Internet of things gets even techier. *China Daily*. Retrieved from http://www.chinadaily.com.cn/china/2017-09/11/content_31833613.htm

Meola, A. (2016, December 19). What is the Internet of Things (IoT)? *Business Insider*. Retrieved from http://www.businessinsider.com/what-is-the-internet-of-things-definition-2016-8

Nash, J.B. (1932). Spectatoritus. New York: Sears Publishing Company. Retrieved from http://babel.hathitrust.org/cgi/pt?id=mdp.39015026444193;view=1up;seq=6

Newcomb, A. (2018, January 12). The CES crystal ball gives glimpse of our high tech future. NBCnews.com. Retrieved from https://www.nbcnews.com/mach/tech/ces-crystal-ball-gives-glimpse-our-high-tech-future-ncna835926

Press, G. (2014, June 18). A very short history of the Internet of Things. *Forbes*. Retrieved from http://www.forbes.com/sites/gilpress/2014/06/18/a-very-short-history-of-the-internet-of-things/#524811322df5.

Sargent, M., and Caldwell, S. (2017, May 27). Homekit: the ultimate guide to Apple home automation. Imore.com. Retrieved from https://www.imore.com/homekit.

Soergel, A. (2018, January 9). 5G: The Coming Key to Technology's Future. *U.S. News & World Report*. Retrieved from https://www.usnews.com/news/economy/articles/2018-01-09/5g-the-coming-key-to-technologys-smart-future

Sorrell, S. Juniper Research. "Smart Homes—It's an Internet of Things Thing." *Smart Home Ecosystems & the Internet of Things Strategies & Forecasts 2014-2018*. Hampshire, U.K. 2014. Print.

Swanner, N. (June 8, 2017). Can Homepod make Homekit a winner for Apple? Dice.com. Retrieved from https://insights.dice.com/2017/06/08/can-homepod-make-homekit-winner-apple/

Tata Consultancy Services (2015). Internet of things: The complete reimaginative force. *TCS Global Trend Study*. White paper. Retrieved from http://sites.tcs.com/internet-of-things/#.

Weinreich, A. (2017, December 18). The Future Of The Smart Home: Smart Homes & IoT: A Century In The Making. Forbes.com. Retrieved from https://www.forbes.com/sites/andrewweinreich/2017/12/18/the-future-of-the-smart-home-smart-homes-iot-a-century-in-the-making/2/#19c7c9f17b70

13

Automotive Telematics

Denise Belafonte-Young M.F.A.[*]

Overview

Smart Cars, Connected Cars, Electric Cars, Driverless Cars... these are some of the major innovations the automotive industry has embarked upon in recent years. Automotive Telematics is a catch-all term for the wide range of applications combining computers, networking, and sensors in automobiles. For example, by consolidating a GPS framework with on-board diagnostics it is conceivable to record and outline precisely where an auto is and how quickly it is traveling, and then cross-reference that with how an auto is operating mechanically. Infotainment, GPS, navigation systems, and other OEMS (original equipment manufacturer) structures help navigate your vehicle and improve diagnostics making vehicle breakdowns and repairs less likely (Laukkonen, 2016).

Connected cars have expanded to fleet and public transportation, which can send information from a vehicle to a central command center. Communication is enhanced using sensors and heightened trackside wireless networks (Carter, 2012). Insurance companies can use this technology to gather evidence of your driving habits, which could lead to safe driver discounts.

Driverless cars will be an interesting challenge to many related industries, including auto insurance companies, and could revolutionize services ranging from driverless pizza delivery and car rental services to Ubers, Lyfts, and even mass transit.

Introduction

Automotive telematics can be defined as "the blending of computers and wireless telecommunications technologies" (Rouse, 2007, para. 1), enabling drivers to get information about the location, movement, and state of their vehicles. Telematics doesn't necessarily need to incorporate two-way communication, yet a large number of the newly developed tools do (Howard, 2015).

Telematics includes wireless data communication, which opens up a huge range of possibilities. To provide the above services, telematics products may include GPS (Global Position System), inter-vehicle Wi-Fi connections, digital audio and video solutions, wireless telecommunication modules, car navigation systems, and other applications (Cho, Bae, Chu, & Suh, 2006).

Background

Ford Motor began a manufacturing revolution with mass production assembly lines in the early 20th century, and today it is still one of the world's largest automakers (Ford Motor Company, 2014). The history of "user" telematics can revert back to Henry Ford's idea in 1903 to create easy transportation for everyday people. Automobiles evolved from strictly a means of

[*] Associate Professor, Lynn University, Boca Raton, Florida

transportation to luxury items as time went on. The development of in-vehicle telecommunications, entertainment, and "infotainment" were part of that revolution and are the landmarks of today's automotive environment.

The Birth of the Car Radio

The first technological breakthrough in automotive electronic devices was the car radio. According to Charlotte Gray (n.d.) "The first radios appeared in cars in the 1920s, but it wasn't until the 1930s that most cars contained AM radios (para. 3)." William Lear, who created the Learjet, also created the first mass market car radio. The first FM car radio was created in 1952 by Blaupunkt, a pioneer in audio equipment and systems. By 1953, the blended AM/FM car radio developed. The first eight-track tape players appeared in cars in 1965, and from 1970 through 1977 cassette tape players made their way into automobiles. By 1984 the first automobile CD player was introduced. The next big innovation was the introduction of DVD entertainment systems, which allowed audio entertainment options to branch into visual displays by 2002 (Gray, n.d.).

Table 13.1
Evolution of Automotive Telematics

In a GSMA study, it was surmised that the evolution of automotive telematics includes:

Telematics 1.0	Hands-free calling and screen-based navigation
Telematics 2.0	Portable navigation and satellite radio
Telematics 3.0	Introduction of comprehensive connectivity to the vehicle
Telematics 4.0	Seamless integration of mobility and the Web—the biggest opportunity yet

Source: SBD & GSMA, 2012

Eventually, as the digital age emerged, terrestrial radio was joined by new ways of listening. MP3, smartphone connections, SiriusXM satellite radio, Bluetooth, HDMI input, Wi-Fi, and voice activation technology are common features in new automobiles.

The Influence of Telecommunications

The history of car phones began in 1946, when a driver extracted an earpiece from under his console

and completed a call. The first wireless service was created by AT&T providing service within 100 cities and highway strips. Professional and industrial users such as utility workers, fleet operations, and news personnel were the first customers (AT&T, 2014). Other uses of technology in cars followed, including:

- Power steering and brake systems
- The "idiot" light (dashboard warning lights)
- Digital and computerized fuel injection, odometer systems
- Cruise control
- Backup cams
- Innovations in vehicle diagnostics: Onstar, FordSYNC etc.

Coe, Prime & Jest (2013) created an evolutionary timeline, depicted in Table 13.2. These innovations in telecommunications and "social" connectivity have expanded the definition of what "driving" is today.

Table 13.2
Developments in Automotive Telematics

Year	Development
1998	The first hands-free car gateways were introduced
2000	The first GSM & GPS systems were brought to market
2002	Bluetooth hand free voice gateways with advanced voice integration features
2003	Integrated GSM phone with Bluetooth
2007	Multimedia handset integration is introduced
2009	Fully-integrated mobile navigation using a car GSM system
2010	3G multimedia car entertainment system
2011	Telematics and Infotainment systems introduced based on Linux

Source: Coe, Prime, & Jest, 2013

Recent Developments

Telematics has advanced significantly since 2010. In March 2013, a group of Ford engineers boldly predicted, "The automobile of the future is electric. It is connected. It is smart." (Phillips et al, 2013; p. 4). Most automobile manufacturers have since begun experimenting with driverless cars. Tesla was a front-runner in the testing of driverless vehicles, along with Google and Apple. But the industry giants like Ford and General Motors are adding a competitive edge in development and testing. Ford Motor Company will increase its investment in electric vehicles to $11 billion by 2022 and plans to develop 40 different models of hybrid and fully electric vehicles by that date (Reuters, 2018).

Meanwhile, Ford plans to begin its market test of autonomous cars in 2021. CEO Jim Hackett says "autonomous vehicles could lead Ford into new businesses. For example, Ford said in August 2017 it would partner with Domino's Pizza to test its autonomous technology on pizza delivery routes in Michigan" (Ferris, 2017, para. 5).

GPS and Navigation Systems

Built-in connectivity and GPS technologies have paved the way to hands-free safety, vehicle tracking, and diagnostic features. General Motor's OnStar system, was the zenith for subscription services. GM OnStar telematics unit has grown into a "profit margin superstar" (Guilford, 2013). GM also has an in-dash GPS system that uses information from a built-in hard drive, which can also be used to store music (Laukkonen, 2016). Each auto manufacturer has developed a system that competes to attract the loyal consumer. Besides the GM On-Star system, other services include: Mercedes Benz mbrace, BMW iDrive Vehicle Control System, Lexus Enform, Toyota Safety Connect, Ford Sync, Hyundai BlueLink, Infiniti Connection, and the Honda Link. Each of these systems is used for connectivity, navigation, and diagnostic analysis. GM has been the main supplier of telematics for over 10 years, offering the innovation as an indispensable piece of OEM embedded telematics systems in North America and China. The website Edmunds.com has sorted out the common structures on these devices, which include:

- Automatic Collision Notification
- Concierge Services
- Crisis Assistance
- Customer Relations/Vehicle Info

- Dealer Service Contact, Destination Download
- Destination Guidance Emergency Services
- Fuel/Price Finder
- Hands-Free Calling
- Local Search
- Location Sharing POI Communication
- POI Search
- Remote Door Lock
- Remote Door Unlock
- Remote Horn/Lights,
- Roadside Assistance
- Sent-To Navigation Function
- Sports/News Information Stock Information
- Stolen Vehicle Tracking/Assistance
- Text Message/Memo Display
- Traffic Information
- Vehicle Alarm Notification
- Vehicle Alerts (speed, location, etc.)
- Vehicle Diagnostics
- Vehicle Locator
- Weather Information.

(Edmunds.com., n.d., par. 1)

GPS can be integrated with built-in navigation and full-fledged *carputers*, or in less expensive, portable units. Basically, with navigation apps on smartphones, tablets, and stand-alone devices, built-ins have significant competition (Laukkonen, 2016).

Safety Features, Advanced Driver Assistance Systems (ADAS) and Vehicle Connectivity

According to the *Journal of Intelligent Manufacturing* in 2012, driver assistance was advancing through camera-based infrastructures to enhance road safety and accident avoidance. These in-vehicle display systems provided visible information on the surrounding environment, specifically to display "blind-spots" and back-up perception. Vision-based integrated ADAS systems are inevitable to standard current automobile feature inclusion. (Akhlaq, Sheltami, Helgeson & Shakshuki, 2012). In 2014, the National Highway Transportation Association (NHTSA) set a regulation for a 2018 deadline for camera backup systems for rearview safety. More and more manufacturers currently offer the system as a standard or optional feature as a safety precaution.

The 2002 Nissan's Infiniti Q45 was the first model to incorporate a rear-view cam. Nissan continues as a leader in this initiative offering an around-view monitoring system, and the 2015 Ford F150 truck also featured four cameras. Tesla experimented with replacing side view mirrors with cameras to enhance vehicle aerodynamics and increase fuel efficiency (Rafter, 2014).

Figure 13.1
Advanced Driver Assistance Systems (ADAS)

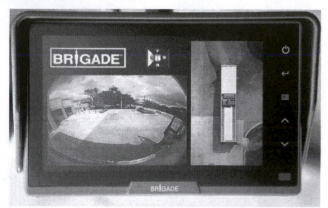

Source: Carline Jean/Sun Sentinel

Studies throughout the 1980s kept tabs on informatics and the development of telecommunications to find new applications for road safety. Through these studies a program known as DRIVE (Dedicated Road Infrastructure for Vehicle Safety in Europe) was started in 1988 to conduct research (CORDIS, n.d.). What auto manufacturers consider "connected vehicles" predominately use the sources of cellular and satellite transmission. One of OnStar's best-known features is automatic crash notification (ACN) which is reported by the vehicle's sensors. A representative, through a voice link, calls 911 and then can correspond with the occupants to keep in contact for reassurance until help arrives. OnStar also sends vehicle monthly diagnostic reports and can issue other reports such as weather, traffic, sports scores, movie times, and nearby amenities (Howard, 2015).

New federal regulations mandated car-to-car communication (for safety) (University of Michigan Transportation Research Institute, 2014). Central car computers as stated are the equivalent of airplane "black boxes" that can read back speed and a lot of other factors when a car is in a collision. Conversely there are a variety of privacy issues here. Questions may arise on who owns the data and whether drivers automatically give up this data when they take their car in for servicing and system services take place. The U.S. government is moving on new rules requiring future vehicles come with Wi-Fi technology necessary for cars to communicate with one another. The goal is to prevent accidents, but networked cars could do much more (Fitchard, 2014).

Driverless, Electric, and Smart Cars

The U.S. Department of Transportation is researching the introduction of driverless cars onto American roadways, soliciting input from researchers and manufacturers. Some of these current endeavors include: Ford plans to bring their autonomous vehicles to a test market in 2018. CEO Jim Hackett explains that the $1 billion investment will help the plan to launch these vehicles by 2021. As mentioned earlier, Hackett is enthusiastic this could lead to joint ventures beginning with a partnership with Domino's Pizza to test the technology on pizza delivery routes in Michigan (Ferris, 2017). Another interesting partnership is the cooperative effort between Lyft and Waymo, the driverless-car division of Google (Isaac, 2017).

The competition is fierce as Lyft has been a second to Uber in the car service market. As of early 2018, Uber was testing its Volvo self-driving SUVs picking up and dropping off passengers in Tempe, Arizona and Pittsburg, Pennsylvania, despite an accident in Arizona, which led to a temporary suspension of their "autonomous driving pilots" (Hawkins, 2017).

Target markets for driverless vehicles include commercial vehicles, buses, and taxis. Driverless cargo units and storage trucks are on the horizon. Imagine a world with safer truck travel, with less fatigue, night-vision hazards, and other human mistakes.

Smart cars have been in the works for years. 2015's Consumer Electronics Show revealed the Mercedes driverless car called the Mercedes Benz F015 Luxury in Motion. The car was designed like a mobile office. It included a steering wheel for alternative human driving. With a bullet-like design, the car is equipped with a built-in laser projector, which beams a light at the crosswalk to alert pedestrians. Laser headlights were another concept initiated by BMW (Martindale, 2015).

Apple, Google, tech companies, and auto manufacturers, continue to expand autonomous vehicle research and prototype releases in America. The European Union had taken research a step further by looking at the roadways and infrastructures ability to correspond with the smart cars. The "smart" roads would be part of a design to offer vehicles "360-degree awareness" through sensors, transmitters, and processors allowing vehicle-to-vehicle correspondence, and ways to be informed of pedestrians and hazardous conditions, carrying the term "vehicle-to-everything." Traffic lights and signs would be equipped to gather real-time data.

After millions of miles driven by self-driving cars, a tragic milestone was achieved in March 2018, when an autonomous Uber vehicle struck and killed a pedestrian on an Arizona highway (Griggs & Wakabayashi, 2018). Given the number of deaths related to automobile accidents, it is probably not reasonable to expect zero accidents with autonomous vehicles, but every accident like this one increases barriers to widespread use.

As the number of million miles driven by driverless cars increases, the number of accidents involving these cars is also increasing. A Tesla Model S on "Autopilot" smashed into a stopped fire truck in Culver City, California. (Stewart, 2018). On the other hand, author Shelley Palmer (2017) notes "Humans are very dangerous" and the self-driving cars and trucks will be safer! (para. 3).

Another concern being addressed is the risk of hackers taking control of a car. Consider the danger of cars shutting down at 60 mph, brakes becoming disabled, and remote commands being sent to cause collisions—all initiated by hackers that can virtually make the world a reality video game (Halsey, 2015). But incidents and precautions such as these are not deterring the industry from advancement and enhancement. Since November 2015, the Toyota Research Institute in association with Jaybridge Robotics continues to work on a car that is less capable of involvement in a crash, which is primarily geared to those who are unable to drive such as the elderly and those with special needs (Toyota Research Institute, 2016).

Trendsetting automaker Tesla is a factor to watch. Elon Musk, Tesla's founder, assured the industry that his electric car company, which currently offers partial autopilot features, is aggressively looking for software engineers to help pave the way to the pool of driverless contenders. On that note, in an article titled "People Keep Confusing Their Teslas for Self-Driving Cars," author Jack Stewart (2018) notes:

> "We are entering a dangerous period in the development of self-driving cars. Today, you can buy a Cadillac, Volvo, Tesla, Audi, and even a Nissan that will do some of your driving for you, as long as you stay on top of things" (para. 1).

Infotainment, Entertainment and Internet in Cars

In-car technology is thriving with manufacturer's inclusion of devices and tools that enhance entertainment and infotainment. As "connected" cars, these gadgets achieve a variety of roles to link drivers to their social media, music, movies and more. Filev, Lu, & Hrovat (2013) describe some of the ways that telematics have changed the world of infotainment:

- *Cloud Computing*—Integrated wireless technology structure enhances driving experiences.

- *The Social Car*—Tweeting at a red light? Traffic delays? Social media is playing a big role in sharing information among drivers, but concerns about distracted driving need to be alleviated first. Many cities and states have passed laws regulating the use of electronic devices while driving.

- *Direct Streaming*—Services such as Sky On Demand and Netflix, allow passengers to download and stream movies in their vehicles. DVD technology is a thing of the past.

- *Voice Recognition*—With the development of cellphone and tablet voice recognition systems, automobiles will use the same technology to avoid driver button pushing and touch display distractions.

Your infotainment system can help through harrowing times of being stuck in traffic, or long travel. The system is progressively adept of delivering options such as:

- Post-secondary learning from your vehicle
- Ebooks being read aloud while you are driving
- Listening to your favorite newspapers read aloud
- Feedback on your driving
- Feedback on road conditions
- Warnings and updates about traffic

In a 2014 study conducted by the Nissan Corporation, consumers under the age of 35 reported that owning an automobile was more of a luxury than a necessity. The discoveries concluded that this generation known as the "Share Natives" mandated "connectivity" and ways to continue their share time. Ideally, the "Share Natives" expected more voice controls, motion sensors, and large charging stations (Euromonitor International, 2015). Infotainment models can help the driving experience with model services such as Pay-Per-View movies, television subscriptions, in-car games for passengers, vehicle-based games, cloud syncing with your music files, and digital access your computer in your office or home (Prime, 2013).

As discussed in Chapter 22, 5G networking will have its commercial launch by 2019. The apparent increase in numbers of connected cars, and the coming wave of autonomous vehicles, constitutes a growing field of connectivity. It is expected that by 2024, three quarters of new vehicles will feature embedded cellular connectivity (Novak, 2018).

Apple CarPlay and Android Auto continue their competition to populate the dashboard as Android offers Google Play Music supporting third-party music apps like *Spotify*, *Tune-In Radio*, and *Radioplayer* (exclusive to users in the United Kingdom), versus Apple CarPlay's iTunes which hosts such apps as *Rdio*, *Dash Radio*, *Stitcher*, *Overcast*, and *Audible*. (Ranosa, 2016).

Even with all of the new ways of listening to music and other audio sources in cars, AM/FM radio continues to dominate: The 2017 Edison Media and Triton Digital Infinite Dial study revealed that a dominant 82% of consumers aged 18+ who have driven or ridden in a car in the last month say they listen to AM/FM radio in the car—a statistic that has shifted little since 2014. (Taylor, 2017, para. 6)

Other entertainment options that are being considered for in-vehicle go beyond audio and radio, including in-vehicle gaming consoles such as the Xbox, PlayStation, and other systems as well as Internet access. Consider the opportunity offered for media consumption if self-driving cars become a reality and users can spend commuting time with any technology that could be built into an automobile.

Current Status

The growing population in major cities and dwellings will necessitate automobile "intelligence" to help navigate through crowded environments (Filev, Lu & Hrovat, 2013). Drivers navigating their way in such vehicles will also use personalized information channels to find the best services and sources. These needs have led to further advancement of automotive controls and connectivity, with Autodealer Monthly identifying 2017 as "the year of Telematics". Writer Mike Esposito (2017) stated:

> "Welcome to the new world of telematics. For years, we have known that connected cars were coming, but I believe 2017 marks their official arrival. Percentage estimates of vehicles with installed telematics that will be sold in 2017 range from 60% to 80%. For now, the market for connected car packages continues to center on premium vehicles. By 2022, however, 75% of connected car packages will be sold as part of smaller, less expensive cars" (para. 2).

The most common telematics systems include:

- Emergency crash notifications
- Emergency or "Good Samaritan" assistance via a "Help" or "SOS" button
- Roadside assistance with embedded GPS
- Vehicle diagnostics and health report emailed to both the customer and the dealer
- Remote door unlock
- Navigation requests
- Traveler services such as gas price finder or local restaurant search
- Streaming sports/news/stock information
- Streaming music or media via Pandora and other providers

- Text message and email display
- Traffic information and rerouting
- Weather information
- Vehicle speed or location alerts
- Stolen vehicle tracking and remote shutdown
- Vehicle software updates

(Esposito, 2017)

A growing concern is that these newer infotainment systems are increasing distractions behind the wheel. The AAA Foundation for Traffic Safety finds:

"…drivers using in-vehicle technologies like voice-based and touch screen features were visually and mentally distracted for more than 40 seconds when completing tasks like programming navigation or sending a text message. Removing eyes from the road for just two seconds doubles the risk for a crash, according to previous research" (As cited in Johnson, 2017, para. 1).

It is expected that approximately 104 million vehicles will contain some form of connectivity with integrated telematics reaching 88% of all vehicles by 2025. Many solutions could lead to safer roads, including smartphone technology and new laws, structures, and regulations for drivers.

One report forecasts the global commercial telematics market to grow from US $20 billion in 2015 to US $47 billion by 2020. Europe is expected to be the largest market in terms of market size, while Middle East & Africa and Asia-Pacific (APAC) are expected to experience an increase in market traction during the forecast period. Latin America is expected to experience a high growth rate in this market" (Commercial Telematics Market, 2015).

As discussed in Chapters 21 and 22, the world will soon see a new generation of wireless communication known as 5G replace previous generations of wireless networking (2G, 3G, 4G and Wi-Fi) to permit more communication. With its key differentiator being a higher capacity network, 5G will act as an enabling factor for self-governing vehicles, machine-to-machine (and what's more, machine-to-framework) administrations, and even the "Internet of Things." As 5G will be an integral part of machine learning and other sensor functions, it will also connect wearable

device technology as well as smart glasses and watches helping integrate augmented and virtual reality in features of cars by 2020 (Gould, 2015).

Ford Motor Company is expanding its Smart Car initiative to China with an investment of $1.8 billion in hopes that the technology will stand out in the country with enhancement of their "smartphone connectivity, autonomous driving" and other advancements in the Chinese market. The initiative will lend itself to building growth in running plants which currently exist in China and incorporating the technology (Yu, 2015). Ford's "MyFord Mobile" is also developing research in "wearable" and "health data" informatics to driver-assist technology. With a surge in smart watch technology, applications can monitor a driver's tiredness, blood-pressure, glucose level, and Ford drivers can remotely start, lock, un-lock and locate their vehicle using voice-activation.

Factors to Watch

As mentioned previously, Tesla is taking the automotive industry by storm with creative innovations. Touch screen monitors are prominent in the automotive evolving technology with horizontal and vertical displays. Tesla has initiated a 17" vertical screen developed by their engineers. (Other companies use 10-12" units.) The high-resolution 3D display system uses the Tegra VCM, "for drivers, the system provides larger, more readable maps and a beautifully rendered instrument cluster that can be personalized from the multifunction steering wheel" (IVI in new Tesla, 2015, para. 2).

Cellular companies like Verizon are offering services for fleet operations for businesses. According to Fleetmatics.com, fleet managers can:

- Access the dashboard, live map, reports, alerts
- Find the closest vehicles for urgent dispatch
- View animated vehicle journeys in route replay
- Display Geofences on a map

Filev, Lu, & Hrovat (2013) state that the boom in population in major cities and dwellings will necessitate automobile "intelligence" to help navigate through such crowded environments. Drivers navigating their way in such vehicles will also be used as personalized information channels to find the best services and sources for the driver. Such needs will lead to further advancement of automotive controls.

The United States government has created regulations for future vehicles, mandating them to be equipped with Wi-Fi technology. The National Highway Traffic and Safety Administration (NHTSA) sees this automotive connectivity as a safety technology similar to seatbelt and airbag laws. Cars can tell each other when a driver is "slamming on brakes," making a turn, or signaling for a lane change. Other drivers—or the cars themselves—can react to these cues to help avoid potential accidents. Driverless cars will in turn be equipped to communicate directly with each other, bypassing human driver involvement (Fitchard, 2014). The NHTSA has reported "Today's electronics, sensors, and computing power enable the deployment of safety technologies, such as forward-collision warning, automatic-emergency braking, and vehicle-to-vehicle technologies, which can keep drivers from crashing in the first place" (NHTSA and Vehicle Cybersecurity, 2015, para. 1). These measures have taken a new toll in the monitoring of cybersecurity communication and development for all vehicle-to-vehicle and vehicle-to-infrastructure applications.

Lastly, the technology behind autonomous vehicles is migrating into other forms of larger transportation vehicles including trucks and ships. Uber is working on a trucking startup called "Otto", whose founder Anthony Levandowski was a former Google employee. Uber performed its first self-driving truck delivery, a 120- beer delivery in October 2016 (Anand, 2017). Self-driving trucks could have a tremendous impact on the shipping industry and the employment of long- and short-haul truckers.

Another joint venture between Rolls-Royce and Google will see ships self-learning because of cutting-edge machine learning algorithms. It will likewise bring the organization's vision of a completely self-sufficient ship setting sail by 2020 coming soon (Marr, 2017). Volvo Cars said it has consented to supply Uber Technologies Inc. with an armada of 24,000 self-driving cabs starting in 2019—one of the first and greatest business orders for such vehicles (Boston, 2017). Furthermore, Tesla is developing an electric big rig truck that can travel 500 miles on just one charge (Fingas, 2017).

In summary, automotive telematics is the intersection of two major trends—mobility and data analytics. Dhawan (2013) explains the relationships and alliances of car manufacturers and other technological entities, "Automobile companies will start collaborating with data companies; mobile device makers will partner with healthcare institutions. Industry lines will blur and experts in one field will soon be experts in another. Not only will technology change as a result, but also our world will change" (para. 6).

Careers and Getting a Job

There are many opportunities to work in the automotive telematics industry. With certificate programs, online learning, BMS Education, CMMI Training, ASE certification, prospects are great for a future in the field. Careerbuilder.com's Automotive Technology positions include: Telematics Engineers, Consultants, Sales representatives, Software/Hardware Developers, Fleet Managers, Infotainment Engineers, Electronic Technicians, Vehicle Diagnostics Specialist, and many more. The possibilities are endless as we see the scope of automotive technology continuing to evolve for years and decades to come.

Bibliography

Akhlaq, M., Sheltami, T. R., Helgeson, B., & Shakshuki, E. M. (2012). Designing an integrated driver assistance system using image sensors. *Journal of Intelligent Manufacturing, 23*(6), 2109-2132. doi:http://dx.doi.org/10.1007/s10845-011-0618-1

Anand, P. (2017, June 1). Uber Rival Waymo Is Testing Self-Driving Trucks Too: The company is expanding its self-driving car efforts. *BuzzFeed News*. Retrieved from https://www.buzzfeed.com/priya/googles-waymo-is-exploring-self-driving-trucks.

AT&T. (2014). 1946: First mobile telephone call. Retrieved from http://www.corp.att.com/attlabs/reputation/timeline/46 mobile.html

Automotive partners. (2016). *SiriusXM*. Retrieved from https://www.siriusxm.com/automakers/vehicle

Boston, W. (2017, November 20). Volvo promises Uber fleet of self-driving taxis by 2019 agreement represents one of the most concrete deals between two big players in the auto field. *Wall Street Journal*. Retrieved from https://www.wsj.com/articles/volvo-promises-uber-fleet-of-self-driving-taxis-by-2019-1511184730.

Carter, J. (2012, June 27). Telematics: What you need to know. *TechRadar*. Retrieved from http://www.techradar.com/news/car-tech/telematics-what-you-need-to-know-1087104

Cho, K. Y., Bae, C. H., Chu, Y., & Suh, M. W. (2006). Overview of telematics: A system architecture approach. *International Journal of Automotive Technology, 7*(4), 509-517.

Coe, J., Prime, R., & Jest, R. (2013, May 30). Telematics history and future predictions. *Telematics.com*. Retrieved from http://www.telematics.com/guides/telematics-history-future-predictions/

Commercial telematics market worth $47.58 billion USD by 2020 [Press release]. (2015, September). *Markets and Markets*. Retrieved from http://www.marketsandmarkets.com/PressReleases/commercial-telematics.asp

CORDIS. (n.d.). History of the deployment of transport telematics. Retrieved from http://cordis.europa.eu/telematics/tap_transport/intro/benefits/history.htm

Dhawan, S. (2013, December 23). Three sectors to drive innovation in 2014. *Harman*. Retrieved from http://www.symphony-teleca.com/company/newsroom/blog-old/the-st-blog/2014/jan/three-sectors-to-drive-innovation-in-2014/

Edmunds.com. (n.d.). Telematics chart. Retrieved from http://www.edmunds.com/car-technology/telematics.html

Esposito, M. (2017, January). Is 2017 the year of telematics? Dealers and their sales and service teams must prepare for a watershed moment in vehicle technology this year. *Auto Dealer Today*. Retrieved from http://www.autodealer-monthly.com/channel/software-technology/article/story/2017/01/is-2017-the-year-of-telematics.aspx

Euromonitor International. (2015, December). A new world of intelligent tech: The smart phone, home and car. Retrieved from http://www.euromonitor.com/

Ferris, R. (2017, October 26). Ford plans to bring autonomous vehicles to a test market in 2018. *CNBC*. Retrieved from https://www.cnbc.com/2017/10/26/ford-plans-to-bring-autonomous-vehicles-to-a-test-market-in-2018.html

Filev, D., Lu, J., & Hrovat, D. (2013). Future mobility: Integrating vehicle control with cloud computing. *Mechanical Engineering, 135*(3), S18-S24. Retrieved from http://search.proquest.com/docview/1316626710?accountid=36334

Fingas, J. (2017, December 26). Elon Musk vows to build Tesla pickup truck 'right after' Model Y: He's also hinting at a number of feature updates for Tesla EVs. *Engadget*. Retrieved from https://www.engadget.com/2017/12/26/elon-musk-promises-to-make-tesla-pickup-truck-after-model-y/

Fitchard, K. (2014, February 3). The networked car is no longer just an idea, it will be mandated in future vehicles. *Gigaom*. Retrieved from http://gigaom.com/2014/02/03/the-networked-car-is-no-longer-just-an-idea-it-will-be-mandated-in-future-vehicles/

Ford Motor Company. (2014). Heritage. http://corporate.ford.com/our-company/heritage

Gould, L. S. (2015). Telematics starts with chips. Automotive Design & Production, 127(12), 52-55. Retrieved from https://lynn-lang.student.lynn.edu/login?url=http://search.ebscohost.com/login.aspx?direct=true&db=bth&AN=111438437&site=ehost-live

Gray, C. (n.d.). Technology in automobiles. *R6 Mobile*. Retrieved from http://support.r6mobile.com/index.php/knowledge-base/152-technology-in-automobiles

Griggs, T. & Wakabayashi, D. (2018, March 20). How a self-driving Uber killed a pedestrian in Arizona. *New York Times*. Retrieved from https://www.nytimes.com/interactive/2018/03/20/us/self-driving-uber-pedestrian-killed.html

Guilford, D. (2013, May 27). Not satisfied with OnStar's steady profits, GM wants to create a global 4G powerhouse. *Automotive News*. Retrieved from http://www.autonews.com/article/20130527/OEM/305279958/not-satisfied-with-onstars-steady-profits-gm-wants-to-create-a-global-

Halsey, A. (2015, July 20). Learning curve for smart cars. *Chicago Tribune*. Retrieved from https://lynn-lang.student.lynn.edu/login?url=http://search.proquest.com/docview/1697182214?accountid=36334

Hawkins, A. J. (2017, March 27). Uber's self-driving cars are back on the road after Arizona accident: Cops say the human, not the car, was at fault. *The Verge*. Retrieved from https://www.theverge.com/2017/3/27/15077154/uber-reactivating-self-driving-car-pilot-tempe-pittsburgh-crash

Howard, B. (2015, March 13). What is vehicle telematics? *ExtremeTech*. Retrieved from http://www.extremetech.com/extreme/201026-what-is-vehicle-telematics

Isaac, M. (2017, May 14). Lyft and Waymo reach deal to collaborate on self-driving cars. *New York Times*. Retrieved from https://www.nytimes.com/2017/05/14/technology/lyft-waymo-self-driving-cars.html?_r=1

IVI in new Tesla Motors electric sedan powered by NVIDIA. (2015). *Telematics Wire*. Retrieved from http://telematicswire.net/ivi-in-new-tesla-motors-electric-sedan-powered-by-nvidia/

Johnson, T. (2017, October 5). New vehicle infotainment systems create increased distractions behind the wheel. *AAA NewsRoom*. Retrieved from http://newsroom.aaa.com/2017/10/new-vehicle-infotainment-systems-create-increased-distractions-behind-wheel/

Laukkonen, J. (2016, October 24). OEM infotainment systems: Navigation and beyond. *Livewire*. Retrieved from https://www.lifewire.com/oem-infotainment-systems-navigation-534746

Marr, B. (2017, October 23). The little black book of billionaire secrets Rolls-Royce And Google partner to create smarter, autonomous ships based on AI and machine learning. Forbes. Retrieved from https://www.forbes.com/sites/bernard-marr/2017/10/23/rolls-royce-and-google-partner-to-create-smarter-autonomous-ships-based-on-ai-and-machine-learning/#79472166dfe9

Martindale, J. (2015, January 16). The car tech of CES 2015. *Telematics.com.* Retrieved from http://www.telematics.com/the-car-tech-of-ces-2015

Novak, D. (2018, January 9). CES 2018: 3 things to watch for. *Qualcomm*. Retrieved from https://www.qualcomm.com/news/onq/2018/01/09/ces-2018-3-things-watch

NHTSA and vehicle cybersecurity. (2015). *National Highway Traffic Safety Administration, U.S. Department of Transportation.* Retrieved from http://www.nhtsa.gov/About+NHTSA/Speeches,+Press+Events+&+Testimonies/NHTSA+and+Vehicle+Cybersecurity

Palmer, S. (2017, April 16). Can self-driving cars ever really be safe? Retrieved from https://www.shelly-palmer.com/2017/04/can-self-driving-cars-ever-really-safe/?utm_source=Daily%2520Email&utm_medium=email&utm_campaign=170430

Phillips, A. M., McGee, R. A., Kristinsson, J. G., & Hai, Y. (2013). Smart, connected and electric: The future of the automobile. *Mechanical Engineering, 135*(3), 4-9. Retrieved from http://search.ebscohost.com/login.aspx?direct=true&db=bth&AN=90499459&site=ehost-live

Prime, R. (2013, May 30). Telematics history and future predictions. *Telematics.com.* Retrieved from http://www.telematics.com/telematics-history-future-predictions/

Rafter, M. V. (2014, May 19). 8 things you need to know about back-up cameras. *Edmunds.com.* Retrieved from http://www.edmunds.com/car-technology/8-things-you-need-to-know-about-back-up-cameras.html

Ranosa, T. (2016, January 5). Apple CarPlay vs. Android Auto: Should Google or Apple take over your dashboard? *Tech Times*. Retrieved from http://www.techtimes.com/articles/121202/20160105/apple-carplay-vs-android-auto-should-google-or-apple-take-over-your-dashboard.htm

Reuters. (2018, January 15). Ford plans $11 billion investment, 40 electrified vehicles by 2022. *CNBC*. Retrieved from https://www.cnbc.com/2018/01/15/ford-plans-11-billion-investment-in-electric-vehicles-by-2022.html?utm_source=Daily+Email&utm_campaign=99b68bcf73-EMAIL_CAMPAIGN_2018_01_15&utm_medium=email&utm_term=0_03a4a88021-99b68bcf73-248596253

Rouse, M. (2007). Telematics. *TechTarget*. Retrieved from http://whatis.techtarget.com/contributor/Margaret-Rouse

SBD & GSMA. (2012, February). 2025 every car connected: Forecasting the growth and opportunity. Retrieved from http://www.gsma.com/connectedliving/wp-content/uploads/2012/03/gsma2025everycarconnected.pdf

Stewart, J. (2018). People keep confusing their Teslas for self -driving cars. *Wired*. Retrieved from https://www.wired.com/story/tesla-autopilot-crash-dui/

Taylor, C. (2017, May 25). The connected car: Radio evolves to keep pace in the dashboard. *Inside radio*. Retrieved from http://www.insideradio.com/the-connected-car-radio-evolves-to-keep-pace-in-the/article_d628a216-411e-11e7-be66-e73fb8d86462.html

Toyota Research Institute expands autonomous vehicle development team [Press release]. (2016, March 9). *Toyota Motor Corporation*. Retrieved from http://toyotanews.pressroom.toyota.com/releases/tri-expands-autonomous-vehicle-team.htm

University of Michigan Transportation Research Institute. (2014, February 5). USDOT gives green light to connected-vehicle technology. Retrieved from http://umtri.umich.edu/what-were-doing/news/usdot-gives-green-light-connected-vehicle-technology

Yu, R. (2015, October 13). Ford bets on smart cars in China: Auto maker hopes $1.8 billion in R&D will yield features tailored for Chinese drivers. *Wall Street Journal*. Retrieved from https://lynn-lang.student.lynn.edu/login?url=http://search.proquest.com/docview/1721535771?accountid=36334

14

Video Games

Isaac Pletcher, M.F.A.[*]

Overview

A video game is an interactive digital entertainment that you "play" on some type of electronic device. The medium has existed in some form since the 1950's and, since that time, it has become a major worldwide industry, spawning technology battles and delivering new iterations that have become ubiquitous parts of modern life. This chapter will look at the development of video games from the beginning, through the golden age, and into the era of the console wars. Further, it will point out the dominant companies and gear in the industry, and track how the current structures for pricing and development have led to the rise of the microtransaction and the potential danger in that business model.

Introduction

On a recent London Underground trip from Heathrow to Victoria this author was surrounded by commuters using mobile phones. However, this wasn't so remarkable as the fact that out of eight visible screens, seven were playing video games. *Angry Birds, Two Dots, Super Mario Run, Sonic Dash, Snake VS Block, Words with Friends,* and *Bowmasters* were immediately evident. One is hesitant to speculate on the motivations of those in the past, but in 1889, could Kyoto playing card manufacturer Fusajiro Yamauchi have foreseen the cultural ubiquity of the entertainment mode his company, Yamauchi Nintendo & Co.—since 1963, Nintendo Co. Ltd.—would help to create and popularize? While Nintendo may be one of the most recognizable names in the video game industry, it is by no means the only player in a market that has seen its share of stargazers, smash hits, and scandal.

Although not as recognizable a name as *Mario* or *Pac-Man, Tennis for Two* may hold the distinction of being the first true video game. Developed in 1958 by Willy Higinbotham of Brookhaven National Laboratory, the interactive game—a tennis match between two players controlled with a single knob, was programmed on an oscilloscope and presented at an open tour day of the lab. Despite the massive popularity of the game Higinbotham would later claim, "It never occurred to me that I was doing anything very exciting. The long line of people I thought was not because this was so great but because all the rest of the things were so dull" (Tretkoff and Ramlagan, 2008). Over the next two years, his game would get a bigger screen in addition to two upgrades that allowed players to simulate the gravity of the moon and Jupiter. While Higinbotham was simply trying to make something amusing for lab visitors, historian Steven Kent (2010) has stated that early iterations of games leading to the marriage of video display and interactive controls

---

[*] Assistant Professor of Film/TV at Louisiana State University, Baton Rouge, Louisiana

were obvious evolutions from the "well-established amusement industry." He even goes so far as to trace these amusements back to the 17th Century game of Bagatelle. Whether this legacy is accurate is debatable. However, what is not at question is the entertaining objective of video games.

Shigeru Miyamoto, creator of characters including Mario and Donkey Kong, has stated that, "the obvious objective of video games is to entertain people by surprising them with new experiences" (Sayre, 2007). This chapter will serve as a brief introduction to the history, current events, and future trends of video games, providing a glimpse into the ecosystem of one of the most pervasive entertainment vehicles of the 20th and 21st Centuries.

Background

While Higinbotham's contributions to video games have been mentioned, it is difficult to say exactly when video games began. Egenfeldt-Nielsen, Smith, and Tosca (2016) have pragmatically pointed out that, "No trumpets sounded at the birth of video games, so we must choose what constitutes the beginning." Some trace the history of video games back to the playing cards sold by Yamauchi. However, it seems unfair to draw no distinctions between the corporate entity and the technological product itself. Kent (2010) is careful to point out the evolutionary forebears of modern game consoles as novelty paddle and mechanical games from the 1940's and 50's that "simulated horse racing, hunting, and Western gunfights." But once the television became popular, the race to put these mechanical games into an electronic version began.

In 1952 Sandy Douglas, a Cambridge PhD candidate developed an electronic version of *Noughts and Crosses* (Tic-Tac-Toe*)* played against Cambridge's first programmable computer, the Electronic Delay Storage Automatic Calculator (EDSAC). The player used a telephone dial to input moves and the game results were displayed on a cathode-ray-tube. While it was the first graphics-based computer game, the presence of programmed punch cards gives ammunition to the argument against it being the first true computer game in the sense of *Tennis for Two*. Once Higinbotham's game showed the possibility of controllable

video entertainment, a number of early gaming luminaries started making their marks.

In 1961, MIT student Steve Russell developed *Spacewar*, which started life as a hack for the PDP-1 computer. Two other students, Alan Kotok and Bob Sanders developed remote controls that could be wired into the computer to control the game functions, and thus, the gamepad was born. By the early 1960s then, the display, ability to play a game, and ability to control the game functions had been achieved. However, all these developments had happened in academic or government environments, and nobody had yet worked out the monetization of the video game.

By 1966 Ralph Baer, an employee at defense contractors Sanders Associates Inc., began work on game systems with colleagues Bill Harrison and Bill Rusch. Their breakthrough came in the form of the "'Brown Box' a prototype for the first multiplayer, multi game system" (The Father of the Video Game, n.d.). Essentially, the trio had managed to create a video game system whereby the user could change between games and the player had a way to connect it to their home television. American company Magnavox saw the value of their system and licensed it, marketing it in May of 1972 as the *Magnavox Odyssey*. Though the console wars of the 1990s and 2000s were still a generation off, once Magnavox put their system on the market, Atari, Coleco, URL, GHP, Fairchild, and others followed suit within four years.

Once the video game console existed, cabinet games quickly emerged, and developers began to look for a game that would draw attention to their product in the vanguard of this emerging market both in the home and arcades. However, controversy followed very quickly. In the same year the *Odyssey* was released, Nolan Bushnell's *Atari* system debuted the game *Pong*, leading to an infringement lawsuit that was settled out of court. By 1976, the first flare up concerning video game violence erupted over a game called *Death Race 98*.

California company Exidy had developed *Death Race 98 (DR98)* based on the platform of their own game, *Demolition Derby (DD)*, which had come out in 1975 and was the very first driving game controlled with a steering wheel. The objective of DD had been to drive cars around a virtual landscape, smashing them to bits against each other. Yet the more-grisly objective of DR98 was to drive a course and run down humanoid figures called "goblins" that made a shriek when hit and were marked by a cross. An Associated Press reporter saw the game in Seattle and wrote a story that was picked up by "all major newspapers, plus *Newsweek, Playboy, National Enquirer, National Observer, Midnight,* the German magazine *Stern,* and many more" (Smith, 2013). In addition, the game was censured by shows on all three major American networks, the CBC, and the BBC. However, the notoriety seemed to set the precedent of controversy boosting sales of a game and as Exidy's marketing director, Paul Jacobs remembered, the company "laughed all the way to the bank" (Smith, 2013). It seemed that within the first few years of video games actually existing, they were starting to earn a controversial reputation.

Through the late 1970's and into the 1980's, a series of games and systems came to prominence with (then) blazing speed in what a number of video game historians have dubbed the "second generation" of video games—the first generation being games that defined both the form and the industry. Kent (2010) has called it simply, "The Golden Age."

Table 14.1

The Golden Age of Video Games

1978		
	Space Invaders	Taito Corporation
	Computer Othello	Nintendo Co. Ltd.
1979		
	Asteroids	Atari, Inc.
	Galaxian	Namco, Ltd.
	Cosmos Handheld System	Atari, Inc.
	Microvision Handheld System	Milton Bradley
1980		
	Pac-Man	Namco, Ltd.
	Game & Watch Handheld System	Nintendo Co. Ltd.
1981		
	Frogger	Konami Holdings Corp.
	Ms. Pac-Man	Midway Games, Inc.
	Donkey Kong	Nintendo Co. Ltd.
1982		
	Burgertime	Data East Corp.
	Q*bert	D. Gottlieb & Co.
	Commodore 64 Home Computer	Commodore Intl.
1983		
	Super Mario Brothers	Nintendo Co. Ltd.

Source: Steven Kent

This "Golden Age" came to an end in the United States in1983 as the game/console market crashed and fell into a deep depression due to a combination of "oversaturated market, a decline in quality, loss of publishing control, and the competition of Commodore, PC, and Apple computers" (Cohen, 2016c). In Japan, however, a different landscape meant the home console market had not felt the same effects and this allowed Nintendo to release the *Famicom* (short for Family Computer) *Home Entertainment System*. Atari initially had the chance to license *Famicom* for American release, but when the deal fell through, Nintendo decided to set up shop itself in the U.S. In 1985 Nintendo of America, which had been established in 1980, began selling the system in North America as the *Nintendo Entertainment System* (NES). The wild success of the system touched off the third generation of video games and cemented Nintendo as an industry titan, spelling doom for Atari and forcing other companies into breakneck development to keep pace.

This mad dash to maintain market superiority led to the Console Wars—in essence, a video game arms race. System technologies were being developed not just for companies to match their competitors but also to provide gamers with more sophisticated experiences. Technologists Mike Minotti and Jordan Minor have traced the first five iterations of console wars, which started in 1989 and have lasted to the present time, pointing out the following winners and losers:

The three industry players have clearly come to be Nintendo, Microsoft, and Sony. Whether this dynamic can be maintained is yet to be seen, especially as all three of these companies have consoles that compete with PCs as a platform. However, the consoles are clearly in the forefront, as the advantage to PC gaming is mainly the potential for improved graphics depending on the machine's card. In fact, the number of games shared between PC and console players is vast, with only a handful of titles like *World of Warcraft* and *League of Legends* being exclusive to the PC platform.

Table 14.2
The Console Wars Age of Video Games

1989 - 1994 (The 16 bit graphic revolution)

Super Nintendo Entertainment System	49.10M units sold
vs	
Sega Genesis	30.75M units sold

1995 - 1999 (Video games now in 3-D)

Sony PlayStation	102.49M units sold
vs	
Nintendo 64	32.93M units sold
vs	
Sega Saturn	9.26M units sold

2000 - 2006 (Sega bows out, Microsoft enters the ring)

Sony PlayStation 2	155M units sold
vs	
Microsoft Xbox	24M units sold
vs	
Nintendo GameCube	21.74M units sold
vs	
Sega Dreamcast	9.13M units sold

2007 - 2013 (Nintendo reclaims the crown)

Nintendo Wii	101.63M units sold
vs	
Microsoft Xbox 360	84M units sold
vs	
Sony PlayStation 3	80M units sold

2013 - Present (4K Gameplay)

Sony PlayStation 4	70.6M units sold*
vs	
Microsoft Xbox One	19M units sold**
vs	
Nintendo Wii U	13.56M units sold*
vs	
Nintendo Switch	10M units sold*^

*Most current data available as of the writing of this chapter

**This is an estimate by EA CFO Blake Jorgensen. Actual sales figures not reported by Microsoft

^New entrant. More details in the *Current Status* section of this chapter

Source: Steven Kent

Recent Developments

In April 2017, analytics firm NewZoo released an update of its *Global Games Market Report* (McDonald, 2017). In it they estimated that worldwide, gamers would spend $108.9 billion on games in 2017. While trends in the market are up—7.8% from 2016—the most telling figure for the gaming ecosystem is that the most lucrative segment of gaming is the mobile market.

NewZoo estimated 42% of the gaming market in 2017 will be on smart phones and tablets, and that by 2020, that particular segment will be fully half of the total market value (McDonald, 2017). However, the ubiquity of mobile gaming, evidenced both by the market and by simple observation on the London Underground has not had a completely smooth ascendance.

By 1987 the *Software Publishers Association of America*, a body of congressional lobbyists for the video game industry, had decided to not institute an MPAA style ratings system and instead let developers police themselves and put warnings on their games when they deemed it necessary. Executive Director, Ken Wasch said, "Adult computer software is nothing to worry about. It's not an issue that the government wants to spend any time with…They just got done with a big witchhunt in the music recording industry, and they got absolutely nowhere…" (Williams, 1988). This laissez-faire attitude, however, was not enough for certain members of the U.S. Congress and by 1993, senators Joe Lieberman and Herb Kohl told the association that if they did not set up a rigorous self-regulatory body that within one year the industry as a whole would be subject to governmental regulation.

This controversy prompted the formation of the Entertainment Software Ratings Board (ESRB). While every game released on main consumer consoles since the 1990's has received an ESRB rating, these ratings appeared relatively recently (since 2015) on the Google Play Store and are still absent on the Apple App Store, which still prefers to use its own internal review process. Unfortunately, as CNET's Josh Lowensohn (2011) pointed out, "the differences in these mobile-content rating strategies makes for cases where one game may be rated for a different age group between two platforms, even if the content is the same."

Since 2016, the weakness of these ratings systems has been highlighted not necessarily in the disagreement of ratings across platforms but rather in the failure of ratings boards to account for user generated content, especially with the amount of online connectivity the current generation of gaming platforms aggressively pursues. User generated content is nothing new for games—however, the processing power of both gaming consoles and PCs has allowed gamers to modify copyrighted material so liberally that open source gaming has started to emerge as a viable market option. In fact, the ability to modify a game and share the result is the reason many players choose PCs over consoles. Letzter (2015) has defined "modding" as fans diving "into the back-end of their favorite games to fix bugs, update graphics, or introduce new elements. Sometimes, fans create new games altogether." Open source refers to games that are developed using software with source code the creators have made publicly available so users can adapt it to their preferences. Open sourcing has even been embraced by a number of companies with older titles. This has come to be known as source porting. ID Software's *Doom* and Maxis' *Sim City* are two games that have had their source coding made public for fans to find nostalgic joy in replaying those decades-old games and modifying them for current preferences. Once can think of a number of viable marketing reasons why some companies are choosing to do this. Yet in terms of viable marketing, this pales in comparison to what has become known as the microtransaction.

Since 2013, three of the most ubiquitous developments in the gaming industry were the, "distribution revolution…the rise of social games…and the body as interface" (Egenfeldt-Nielsen, et al., 2016). Since that writing, the body as interface has progressed with great strides and is discussed at length in the chapter on Virtual Reality (Chapter 15). Social games like *Farmville* and *Words with Friends,* which seem to be the distant descendants of the MMO (massive multiplayer online) and MMORPGs (massive multiplayer online role-playing games) are still played but with the wide interconnectivity of online objects (Chapter 12 on the Internet of Things), it seems almost a foregone conclusion that almost any game at this point has the potential to be a social game.

At the same time the distribution models and methods have seen acute changes. The method of packaging a video game on a solid-state object and selling it as a physical product in a store is quickly becoming passé. There are now app stores that sell games for both Apple and Android devices as well as specific app stores for Sony PlayStation and Microsoft Xbox users. In addition, if a gamer is using a Mac or PC, "platforms such as Steam offer direct purchase, data storage, downloadable content (DLC), and other

community features" (Egenfeldt-Nielsen, et al., 2016). In the last quarter of 2015, Electronic Arts (EA) made more than 40% of their revenue from DLC. The analysts at NewZoo show the 2017 year-end revenue from boxed games to be less than 20% of the global industry market.

As the video game world shifts towards electronic purchases and games played on phones and tablets, developers and game conglomerates can no longer make the same revenues with boxed games. Therefore, since 2015, game developers have not produced nearly as many boxed units of game titles and instead are opting for DLC and microtransactions. Monetization service company Digital River released a report in early 2017 which states that because of the digitalization of video games, the average gamer is expecting more from the developer for less money:

"Consumers are less willing to pay $60 for a boxed game and instead choose titles with a steady stream of new content. Publishers seek to meet these expectations and have adopted a 'games as service' model, releasing fewer titles over time while keeping players engaged longer with regular updates and add-ons" (SuperData Research Holding, 2017).

Of course, if the game is cheap or "free-to-play" (costing nothing for the download and to start gameplay), these updates and add-ons are what cost money, and players must pay for them or risk never being able to finish the game or see the premium content. These are the microtransactions that add up and, according to some estimates, have almost tripled the game industry's value. After all, why charge $60 for a game once when over the development life of a game, a player is willing to pay upwards of $100-$500? This model has gained traction and, as journalist Dustin Bailey (2017) mentions, while these "modern business practices are failing in the court of public opinion, they are profitable. Extremely, *obscenely* profitable." It is this very profitability that ultimately drives the developments in the industry. In 2017, using these models for distribution and microtransactions, three of the world's largest game developers, Activision Blizzard Inc., Take-Two Interactive Software Inc., and Electronic Arts have seen revenue growth of 70%, 112%, and 51%, respectively. These numbers indicate that the video game industry is learning how to grow with consumer demand and technology, but whether these models are sustainable is going to be interesting to watch over the next few years.

Finally, many recent developments that impact the video game industry are discussed in other chapters. Studying the chapters on Virtual Reality (Chapter 15) and Internet of Things (Chapter 12) will provide a substantial background on those developments and a hint as to where the playability and connectivity of video games is headed.

Current Status

According to the *Global Games Market Report*, digital games—mobile and downloaded games—make up 87% of the current market figures. It is also interesting to note that the largest gaming region is Asia-Pacific, which generates nearly half—47% of global revenue (McDonald, 2017). Chinese gamers alone are responsible for a full 25% of the total market. North America is the second largest revenue-generating region, and it is interesting to note that the majority of growth has come and will continue to be from smartphone gaming. McDonald (2017) notes that, "Growth is fueled by a combination of a higher share of spenders as well as average spend per paying gamer." This increased spending is the premium content paid for with microtransactions.

In terms of gaming platforms, we are still in the fifth iteration of the console wars, waiting for new generations of the PlayStation and Xbox (which industry pundits predict will not happen this decade). However, in March 2017, Nintendo launched their new platform, *Nintendo Switch*. The Switch attempts to bridge gaps by being a "handheld/living room box hybrid…It became the console of choice for a traveling companion and touting to parties" (Corriea, 2017). While the sales figures for Switch seem to be relatively low, it must be remembered that these 10 million+ units sold within eight months, a quicker pace than even the massively successful PlayStation 4. Industry analyst Ben Gilbert (2017) has mentioned that, "demand has been so high, in fact, that Nintendo had to revise its sales projections for first-year sales of the Switch…Nintendo expects to sell over 14 million by March 2018."

Of the three main gaming platforms—console, mobile, and PC—the only one that is currently losing numbers is PC gaming. Worldwide, PC gaming accounted for $29.4 billion, which is 27% of total 2017 revenue, down 2.3% from 2016. Companies like Facebook and Zynga, who track the analytics of their games in both PC and Mobile markets reported 2017 to be the worst year for PC browser specific gaming since 2011, with Zynga hit especially hard, with a 30% decrease in revenue. NewZoo expects this slide to continue, with PC gaming in toto barely reaching $24 billion in 2019.

Overall, NewZoo numbers gamers at 2.2 billion worldwide. Their analysis projects that the upward trends will continue, projecting industry revenues of over $128 billion by 2020. Digital River is quick to add that because the digital marketplace is only going to become more robust that publishers must be more concerned than ever about fraud. Currently, when a game is bought online the buyer is given a digital key that unlocks gameplay. Because it is easy enough to fraudulently buy the key, it is also easy to sell the ill-gotten key on a gray market. They point out that the five top grossing games of 2017 are all free-to-play, and, if the revenue projections hold, by 2020 that particular market will increase by 47%. Unfortunately, the security of all the in-game items of all those players may well be in doubt. Therefore, it seems incumbent on developers and security personnel to make sure that the security of the system remains consistent with the playing demands as the market increases.

Factors to Watch

If there is a potential pitfall to the growth of the video game industry it may be in the pricing structure itself. NBC's Tom Loftus (2003) explained that when the next generation of game consoles are in development, the current generation is in the downward part of its life cycle. Historically, it made sense that developers let the current generation of games fade away without updates when the new consoles were on the horizon. Yet because of the current "games as service" model this is not happening. The games keep getting updated and fans keep making their own updates when developers stop. A major title game that retailed between $40-$65 in 2015 still retails for between

$40-$50. A standard PlayStation game cost less than $1 million to develop in 1998, but now can cost upward of $200 million while the life cycle of the game needs to be far longer. However, the game costs to the consumers aren't changing, unless the games rely on microtransactions, which are becoming more and more unpopular.

In August 2017, Warner Brothers Interactive launched the game *Middle Earth: Shadow of War* at a retail price (download or disc) of $59.99. Soon after the launch, they announced that loot boxes could be bought for the game with premium items for in game or real-world money. This conflation of a full-price game and paid premium content got worse when EA launched a beta version of *Star Wars Battlefront II* in October and players realized this game was also built to utilize the same loot box structure. After the public outcry and a call from licensing corporate overlord Disney, EA removed the microtransactions altogether. The ire of the gamers lies in the fact that, as the BBC put it, loot boxes "amount to a covert lottery" ("Apple Changes Rules", 2017).

The backlash against this practice became so intense that by December 2017 certain countries (Belgium) banned games with loot boxes, and after pressure from the U.S. Congress, Apple declared that any developer with a game on its app store must post the odds of getting certain items. Because of the closed nature of loot boxes and other styles of microtransactions, it will be worth watching how the video game industry decides to move forward, especially in premium games that cost far more than two or three dollars on an app store.

Because of the fast-paced nature of game and system development, the greatest changes that happen year to year come from the games themselves. As of 2018, the consoles are all 128 bit and work with the most sophisticated graphics cards. However, that does not stop developers from doing all they can to test those limits. The only disadvantage of the Nintendo Switch is that its hard drive is only 32 Gbs. Nintendo had announced a 64 Gb hard drive card that was slated for release in March 2018 but now has been moved back to November. Unfortunately for many players, some of the current titles severely tax that hard drive size. This leads to further gamer ingenuity

to find workarounds. It will be telling to see if Nintendo leans on the admittedly very good game lineup for the Switch to keep on top of the current market or if gamers get tired of the limitations of the system and revert back to other titles on PlayStation and Xbox.

It is also worth noting the potential for communication that now exists between developers and gamers. That *Battlefront* was pulled and changed so quickly after public outrage is interesting. Gone are the days when developers like John Romero of Ion Storm can tease a game like *Daikatana* with one release date only to repeatedly renege and release the game three years later. The checks and balances that have been brought to bear by our growing interconnectedness may prove to be advantageous to the gamer, but will we eventually see a push back by developers? What is clear is the health of the industry and the variety of gamers, whether one builds whole platforms and games on their own or simply uses the game as a way to pass time.

Getting a Job

Until recently, the most popular way to break into the video game industry was to start work as a game tester for any independent or big-name developer that would hire you. Author Michael Cheary (2015) explains testing: "It will require you to play the same level, sometimes for days at a time, in an effort to find every possible bug before it can go live." While this is still the starting point for many, the open communication that has come about between developers and the gaming public has opened up a number of other avenues to start careers. Learning how to modify existing games and creating new levels that can be shared on sites like Steam can show developers that you already have some coding and programming skill.

Another tactic used by some is to start a gaming blog. Two of the most influential names in the gaming world are Jerry Holkins and Mike Krahulik. Their gaming webcomic/blog, *Penny Arcade*, debuted in 1998, and their dedication has been noticed by industry figures and earned them invitations to help develop games. Several Universities across the U.S. also have degrees in video game design and development.

Finally, the standard path of interning for a game developer and working your way up the ladder is still perfectly viable. If you have a skill like programming, art, or sound design, there are a number of companies that want to give you unpaid work that might lead to something that is paid. No matter what path you take remember to be positive, work hard, and network all you can.

Bibliography

Apple changes rules on app 'loot boxes'. (2017, December 22). Retrieved from http://www.bbc.com/news/technology-42441608.

Bailey, D. (2017, October 11). The game industry's value has tripled because of DLC and microtransactions. Retrieved from https://www.pcgamesn.com/games-as-service-digital-river-report.

Bay, J. (2016, March 29). 10 Proven Ways to Break Into the Video Game Industry. Retrieved from http://www.gameindustrycareerguide.com/how-to-break-into-video-game-industry/.

Chalk, A. (2007, July 20). Inappropriate Content: A Brief History of Videogame Ratings and the ESRB. Retrieved from http://www.escapistmagazine.com/articles/view/video-games/columns/the-needles/1300-Inappropriate-Content-A-Brief-History-of-Videogame-Ratings-and-t.

Cheary, M. (2015, December 14). How to: Get a job in the games industry. Retrieved from https://www.reed.co.uk/career-advice/how-to-get-a-job-in-the-games-industry/.

Cohen, D. (2016a, October 19). OXO aka Noughts and Crosses - The First Video Game. Retrieved from https://www.lifewire.com/oxo-aka-noughts-and-crosses-729624.

Cohen, D. (2016b, October 19). The History of Classic Video Games - The Second Generation. Retrieved from https://www.lifewire.com/the-second-generation-729748.

Cohen, D. (2016c, October 19). The History of Classic Video Games - The Industry Crash and Rebirth. Retrieved from https://www.lifewire.com/the-industry-crash-and-rebirth-729749.

Corriea, A. (2017, December 06). 2017 Year in Review: Video Games. Retrieved from http://fandom.wikia.com/articles/year-review-2017-video-games.

Crider, M. (2017, December 24). The Best Modern, Open Source Ports of Classic Games. Retrieved from https://www.howtogeek.com/335259/the-best-modern-open-source-ports-of-classic-games/.

Egenfeldt-Nielsen, S., Smith, J. H., & Tosca, S. P. (2016). *Understanding video games: the essential introduction.* New York, NY: Routledge.

Gilbert, B. (2017, December 12). Nintendo has already sold 10 million Switch consoles just 9 months after launch. Retrieved from http://uk.businessinsider.com/nintendo-switch-sales-top-10-million-in-just-9-months-2017-12?r=US&IR=T.

Kent, S. L. (2010). *The Ultimate History of Video Games: from Pong to Pokémon and beyond... the story behind the craze that touched our lives and changed the world.* New York, NY: Three Rivers Press.

Kolakowski, M. (2017, October 16). Why Video Game Stocks' Hot Streak May Get Hotter. Retrieved from https://www.investopedia.com/news/why-video-game-stocks-hot-streak-may-get-hotter/.

Letzter, R. (2015, July 20). Online communities are changing video games to make them better, weirder, and much more wonderful. Retrieved from http://uk.businessinsider.com/video-game-modding-2015-7.

Loftus, T. (2003, June 10). Top video games may soon cost more. Retrieved from http://www.nbcnews.com/id/3078404/ns/technology_and_science-games/t/top-video-games-may-soon-cost-more/#.WkTY7raZPOQ.

Lowensohn, J. (2011, November 29). Apple, Google kneecap 'universal' content rating for apps. Retrieved https://www.cnet.com/news/apple-google-kneecap-universal-content-rating-for-apps/.

McDonald, E. (2017, April 20). The Global Games Market 2017 | Per Region & Segment. Retrieved from https://newzoo.com/insights/articles/the-global-games-market-will-reach-108-9-billion-in-2017-with-mobile-taking-42/

Minor, J. (2013, November 11). Console Wars: A History of Violence. Retrieved from http://uk.pcmag.com/game-systems-reviews/11991/feature/console-wars-a-history-of-violence.

Minotti, M. (2014, August 20). Here's who won each console war. Retrieved from https://venturebeat.com/2014/08/20/heres-who-won-each-console-war/.

Nintendo History. (n.d.). Retrieved from https://www.nintendo.co.uk/Corporate/Nintendo-History/Nintendo-History-625945.html.

Owen, P. (2016, March 09). What Is A Video Game? A Short Explainer. Retrieved from https://www.thewrap.com/what-is-a-video-game-a-short-explainer/.

Sayre, C. (2007, July 19). 10 Questions for Shigeru Miyamoto. Time, Inc. Retrieved from http://content.time.com/time/magazine/article/0,9171,1645158-1,00.html.

Smith, K. (2013, May 24). The Golden Age Arcade Historian. Retrieved from http://allincolorforaquarter.blogspot.co.uk/2013/05/the-ultimate-so-far-history-of-exidy_24.html.

Spohn, D. (2017, August 14). PC vs. Console for Online Gaming. Retrieved from https://www.lifewire.com/pc-vs-console-1983594.

SuperData Research Holding, Inc. 2017. *Defend Your Kingdom: What Game Publishers Need to Know About Monetization & Fraud.* New York: Digital River.

Taylor, H. (2017, October 10). Games as a service has "tripled the industry's value." Retrieved from http://www.gamesindustry.biz/articles/2017-10-10-games-as-a-service-has-tripled-the-industrys-value.

The Father of the Video Game: The Ralph Baer Prototypes and Electronic Games. (n.d.). Retrieved from http://americanhistory.si.edu/collections/object-groups/the-father-of-the-video-game-the-ralph-baer-prototypes-and-electronic-games.

Tretkoff, E., & Ramlagan, N. (2008, October). October 1958: Physicist Invents First Video Game (A. Chodos & J. Ouellette, Eds.). Retrieved from https://www.aps.org/publications/apsnews/200810/physicshistory.cfm.

Walton - Jan 29, 2016 2:30 pm UTC, M. (2016, January 29). EA lets slip lifetime Xbox One and PS4 consoles sales. Retrieved from https://arstechnica.com/gaming/2016/01/ea-lets-slip-lifetime-xbox-one-and-ps4-consoles-sales/.

What is open gaming? (n.d.). Retrieved from https://opensource.com/resources/what-open-gaming.

Williams, J. (1988, January). Goodbye "G" Ratings (The Perils of "Larry"). *Computer Gaming World, 43,* 48-50.

Winter, D. (n.d.). Pong Story. Retrieved from http://www.pong-story.com/intro.htm.

15

Virtual & Augmented Reality

Rebecca Ormond[*]

Overview

Virtual and Augmented Reality (VR/AR) systems have been used in professional applications for decades. However, when Oculus released the Oculus Rift SDK 1 system in 2012 and Facebook purchased Oculus for $2 billion in 2014, a consumer-entertainment industry was born. Although market predictions for VR and AR have turned out to be exaggerated in the past, John Riccitellio, CEO of Unity Technologies, in his Google I/O 2017 developers' conference presentation reminded developers: "The industry achieved $2.5 Billion from a near zero start in 2015, that is stupendous!" (Riccitellio, 2017).

Early inhibitors to consumer adoption, such as latency issues that caused VR Sickness, high costs and proprietary distribution platforms are slowly being resolved. As of January 2018, improvements to VR/AR technologies, supported by substantial industry and government financing, are emerging at a never before seen rate, which is predicted to increase content development and consumer adoption rates.

This chapter explores the issues inhibiting and enabling the trajectory of VR and AR from an early innovator or adopter stage into a mass market or majority adopter stage.

Introduction

Virtual Reality (VR) and Augmented Reality (AR), as well as their related subcategories, Mixed Reality (MR) and Augmented Virtuality (AV), are categorized by software, hardware and content. VR artist and author Celine Tricart (2018) describes "Virtual Reality" content as "an ensemble of visuals, sounds and other sensations that replicate a real environment or create an imaginary one," while AR content "overlays digital imagery onto the real world." When VR and AR are used together, we call it "mixed reality." The final subcategory is AV, which includes interaction with physical objects via Haptics, which refers to a sense of "touch" usually via a controller or input device as "many haptic systems also serve as input devices" (Jerald, 2016). Finally, we "use the term 'VR' when the content is watched in a VR headset," but call it "360° video when it is watched on a flat screen using a VR player" (Tricart, 2018). VR or AR will be used here, with the understanding that these include their subcategories of MR, AV and 360° video (also referred to as "Magic Window" in the industry).

[*] Assistant Professor, Department of Media Arts, Design, and Technology, California State University, Chico

Typically, VR is watched on a stereoscopic (a distinct image fed to each eye) head mounted display (HMD) that employs inertial measurement unit tracking (IMU) that "combine gyroscope, accelerometer and/or magnometer hardware, similar to that found in smartphones today, to precisely measure changes in rotation" (Parisi, 2016). HMDs fall into three broad categories: smartphone-based, tethered (requiring a computer and software development kit or SDK), and all-in-one (also referred to as stand-alone, similar in function to a tethered HMD, but not tethered to a computer). Tethered and all-in-one HMDs allow for Active Haptics, 6 Degrees of Freedom (DOF), which means you can move around, and Agency, which means you can guide the experience. Playstation VR or Oculus Rift are examples of tethered HMDs that come with controllers (Haptics).

The all-in-one HMD is relatively new, with many, including Oculus Go, Vive Focus, and others (Conditt, 2017) announced for release in 2018. Smartphone based HMDs use a smartphone compatible with the HMD, but have only 3 DOF and no Agency, which translates into limited movement and user control of the experience. Google Cardboard and Samsung Gear VR are both smartphone HMDs. VR can also be projected with a Cave Automatic Virtual Environment (CAVE) system, "a visualization tool that combines high-resolution, stereoscopic projection and 3D computer graphics that create the illusion of a complete sense of presence in a virtual environment" (Beusterien, 2006). What all VR systems have in common is that they block out the real world and are therefore considered immersive.

AR is viewed through glasses or smart devices such as a smartphone (not housed in an HMD), which do not occlude the real world as seen by the wearer and are therefore not immersive. As of this writing, the number of AR glasses in use is limited, with the notable exception of the 2015 release of the Microsoft HoloLens, which uses "a high-definition stereoscopic 3D optical see-through head-mounted display (and) uses holographic waveguide elements in front of each eye to present the user with high-definition holograms spatially correlated and stabilized within your physical surroundings" (Aukstakalnis, 2017). However, the popularity of *Pokémon Go* proves that AR can be engaging, even through a smartphone.

AR is projected to be a major market in the future, and there are a number of recent software development kits (SDKs) that should drive content creation. "A software development kit (or SDK) is a set of software development tools that allows the creation of applications for a certain software package, software framework, hardware platform, computer system, video game console, operating system, or similar development platform" (Tricart, 2018).

VR content can be "live" action, computer generated (CG) or both. Live action VR content is captured via synchronized cameras or a camera system using multiple lenses and sensors, with individual "images" being "stitched" together to form an equi-rectangular or "latlong" image (similar to how a spherical globe is represented on a flat map). VR experiences are usually posted in a non-linear edit system (NLE) or Game Engine and Digital Audio Workstation (DAW). VR content is also streamed live for sporting events (usually viewed as 360° video) though this application has required specialized high end professional VR cameras such as the Ozo 360° stereoscopic camera (Kharpal, 2015).

VR games are built in game engines, for example, *Unity 3D* or *Unreal*. Since game engines include "video" importing and exporting capabilities, VR content is also created in game engines. These game engines are free and have been very popular with both professional and amateur content creators, but 3D (stereoscopic) 360° VR cameras have been very expensive, and therefore only affordable for big budget productions

While this chapter primarily addresses entertainment applications, Table 15.1 lists many examples of VR/AR companies and products for non-entertainment industries. These examples were sourced from *Practical augmented reality: A guide to the technologies, applications, and human factors for AR and VR* (Aukstakalnis, 2017) where full descriptions can be found. In particular note that real estate, architecture and flight simulators are robust markets for VR.

Table 15.1

Non-Entertainment VR/AR Companies and Products

Industry	Company/Organization	VR/AR Application
Real Estate	Matthew Hood Real Estate Group & Matterport, Inc.	Real estate viewable as MR on HMD
Automotive Engineering	Ford Motor Co. & NVIS	Design Analysis & Production Simulation
Aerospace Engineering	NASA & Collaborative Human Immersive Laboratories (CHIL)	CHIL's CAVE to assist in engineering
Nuclear Engineering and Manufacturing	Nuclear Advanced Manufacturing Research Center (NAMRC) & Virtalis	Active Cube /Active Wall 3D visualization
Medicine (non-profit)	HelpMeSee (non-profit organization and global campaign) uses a number of organizations including Moog, Inc., InSimo, and Sense Graphics	The Manual Small Incision Cataract Surgery (MSICS) simulator
Medicine (education)	Academic Center for Dentistry & Moog Industrial Group	The Simodont Dental Trainer
Heath (PTSD treatment)	The Geneva Foundation & Virtually Better, Inc., Navel Center	Bravemind: MR using stereoscopic HMD (sensor& hand-held controller)
Health (phobias)	Virtually Better & VRET software suites	MR using stereoscopic HMD (sensor & hand-held haptics)
Medicine (operation room)	Vital Enterprises	Vital Stream (medical sensor & imaging)
Defense (flight)	NASA's Armstrong Flight Research Center & National Test Pilot School	Fused Reality (flight simulator)
Defense (planning and rehearsal)	U.S. Military developed The Dismounted Soldier Training System (DSTS)	Fully portable immersive environment MR training system
Defense (rescue)	U.S. Forest Service with Systems Technology, Inc.	VR Parachute Simulator (PARASIM)
Advanced Cockpit Avionics	US Military & Lockheed Martin	f-35 Joint Strike Fighter Helmet Mounted Display system
Education (vocational training)	Lincoln Electric with VRSIM, Inc.	VTEX (a virtual welding system)
Education (theory, knowledge, acquisition & concept formation)	Multiple educational institutions using multiple Augmented Displays	Multiple Building and Information Modeling (BIM) applications, AR display & interactive lessons
Education (primary)	Multiple Schools with the Google Expeditions Pioneer Program	Google's Expeditions Kit
Big Data	Big Data VR challenge launched by Epic Games & Welcome Trust Biomedical Research Charity	LumaPie (competitor) built a virtual environment where users interact with data
Telerobotics/ telepresence	Developed through a space act agreement between NASA &General Motors	Robonaut 2 (remote robot controlled for use in space via stereoscopic HMD, vest and specialized gloves)
Architecture	VR incorporated into computer-assisted design programs	Allow visualization of final structure before or during construction.
Art	Most VR tools can be repurposed by artists to create works of art	Virtual art extends our understanding of reality

Source: Rebecca Ormond

Background

"Virtual reality has a long and complex historical trajectory, from the use of hallucinogenic plants to visual styles such as Trompe L'Oeil " (Chan, 2014). Even Leonardo da Vinci (1452-1519) "in his *Trattato della Pittuara* (Art of Painting) remarked that a point on a painting plane could never show relief in the same way as a solid object" (Zone, 2007), expressing his frustration with trying to immerse a viewer with only a flat frame.

Then in 1833 Sir Charles Wheaton created the stereoscope, a device that used mirrors to reflect two drawings to the respective left and right eye of a viewer, thus creating a dual-image drawing that appeared to have solidity and dimension (Figure 15.1) Wheaton's device was then adapted into a handheld stereoscopic photo viewer by Oliver Wendall Holmes, and "it was this classic stereoscope that millions used during the golden age of stereography from 1870 to 1920" (Zone, 2007).

Figure 15.1
Charles Wheaton's Stereoscope (1833)

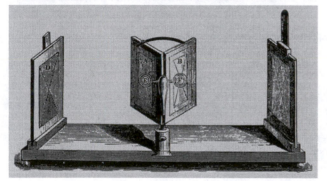

Source: Wikimedia Commons-public domain

"Google's Cardboard as an inexpensive mass-produced VR viewer is almost identical in purpose and design to its analog ancestor, Holmes' patent-free stereoscope, from over 150 years ago" (Tricart, 2018). Later stereoscope was popular in early films. However, by the 1930s, film moved from capturing reality to more complex narrative styles, while stereoscope was still primarily for "reality" gimmick films. It is the limit of stereoscope for story-telling, more so than the need for glasses, that noted stereographer and author

Ray Zone (2007) argues led to its marginalization by the mid-1900s.

Between the mid-1900s and 1960, there were a few "mechanical" VR devices, most notably Edwin Links' flight simulator created in 1928, with 10,000 units being sold principally to the Army Air Corps by 1935 (Jerald, 2016), but the next real big innovation involved the computer.

In 1968 Ivan Sutherland and Bob Sproull constructed what is considered the first true VR head-mounted display, nicknamed the Sword of Damocles, which was "somewhat see through, so the user could see some of the room, making this a very early augmented reality as well" (Tricart, 2018). By the 1980s NASA developed a stereoscopic HMD called Virtual Visual Environment Display (VIVED), and Visual Programming Language (VPL) laboratories developed more HMDs and even an early rudimentary VR glove (Virtual Reality Society, 2017). During this same time VR creators, most notably VPL laboratories, turned their attention to the potential VR entertainment market. By 1995 the *New York Times* reported that Virtuality managing director Jonathan Waldern predicted that the VR market would reach $4 billion by 1998. However, this prediction ended with "Most VR companies, including Virtuality, going out of business by 1998" (Jerald, 2016). At this time limited graphic ability and latency issues plagued VR, almost dooming the technology from the start. Still, "artists, who have always had to think about the interplay between intellectual and physical responses to their work, may play a more pivotal role in the development of VR than technologists" (Laurel, 1995). During this same time, artist Char Davies and her group at SoftImage (CG creator for Hollywood films) created *Osmose*, "an installation artwork that consists of a series of computer-generated spaces that are accessed by an individual user wearing a specialized head mounted display and cyber-vest" (Chan, 2014). *Osmose* is an early example of the "location based" VR experiences that are re-emerging today.

In 2012 Oculus released the Oculus Rift SDK I, a breakthrough tethered stereoscopic HMD (software and hardware) with a head-tracking sensor built into the lightweight headset (Parisi, 2016). Oculus Rift SDK1 and SDK 2 (both tethered HMDs) were released to the public, but between the cost of the hardware and the required computer, the final cost remained well over $1,000 and required that the user have some computer knowledge. AR was also introduced via Google Glass in 2012, but many issues from limited battery life to being "plagued by bugs" and even open hostility towards users (because of privacy concerns) led to them being quickly pulled from the market (Bilton, 2015). Then in 2014 Facebook bought Oculus for $2 billion (Digi-Capital, 2017a) while Google cardboard was introduced at the Google I/O 2014 developers' conference. The Google cardboard SDK was available for both Android and iOS operating system's and could be handmade from simple cardboard instructions available on the web (Newman, 2014). In 2015 Microsoft announced HoloLens, "a high-definition stereoscopic 3D optical see-through head-mounted display which uses holographic waveguide elements in front of each eye to present the user with high-definition holograms spatially correlated and stabilized within your physical surroundings" (Aukstakalnis, 2017). Meanwhile, Valve, which makes Steam VR, a virtual reality system, and HTC collaborated to create HTC Vive, a tethered HMD (PCGamer, 2015). During the same year, Oculus partnered with Samsung on an Android smartphone HMD, the Samsung Gear VR, releasing a smartphone HMD in November 2015 (Ralph & Hollister, 2015) with this lower cost smartphone HMD outselling both Vive and Occlulus Rift tethered HMDs (Figure 15.2).

By 2016, Sony released the Playstation VR, which, while "tethered," could be used with the existing Playstation 4, selling the most "tethered" HMD units in 2016 (Liptak, 2017). During these years, most VR or AR was proprietary, with manufacturers only working with specific smartphones or only supporting their own platforms. The exception was Google Cardboard, which was biggest seller in the VR and AR market, outselling all other HMDs (tethered and smartphone) as well as AR glasses combined, with estimates by 2017 of over 10 Million sold (Vanian, 2017). While it is clear that cost, ease of use, and cross platform compatibility strongly affected consumer adoption, it is most important to note that the overall costs dropped and access to content increased, thereby increasing overall consumer adoption of VR and AR technologies.

Figure 15.2

VR Demonstration using the HTC Vive

Source: H.M.V. Marek, TFI

In North America's Silicon Valley in 2015 alone, "AR/VR startups raised a total of $658 million in equity financing across 126 deals" (Akkas, 2017). In Vancouver, "Creative B.C. invested $641,000 through its interactive fund in 14 different VR companies. By July 2016, there were 168 mixed reality companies, and 5,500 full-time employees in B.C." while "4 billion RMB ($593 million USD) has been invested in VR in China in 2015 and 2016; a large majority funded by the Chinese government" (Akkas, 2017).

In addition, although not directly related to sales at the time, in 2014 the Mozilla team (Mozilla Firefox web browser) and a member of the Google Team First announced "a common API between the two browsers...using this new API, the programmers could write VR code once and run it in both browsers. At that moment WebVR was born" (Parisi, 2016).

Recent Developments

The release of the smartphone HMD drove a rapid increase in sales, including even more newcomers such as the Daydream, a higher end smartphone HMD announced at the Google I/0 16 developers' conference (Holister, 2016). However, when 2016 actual sales numbers were released in 2017, they fell far short of projections for 2016. "Where at the start of the year we thought 2016 could deliver $4.4 billion VR/AR revenue ($3.8 billion VR, $0.6 billion AR), the launch year's issues resulted in only $2.7 billion VR revenue. This was counterbalanced by *Pokémon Go's* outperformance helping AR to an unexpected $1.2 billion revenue, for a total $3.9 billion VR/AR market in 2016" according to the Digi-Capital Corporation, a Swiss company managing equity investments in the entertainment industry (2017a).

AR exceeded projections in 2016 but must overcome some major inhibitors to move from the early adopter stand to a majority stage in the consumer market. According to Digi-Capital (2017b): "5 major challenges must be conquered for them [AR glasses] to work in consumer markets: (1) a device (i.e., an Apple quality device, whether made by Apple or someone else), (2) all-day battery life, (3) mobile connectivity, (4) app ecosystem, and (5) telco cross-subsidization." A major inhibitor to VR growth has been the need for a specific app, headset and browser. In a 2017 interview, Antti Jaderholm, Co-founder and Chief Product Officer at Vizor, said, "We see that currently about 15% of viewers view the experiences in VR. That number has remained stable for a while, haven't seen a big change in the last year or so…85% consume WebVR experiences from our site without any headset" (Bozorgzadeh, 2017). 2017 saw several advances addressing the above inhibitors to consumer adoption.

In 2017 Google announced advances to WebVR and the release of ARcore, while Apple announced ARKit. On the VR hardware side, Oculus announced the Oculus Go, an all in one HMD priced under $200 and HTC announced the Vive Stand Alone (Greenwald, 2017). On the AR hardware side, Magic Leap announced all-in-one consumer AR glasses (Conditt, 2017). Combined, these addressed the major inhibitors to VR and AR adoption.

Figure 15.3
Market Shares 2016

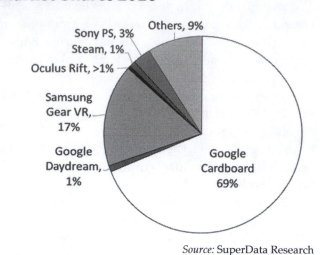

Source: SuperData Research

Figure 15.4
Total Devices Shipped 2016

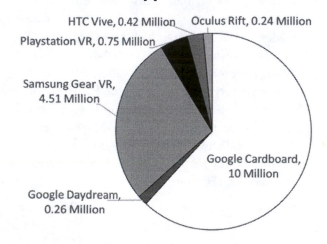

Source: Strategy Analytics

At the *Google I/O 2017* conference, Megan Lindsay, Google's WebVR Project Manager, in her presentation "Building Virtual Reality on the Web," (2017) announced advances to WebVR that "opens VR up … to content from any device through any web browser," then Lindsay went on to say that "content is absolutely critical to the success of any new ecosystem…WebVR opens VR up to the largest developer in the world…web developers." While initially available on Daydream, Google's own mobile VR platform, Google is working cross platform with other HMDs and other browsers such as Mozilla (who initially started WebVR and involved Google) towards all HMDs working on all VR-enabled browsers.

Who are web developers? Everyone, and that is the point. We are seeing an increase in WebVR development tools, for example Amazon's Sumerian (in Beta at this writing) "lets you create and run virtual reality (VR), augmented reality (AR), and 3D applications quickly and easily without requiring any specialized programming or 3D graphics expertise" Amazon Web Services, 2017).

Apple and Google also introduced their AR frameworks ARKit and ARCore, respectively, to allow developers to create mobile AR experiences. "The wide support of mobile AR presents an exciting opportunity for marketers. AR is no longer limited to social media apps like Snapchat; and instead a set of natively integrated tools on both iOS and Android, enabling developers to build cool, engaging applications" (Yao and Elchison, 2017).

SDKs, whether you are developing content on the web, iOS or Android, have many uses including VR, AR or mixed MR (Alexandar, 2017).

It is also important to note that while free game engines such as *Unity 3D* or *Unreal*, which work with 3D 360° video, have been very popular with both professional and amateur content creators, stereoscopic VR cameras were priced only for big budget productions. Affordable live action 3D 360° video VR cameras, such as Human Eye's *Vuze* camera, priced under $1,000 (see figure 15.5) have entered the market, with a new Vuze+ camera to be released in 2018 that even allows for live VR streaming for under $1,200 (Antunes, 2017).

Figure 15.5
The Vuze Camera

Source: Human Eye's Ltd.

VR has evolved from early suggestions that it might be the new "film" to its own category, Cinematic VR Experience, with most major film festivals (Sundance, Tribeca, Sundance, etc.) having a VR category. The Academy of Motion Picture and Sciences awarded its first VR award "in recognition of a visionary and powerful experience in storytelling" to director Alejandro González Iñárritu for his virtual reality installation "Carne y Arena" (Virtually present, Physically invisible) (Robertson, 2017a). Top cinematic VR experience companies include Felix & Paul, who produced among many, Obama's White House tour, "The People's House," which won an Emmy (Hayden, 2017), and companies such as Jaunt VR with multiple Cinematic VR originals (Jaunt Studios, 2017), plus major film and television studio and sports news partnerships.

Finally, distribution has opened up. For example, "Paramount has partnered with MoveoPlus to provide Moveo Virtual Reality Simulators in theaters and theme parks (BusinessWire, 2017). Disney has partnered with Jaunt, a VR company currently boasting roughly $101 million in investment capital, to create content (Ronald, 2017) and the Void, which builds location-based VR experiences" (Fickley-Baker, 2017). Much of this is an extension of the concept of transmedia, i.e., reselling the same story franchise across multiple experiences. As just one example, Star Wars VR experiences such as Star Wars Droid Repair Bay, powered by Steam VR (a virtual reality system) are available for home use, or "bigger" (e.g., on a better HMD, CAVE type environment or simply as an "outing"), at IMAX theatres for about $1 a minute (IMAX VR, 2017). Other Star Wars experiences, such as Star Wars: Secrets of the Empire are also at Void locations (The Void, 2017).

Current Status

Recent developments in hardware (all-in-one displays, including both HMDs and AR glasses) and software (SDKs) are fueling predictions and current sales. According to industry analyst Canalys (2017), 2017 Q3 saw shipments totaling over 1 million for the first time in any quarter, namely companies creating both infrastructure and content.

Factors to Watch

Now that someone has tasted VR, how much more will they want? For example, how many of the 85% of 360° Magic Window viewers will purchase an HMD to watch VR on the web. Of course, it is not as simple as just having the headwear. In looking back historian Ray Zone noted stereoscope wasn't "killed" by the technology, but content, as stereoscopic film was used primarily as "gimmick" or "novelty" (Zone, 2007) Fortunately, VR/AR content creators today are actively exploring how "plot in the classic sense may not be the only type of high-level narrative structure" in VR and AR worlds (Aylett, 1999). Will affordable 3D 360° camera's increase the number of indie-artists making VR experiences as DSLR cameras increased the number of film makers? Certainly, developments including WebVR, ARcore, and Arkit represent a collaborative push by the major companies to make content easier to create. Faster data rates could make a difference, too; for example, Google's Seurat demonstrated at Google I/0 '17, compressed content that used to take an hour to transmit down to 13 milliseconds, noting how this enabled easier "content creation" by VR and AR artists. Now we must watch to see if more content creation will lead to more quality content, enticing both consumer adoption and long-term consumer interest.

We also need to watch emerging potential problems with widespread VR and AR usage. For example, there is some concern over long-term effects of VR and AR on an individual and society. In the medical field, it's been demonstrated that VR is so powerful that it is actually more effective than opioids in pain management (Hoffman, 2004) but the medical field is also studying negative effects. "Neuroscientists at UCLA have been conducting research on how the use of virtual reality systems affect the brain... the hippocampus, a region of the brain involved in mapping and in individual's location in space showed 60% of neurons shut down" (Aukstakalnis, 2017). This is only one study, so it will be important to see the results of more studies. Clearly, more Big Data will be collected if individuals are constantly online via smart glasses, so concerns about Big Data's use—or mis-use—will intensify. There is also a phenomenon known as the "Uncanny Valley" effect, in which CG people appear too "real" resulting in an "evil" quality many viewers find unsettling. Finally, there are some practical questions of safety while using the various headgear. This is not a comprehensive list, and many of these types of concerns may dwindle or be overcome, but when a medium is predicted to be as pervasive in our lives as VR and AR, potential adverse outcomes should be watched.

Getting a Job

VR and AR will exist across many industries and so there are multiple points of entry: programmers, designers, creative coders, cinematographers, animators, visual special effects artists, audio engineers and mixers, editors, special effects editors and many of the same jobs you see in movie credits today. As of this writing, the mode of entry is very closely related to modes of entry into other media areas, such as film, television and gaming. A good place to start is an institution of higher learning where students study the fundamentals of media technology, craft, professional workflows, artistry, and theory. It is important to note that there is a big gap between student skill level and professional production, so internships are also very important as a gateway to employment.

Projecting the Future

"Watching the world through goggles may not be everyone's cup of tea, but analysts at Goldman Sachs recently projected an $80 billion industry by 2025, with $45 billion spent on goggles and other equipment and $35 billion more for the content to play video games or watch sports and entertainment in VR" (Grover, 2017).

We should continue to see improvements including advances to "inside out" tracking where sensors on the headset itself will allow for improved depth and acceleration cues (Robertson, 2017b), and these VR and AR headsets will even likely merge so that "future optical see-through HMDs will be able to make individual pixels opaque so digital imagery can completely occlude all or part of the real world" (Jerald 2016), meaning one set of glasses will work for both immersive and augmented uses. More businesses will experiment with VR, AR, and MR for both marketing and training, although the manner of use will be influenced by the user simplicity of the SDKs. Universities and trade schools will develop more programs to train students who want to work in VR, AR, and MR in both entertainment and commercial environments. It is likely that laws or regulations regarding some uses of VR, AR, and MR will emerge, alongside VR/AR/MR specialists in industries such legal or education.

Bibliography

Akkas, E. (2017). The International Virtual Reality Market: Who is in the Lead? Global Me: Language and Technology. Retrieved from https://www.globalme.net/blog/international-virtual-reality-market.

Alexandar, J. (2017). Filmmakers are Using AR and Apple's ARKit to Create Extravagant Short Films. Polygon. Retrieved from: https://www.polygon.com/2017/9/20/16333518/ar-arkit-apple-arcore-google-movies.

Amazon Web Services. (2017). Amazon Sumerian. Retrieved from https://aws.amazon.com/sumerian/.

Antunes, J (2017) Vuze VR camera will get live streaming... soon. Provideo Coalition. Retrieved from: https://www.provideocoalition.com/vuze-vr-camera-gets-live-streaming/.

Aukstakalnis, S. (2017). *Practical augmented reality: A guide to the technologies, applications, and human factors for AR and VR.* London, England: Pearson Education, Inc.

Aylett, R. (1999). Narrative in virtual environments – towards emergent narrative. (Technical Report FS-99-01). Palo Alto, CA. Retrieved from https://pdfs.semanticscholar.org/6e82/5e0af547f8da6b7ff16d31248be03cd571a4.pdf.

Beusterien, J. (2006). Reading Cervantes: A new virtual reality. *Comparative Literature Studies,* 43(4), 428-440. Retrieved from http://www.jstor.org/stable/25659544.

Bilton, N. (2015). Why Google Glass broke. *New York Times.* Retrieved from https://www.nytimes.com/2015/02/05/style/why-google-glass-broke.html.

Bozorgzadeh, A. (2017). WebVR's Magic Window is the Gateway for Pushing VR to Billions of People. Upload VR. Retrieved from https://uploadvr.com/webvrs-magic-window-gateway-pushing-vr-billions-people/.

Business Wire. (2017). MoveoPlus and Paramount Studios Announce Strategic Partnerships to Redefine the Future of Virtual Reality Entertainment. Retrieved from https://www.businesswire.com/news/home/20170426005274/en/MoveoPlus-Paramount-Studios-Announce-Strategic-Partnership-Redefine.

Canalys. (2017). Media Alert: Virtual Reality Headset Shipments Top 1 Million for the First Time. Newsroom. Retrieved from www.canalys.com/newsroom/media-alert-virtual-reality-headset-shipments-top-1-million-first-time.

Chan, M. (2014). *Virtual reality: Representations in contemporary media.* New York, NY: Bloomsbury Publishing, Inc.

Conditt, J. (2017, December 20). Worlds Collide: VR and AR in 2018. Engadget. Retrieved from https://www.engadget.com/2017/12/20/vr-and-ar-in-2018/).

Digi-Capital. (2017a). After a Mixed Year, Mobile AR to Drive $108 Billion VR/AR Market by 2021. Retrieved from https://www.digi-capital.com/news/2017/01/after-mixed-year-mobile-ar-to-drive-108-billion-vrar-market-by-2021/#.Wlj7O0tG1mo.

Digi-Capital. (2017b). The Four Waves of Augmented Reality (that Apple Owns). Retrieved from https://www.digi-capital.com/news/2017/07/the-four-waves-of-augmented-reality-that-apple-owns/#.Wi63-7Q-fOQ).

Fickley-Baker, J. (2017). Tickets Now Available for Star Wars: Secrets of the Empire Hyper-Reality Experience by ILMxLAB and the VOID. Disney Parks Blog. Retrieved from https://disneyparks.disney.go.com/blog/2017/10/tickets-available-now-for-star-wars-secrets-of-the-empire-hyper-reality-experience-by-ilmxlab-and-the-void/.

Greenwald, W. (2017). The best VR (virtual reality) headsets of 2018. *PCMag.* Retrieved from https://www.pcmag.com/article/342537/the-best-virtual-reality-vr-headsets.

Grover, Ronanld. (2017). Disney and Other media giants are betting VR is the next big play in entertainment. CNBC retrieved from https://www.cnbc.com/2017/06/08/virtual-reality-startup-jaunt-to-shake-up-entertainment-industry.html

Hayden, S. (2017). Felix & Paul's 'The People's House' 360 Video wins an Emmy." Road to VR. Retrieved from https://www.roadtovr.com/felix-pauls-360-video-peoples-house-inside-white-house-wins-emmy/.

Hoffman, H. (2004). Virtual-reality therapy. *Scientific American,* 291(2), 58-65. Retrieved from http://www.jstor.org/stable/26060647.

Holister, S. (2016). Daydream VR is Google's New Headset – and it's Not What You Expected. CNET. Retrieved from https://www.cnet.com/news/daydream-vr-headset-partners-google-cardboard-successor/.

IMAX VR. (2017). Retrieved from https://imaxvr.imax.dom/.

Jaunt Studios. (2017). Global Leaders in Immersive Storytelling. Retrieved from www.jauntvr.com/studio.

Jerald, J. (2016). *The VR book: Human centered design for virtual reality*. New York, NY: Association for Computing Machinery.

Kharpal, A. (2015). This Is Why Nokia Is Betting Big on Virtual Reality. CNBC. Retrieved from https://www.cnbc.com/2015/11/12/nokia-ozo-virtual-reality-big-bet.html.

Laurel, B. (1995). Virtual reality. *Scientific American*, 273(3), 90-90. Retrieved from http://www.jstor.org/stable/24981732.

Lindsay, M. (2017). Building virtual reality on the web. Paper presented at the Google I/O 2017 Developer Festival, May 17-19, 2017, Mountain View, CA.

Liptak, A. (2017) Sony's Playstation VR sales have been stronger than expected. The Verge. Retrieved from https://www.theverge.com/2017/2/26/14745602/sony-playstation-vr-sales-better-than-expected.

Newman, J. (2014). The weirdest thing at Google I/O was this cardboard virtual reality box. *Time*. Retrieved from http://time.com/2925768/the-weirdest-thing-at-google-io-was-this-cardboard-virtual-reality-box/.

Parisi, T. (2016). *Learning virtual reality: Developing immersive experiences and applications for desktop, web and mobile*. Sebastopol, CA: O'Reilly Media, Inc.

PCGamer, SteamVR, Everything you need to know. Retrieved from http://www.pcgamer.com/steamvr-everything-you-need-to-know.

Ralph, N. & Hollister, S. (2015). Samsung's $99 Gear VR Launches in November, Adds Virtual Reality to All New Samsung Phones (hands-on). CNET. Retrieved from https://www.cnet.com/products/samsung-gear-vr-2015/preview/.

Riccitellio, J. (2017). Presention at the Google I/O 2017 Developer Festival, May 17-19, 2017, Mountain View, CA.

Robertson, A. (2017a). Alejandro Gonzales Inarritu's Incredible VR Experience Is Getting a Special Oscar Award. The Verge. Retrieved from https://www.theverge.com/2017/10/27/16562434/alejandro-gonzalez-inarritu-carne-y-arena-oscar-special-award-vr.

Robertson, A. (2017b). Self-tracking Headsets Are 2017's Big VR Trend – but They Might Leave Your Head Spinning. The Verge. Retrieved from https://www.theverge.com/2017/1/12/14223416/vr-headset-inside-out-tracking-intel-qualcomm-microsoft-ces-2017.

Ronald, G. (2017). Disney and Other Media Giants Are Betting VR Is the Next Big Play in Entertainment. CNBC. Retrieved from https://www.cnbc.com/2017/06/08/virtual-reality-startup-jaunt-to-shake-up-entertainment-industry.html.

The Void. (2017). Star Wars: Secrets of the Empire. Retrieved from https://www.thevoid.com/dimensions/starwars/secretsoftheempire/.

Tricart, C. (2018). *Virtual reality filmmaking: Techniques & best practices for VR filmmakers*. New York, NY: Routledge.

Vanian, J (2017) Google has shipped Millions of Carboard Reality devices. *Fortune*. Retrieved from http://fortune.com/2017/03/01/google-cardboard-virtual-reality-shipments/.

Virtual Reality Society. (2017). Profiles. Retrieved from https://www.vrs.org.uk/virtual-reality-profiles/vpl-research.html.

Yao, R. & Elchison, S. (2017). Apple's ARKit vs Google's ARcore: What Brands Need to Know. Geo Marketing. Retrieved from http://www.geomarketing.com/apples-arkit-vs-googles-arcore-what-brands-need-to-know.

Zone, R. (2007). *Stereoscopic Cinema and the Origins of the 3-D Film 1838-1952*, Lexington, KY: University Press of Kentucky

16

Home Video

Matthew J. Haught, Ph.D.*

Overview

Home video technology encompasses a range of signal formats, from broadcast, cable, satellite, Internet protocol television (IPTV), over-the-top media services (OTT), and online video distributors (OVD). The three web-based signal formats have spurred major change to television sets, with Internet connectivity either built in (smart TVs), added via multi-platform streaming connectors (Google Chromecast, Amazon Fire TV Stick), or through an external device (Blu-ray player, gaming system, Apple TV, Roku). The trend of cord-cutting, dropping traditional cable or satellite service, and the rise of popular niche programming by OTT and OVD providers, has disrupted the business model for the entertainment and service provider industries.

Introduction

For decades, home video technology centered on receiving and displaying a signal carried through broadcast, cable, and satellite providers. Second to that purpose was recording content via videocassette recorder (VCR) and digital video recorder (DVR) and displaying content from personal video recorders, on video home system cassette tapes (VHS), laserdiscs, digital video discs (DVDs), and Blu-ray discs. However, the gradual shift toward the Internet and web-based home video has eroded home video's legacy formats, in favor of Internet protocol television (IPTV), over-the-top media services (OTT), and online video distributors (OVD).

Today's home video receivers can receive signals over the air (via an external antenna), via cable, satellite, or the Internet, from DVDs and Blu-rays, and from streaming service providers. Tube-based televisions have largely disappeared from stores, replaced by plasma, OLED, LCD, and LED flat-panel displays.

However, some video gamers have re-embraced tube displays for their better connectivity with classic video game platforms (Newkirk, 2017).

Home video technology exists primarily for entertainment purposes; it also has utility as a channel for information and education. Broadcast television delivers a broad national audience for advertisers, and cable and OVD platforms deliver niche demographics.

Background

Today's home video ecosphere sits at the intersection of three technologies: television, the Internet, and the film projector. Home video encompasses the many paths video content has into users' home, as well as the devices used to consume that content.

Broadcast television began in 1941 when the Federal Communications Commission (FCC) established a standard for signal transmission. Up to that point, Philo T. Farnsworth, an independent researcher, as

---

* Assistant Professor, University of Memphis, Department of Journalism and Strategic Media, Memphis, Tennessee

well as engineers and researchers at RCA and Westinghouse, had been working separately to develop the mechanics of the television set. Farnsworth is credited with inventing the first all-electric television, while RCA developed the cathode ray tube, and Westinghouse developed the iconoscope as a component of early video cameras. These inventions coalesced into the broadcast television system, where an image with video was captured, encoded into an electromagnetic signal, sent through the airwaves, received via antenna, and reproduced on a television screen (Eboch, 2015).

In 1953, issues with limited spectrum created a second tier of broadcast frequencies; the initial space was categorized as very high frequency (VHF), and the additional space as ultra-high frequency (UHF). Consumers were slow to adopt UHF, but eventually, in 1965 the FCC mandated UHF compatibility for television set manufacturers.

As discussed in Chapter 7, cable television originated in 1948 when rural communities in Arkansas, Oregon, and Pennsylvania separately erected community antennae to receive distant broadcast signals. By the late 1950s, these antennae could receive local and distant signals, providing subscribers choices for content. Following a few regulation issues, cable television began to expand rapidly in the 1970s, and some networks began using satellites to send signal to cable systems. In the mid- to late-1990s, cable systems began to transition from antenna-satellite distribution to a fiber optic and coaxial network, capable of supporting television signals, telephone calls, and Internet (CalCable, n.d.). Satellite television began as users, primarily in rural areas, installed large dish receivers to receive signals sent via satellite. In 1990, DirecTV was the first direct broadcast satellite service available to users, followed soon by USSB, Dish, and others. Over the next two decades, the industry consolidated to two primary providers: AT&T's DirecTV and Dish.

Consumer Internet access began in the 1980s through telephone dial-up service. However, by the mid-1990s, the slow speed of telephone modems was evident, and cable companies began to provide Internet access using the same coaxial cables that distributed television signals, supplanting telephone companies and third-party Internet Service Providers (ISPs). As Internet connectivity improved, video began to migrate to the platform.

Finally, the home film projector began as a way to show films and recorded home movies. Its use was expanded with the introduction of the consumer VHS and Betamax VCRs in the 1970s. After VHS won the standard war with Betamax, home video recording became quite popular through the VCR and through portable video camera recorders. The landmark *Sony vs. University City Studios* case in 1984, commonly called the Betamax case, established that home VCR users could record television programming for non-commercial use without infringing on copyright. In the 1990s and 2000s, DVDs replaced VHS as the primary means of storing recorded video, and portable video camera recorders transitioned to flash media for storage. In the mid-2000s, Sony's Blu-ray format beat out Toshiba's HD DVD as the format of choice for high definition video storage. Further, the digital video recorder, available through most cable and satellite systems, replaced the VCR as a means of recording and time-shifting/storing television programming.

Recent Developments

The recent developments in home video extend ongoing evolutions in the industry. Three key areas: Smart TVs, cord cutting via the expansion of OVD and OTT services, and changes to Net Neutrality laws, shape the industry's present state. And, for the foreseeable future, these issues should remain at the forefront of home video.

Smarter TVs

The *2018 Consumer Electronics Show*, as typical, featured a bevy of new television features, increases in size and quality, and incremental improvements to features in standard consumer models. Samsung and LG Electronics both expanded on new OLED and LED display technologies from 2017. LG launched a 65-inch 4K resolution OLED television that rolls down into a small box. Meanwhile, Samsung launched a 146-inch TV with its microLED display; the display's name reflects its size: The Wall (Falcone, 2018).

Aside from the boasts in size and quality, the latest development with television sets themselves are a deeper connection to the Internet and added functionality beyond program viewing. With the Internet of Things (IoT) technology, television sets are just glass to gaze at an online world. Home video has spread beyond the television set: video-capable screens are even being installed in refrigerators. Television systems are working to serve their many purposes, from connected televisions to stream, better refresh rates to game, and integration with personal assistants such as Amazon Echo for convenience (Pierce, 2018).

Cord Cutting, OTT, and OVD

Over-the-top services and online video distributors, such as cable-like providers VerizonFIOS, AT&T U-verse, and Dish's Sling app, plus streaming video services Netflix, Hulu, Amazon Prime, MLB.TV, YouTube Red, WWE Network, CBSAllaccess, and Crunchyroll, are disrupting the traditional television ecosystem. More than half of Wi-Fi homes have at least one OTT service. For many OTT consumers, streaming services have displaced traditional television viewing; users watch OTT content 19 days a month on average, and for 2.2 hours per day. The highest concentration OTT viewing also happens during prime-time hours. Sling leads all platforms in viewing time, with an average of 47 viewing hours per month, with Netflix following at 28 (Perez, 2017). Sling viewers, typically, are using the service as a cable replacement—a skinny way to cut the cord.

Table 16.1

Paying Subscribers, Popular Online Video Apps, 2017

Rank	App	Subscribers (Millions)
1	Netflix	53
2	Amazon	30.8
3	Hulu	18
4	MLB.TV	<2
5	HBO Now	<2
6	Starz	<2
7	YouTube Red	<2
8	Showtime	<2
9	CBS All-Access	<2
10	Sling	<2

SOURCE: Parks Associates, 2017

Traditional television providers can see the writing on the wall. AT&T has purchased DirectTV as a satellite provider and launched DirectTV Now as a streaming service. Comcast has taken its Xfinity network mobile. Dish launched Sling. Comcast, Disney, and Time Warner co-own Hulu. The major brands of pay television are likely to stick around as major brands in streaming television.

Net Neutrality

The 2017 FCC decision to rollback prohibitions on throttling speeds on certain websites or charging users for premium access to websites could have a drastic effect on home video (Kang, 2017). However, the ultimate impact is too early to know. Some fear that, as traditional cable television providers absorb OTT and OVD services, users might see ISPs favor their own streaming service over others. (Net neutrality is discussed in more detail in Chapter 5.)

Current Status

Home video is making a pivot from terrestrial and satellite providers to web-based providers. In 2017, pay TV market penetration was down to 79%, from a high of 88% in 2010 (Leichtman, 2017). Yet, 96.5% of U.S. homes with a television receive a video signal of some form, paid or otherwise, (Nielsen, 2017a). Of the 21% of households that do not have pay television, two-thirds had previously been pay TV subscribers. Cord-cutting is more common among those with household incomes below $50,000 annually. Thus, the cord-cutting phenomenon from traditional pay TV platforms is on the rise and is likely to define the home video industry through 2020.

Home video technology now engages mobile devices alongside television sets. Users, particularly younger users, are watching home video content on tablets and mobile devices increasingly, and soon will do so more often than on a television set (Eck, 2017). Even content creators for home video are shifting; Amazon, Netflix, and Hulu are producing original (critically acclaimed and award-winning) content for their platforms, in addition to content from traditional video creators. And specialized platforms are playing a role to provide content previously inaccessible to consumers; for example, Crunchyroll, an OVD specializing in anime, reports having 1 million paid subscribers and 20 million total users (Orsini, 2017). Specialized web networks with premium ad-free subscriptions or free-to-watch with advertising options are providing less-visible content to their niche audience.

Moreover, creatives developing new shows might first test them on YouTube or another low-end platform, and larger content companies can pluck new talent from these engaged creatives (Eck, 2017). Or, shows on traditional television can extend their brands with web shows.

Consider first the example of World of Wonder. The production company's flagship show, RuPaul's *Drag Race*, airs on VH1 and LOGO. However, because fans of the show are deeply engaged with its content, WOW has multiple active free programs on YouTube. Then, in 2017, it launched WOW Presents Plus, a premium YouTube channel to air additional content from former RuPaul's *Drag Race* contestants. One free show, *UNHhhh*, starring former contestants Trixie Mattel and Katya Zamolodchikova, was picked up and expanded to a full 30-minute program by Viceland Network as *The Trixie and Katya Show* in 2017.

Second, the example of Issa Rae's *The Mis-Adventures of Awkward Black Girl* YouTube series, exemplifies online video as an incubator of new talent. Rae launched the web series in 2011; its first episode, "The Stop Sign," earned 2.4 million views. Rae's success translated to an HBO series, *Insecure*, in 2016, which itself landed her deals for two more series with the premium network (Crucchiola, 2017).

Both these examples show how home video has democratized the television landscape. WOW's programming caters to an LGBT audience, and Issa Rae's work is popular among African-American audience. Networks have typically been hesitant to embrace programming for these limited audiences; however, because of their strong online followings, their creative content found a place on mainstream television.

The power of home video today is the ability reach diverse audiences with diverse content. No longer are audiences limited to the content created by studios and networks, or by the channels offered by their local cable companies. Users are able to find content that suits their tastes, and watch it on whatever device they have, whenever they want, and often, paying to avoid advertising, or watching free content with ads specific to their demographics.

Technology Ecosystem

Hardware:

The hardware for home video encompasses a range of devices, including television sets, computer monitors, tablets, and mobile phones (Eck, 2017). Similarly, these displays have associated terrestrial, satellite, or web-based connectivity through a receiver, and ultimately, the sender mechanism for encoding the video signal.

Television sets continue to evolve through Internet of Things (IoT) technology (Spangler, 2018). What used to be a purely home video terminal now ranks among twenty or so other smart devices in the home capable of displaying video from OVD, as well as monitoring other smart home systems. However, the key feature of a television set is its size. Mobile device screens are rarely more than 12 inches wide, but television screens can measure up to 60 or more inches (Pierce, 2018).

Software:

Software needed for home video viewing is limited. Typically, the viewing device itself has built-in viewing capabilities. However, mobile applications for OVD are common. Netflix, Amazon, and Hulu are the most popular apps, but more niche programming apps follow.

Organizational Infrastructure:

Home video infrastructure can be broken down into two broad categories: Viewing device manufacturers and service providers. Devices include television sets, computer monitors, mobile devices, Blu-ray and DVD players, and smart television connectors. Providers include cable and satellite services, Internet service providers, IPTV services, and OVD.

With 27.2% market share, Samsung leads all LCD television set manufacturers. The second most popular, LG Electronics, has an 11.9% market share, with TCL third at 9% (Statista, 2016). Higher-end Sony and LG organic light emitting diode (OLED) and Samsung quantum dot light emitting diode (QLED) are popular and deliver higher definition, but at higher costs (Katzmaier, 2017). Samsung, Sony, LG, and Panasonic also dominate the list of Blu-ray and DVD player manufacturers.

Top computer manufactures are led in market share by HP (21.8%), Lenovo (20.4%), Dell (15.9%), Apple (7%), and Acer (6.8%); Google's Chromebook is a rising competitor, but still has a minimal market share (Dunn, 2017). Mobile devices are dominated by Apple's iOS and Google's Android OS, together accounting for 99.6% of the market share (Vincent, 2017). Globally, Android serves 81.7% of users, while Apple serves 17.9%. In the U.S., however, the market is more even, with about 53% of users on Android and 44% on iOS (ComScore, 2017). Apple leads smartphone manufacturers with 44.6% of the market; Samsung (28.3%) and LG (10%) follow (ComScore, 2017). Apple (25%) and Samsung (15%) also dominate tablet market share.

Comcast (22.4 million subscribers) is the largest cable provider in the U.S., followed by Charter (17.0 million), which in 2016 absorbed Time Warner Cable. AT&T's Direct TV leads satellite with 21.0 million subscribers, Dish follows with 13.2 million subscribers. Verizon FIOS is the largest IPTV provider with 4.6 million subscribers; AT&T's U-verse has about 4 million (NCTA, 2017; Spangler, 2017). Comcast (25.1 million), Charter (22.6 million), and AT&T (15.6 million) lead Internet Service Providers, with Verizon (7.0 million), Century Link (5.9 million), and Altrice (3.9 million) following (Molla, 2017).

Netflix leads all OVD platforms with 52.8 million subscribers, with Amazon and Hulu boasting high subscriptions. However, some OVD platforms offer free subscriptions. YouTube, for example, has 1 billion active users per month, while YouTube Red has more than 2 million paying subscribers; those 2 million can watch videos on the network ad-free, while the remainder can use the service, but must watch ads before most videos. This free-with-ads/paid-ad-free model operates for several platforms.

Social Systems:

The NCTA—The Internet & Television Association, is the lead trade association for the U.S. broadband and television industries. The NCTA was one of the lead lobbying groups pushing for the FCC's 2017 rollback of Net Neutrality rules. However, the association claims it supports an open Internet, including no blocking of legal content, no throttling, no unfair discrimination, and transparency in customer practices (NCTA, n.d.). The 2017 Net Neutrality rollback removes rules blocking extra charges or slowing of content for specific sites. However, at the time of this writing, lawsuits and legislation could change this rollback, instituting new limits or overriding the change entirely (Kang, 2017).

Individual User:

Service penetration for television remains high; Nielsen estimated there were 119.6 million TV homes in the U.S. for the 2017-18 season, which is about 80% of households (Nielsen, 2017a). Meanwhile, 77% of U.S. adults own a smartphone, 51% own a tablet, and 78% own a desktop or laptop—all capable of home video viewing (Pew, 2017).

As of June 2017, 58.7% (69.5 million) of the 119.6 million TV households have at least one Internet-enabled device capable of streaming content to the television set (Nielsen, 2017b). Users access video content through gaming systems (39.4 million); smart televisions (36.6 million), and multimedia devices (30.6 million) with some overlap (Nielsen, 2017b).

Ultimately, individual user adoption can vary widely, from full cable and satellite with premium channels and multiple OTT, OVD platforms via multiple enabled devices to no television service with only a free OVD subscription on a single web-enabled device.

Figure 16.1

Household Ownership of Enabled Devices (One, Two or Three)

Of the 58.7% of TV Households that have an enabled device...

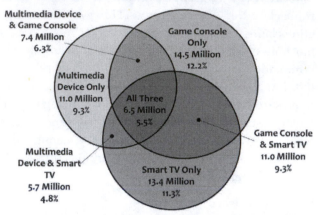

Multimedia Device & Game Console
7.4 Million
6.3%

Game Console Only
14.5 Million
12.2%

Multimedia Device Only
11.0 Million
9.3%

All Three
6.5 Million
5.5%

Game Console & Smart TV
11.0 Million
9.3%

Multimedia Device & Smart TV
5.7 Million
4.8%

Smart TV Only
13.4 Million
11.3%

Source: Nielsen National Panel, 2017

Sales and Subscriptions

In 2017, DVD and Blu-Ray disc sales dropped 14% to $4.7 billion; video rentals from kiosks and stores dropped 17% for a little more than $2 billion. Streaming home video, including OTT and OVD platforms, grew by 30% to $9.5 billion. Pay-per-view and on-demand home video dropped nearly 7% to $1.9 billion. Meanwhile, electronic-sell-through, where users pay for a video and keep it on a hard drive, rose to about $2 billion (Lopez, 2018).

Outside the home, domestic box office spending dropped 2% to $11.1 billion. More often, consumers are opting to stay home and watch new streaming content rather than go to the cinemas and watch a movie (Lopez, 2018).

The number of home cable and satellite subscribers continues to decline as cord-cutters drop cable services in favor of OTT and OVD services; at the end of 2017, just below 80% of U.S. homes had cable television, down from a high of 91% in 2011. Kagan, a research unit in S&P Global Market Intelligence, reports that nearly 4 million homes dropped television in 2018, and it and other researchers expect the slide to continue, likely to about 67% of homes by 2027 (Adgate, 2017).

Despite declines in cable and satellite subscriptions, revenue in the industry is expected to increase.

Kagan expects revenues from residential subscriptions to increase from $108.4 billion in 2016 to $117.7 billion in 2026. Broadband Internet services are largely driving this growth, with an expected 71 million subscribers by 2026 (Burger, 2016).

In advertising, television is no longer king. Brands spent about $71.3 billion on domestic television in 2016, while digital advertising spiked to $72.5 billion. Companies are concentrating television spending on fewer brands and products, and they are even launching new products without national television campaigns. Advertising on digital video platforms is on the rise, however, with digital video hitting a record $9.1 billion in 2016 and mobile video revenue hitting $4.2 billion (Slefo, 2017).

Factors to Watch

As cord-cutting and online transitions escalate, home video could transition further away from its traditional tether to a television set (Eck, 2017). Mobile video consumption is on the rise, but still lags traditional television consumption four-to-one (eMarketer, 2016). As OVD platforms Hulu, YouTube Red, Amazon Prime, Netflix, and others continue to generate original content, expect more mobile video consumption.

The caveat to this mobile pivot comes through the FCC's 2017 decision to overturn Net Neutrality rules that prohibited ISPs from throttling speeds or charging uses more for streaming services. Considering the bulk of ISPs are cable, satellite, or IPTV companies, users who drop mainline video service in favor of OVD platforms outside the ISPs home brand might see slower speeds or higher costs.

Regardless, the ability of OVD and legacy video distributors to deliver an audience to advertisers remains strong. Big data tracking of users will be easier as more make the switch to online video. Thus, brands will be able to target specific users and demographics more easily and will be able to more effectively measure the return on investment for ad spending. Further, demographic-specific creative advertising could be channeled to users on the fragmented OVD platform, all driven by big data.

Home video is a commodity of the media convergence economy. Television set manufacturers have switched to making mobile phones and tablets. Computer manufacturers are making mobile phones and tablets and smart television systems. Movie and television studios, independent content creators, and cable companies have their own streaming platforms or investing in others. Home video is evolving far beyond channels and receivers to a new era of content and glass.

Getting a Job:

Students who are interested in production careers in home video need to learn the current technology for studios, including switchers, cameras, and lighting, as well as editing. Systems maintenance professionals can work outdoors or in close spaces to maintain the substantial infrastructure needed for transmission. Home video systems designers need to study engineering or physics to build knowledge about fiber optics and electricity.

Bibliography

Adgate, B. (2017, December 07). Cord Cutting Is Not Stopping Any Time Soon. Retrieved from https://www.forbes.com/sites/bradadgate/2017/12/07/cord-cutting-is-not-stopping-any-time-soon/#59fab3625ef0.

Burger, A. (2016, August 12). U.S. Cable Industry Revenue Forecast to Hit $141 Billion in 2026. Retrieved from http://www.telecompetitor.com/u-s-cable-industry-revenue-forecast-to-hit-141-billion-in-2026/.

CalCable. (n.d.). History of Cable. Retrieved from https://www.calcable.org/learn/history-of-cable/

ComScore. (2017, April 11). ComScore Reports February 2017 U.S. Smartphone Subscriber Market Share. Retrieved from https://www.prnewswire.com/news-releases/comscore-reports-february-2017-us-smartphone-subscriber-market-share-300437639.html.

Crucchiola, J. (2017, December 18). Issa Rae Is Working on Two New Shows for HBO. Retrieved from http://www.vulture.com/2017/12/issa-rae-is-working-on-two-new-shows-for-hbo.html.

Dunn, J. (2017, April 14). Here are the companies that sell the most PCs worldwide. Retrieved from http://www.businessinsider.com/top-pc-companies-sales-idc-market-share-chart-2017-4.

Eboch, M. M. (2015). *A History of Television*. Minneapolis: ABDO Publishing

Eck, K. (2017, April 20). Is Mobile About to Overtake Traditional TV Viewing? Retrieved from www.adweek.com/tv-video/why-more-people-than-ever-are-watching-tv-on-mobile/.

eMarketer. (2016, April 14). Is TV vs. Digital Video a Close Race? Retrieved from www.emarketer.com/Article/TV-vs-Digital-Video-Close-Race/1013826

Falcone, J. (2018, January 15). CES 2018: The final word. Retrieved January 16, 2018, from https://www.cnet.com/news/ces-2018-the-final-word/.

Kang, C. (2017, December 15). What's Next After the Repeal of Net Neutrality. Retrieved from https://www.nytimes.com/2017/12/15/technology/net-neutrality-repeal.html.

Katzmaier, D. (2017, May 24). QLED vs. OLED: Samsung's TV tech and LG's TV tech are not the same. Retrieved from https://www.cnet.com/news/qled-vs-oled-samsungs-tv-tech-and-lgs-tv-tech-are-not-the-same/

Leichtman. (2017, September 26). 79% of TV Households Subscribe to a Pay-TV-Service. Retrieved from http://www.leichtmanresearch.com/press/092617release.html.

Lopez, R. (2018, January 09). Disc Sales Decline Deepens in Annual Home Entertainment Spending Report. Retrieved from http://variety.com/2018/digital/news/home-entertainment-spending-2017-1202658638/.

Molla, P. K. (2017, April 27). Comcast, the largest broadband company in the U.S., is getting even bigger. Retrieved from https://www.recode.net/2017/4/27/15413870/comcast-broadband-internet-pay-tv-subscribers-q1-2017.

NCTA. (n.d.). ISPs Commit To An Open Internet. Retrieved from https://www.ncta.com/positions/supporting-an-open-internet.

NCTA. (2017). Top 10 Video Subscription Services. Retrieved from https://www.ncta.com/chart/top-10-video-subscription-services.

Neikirk L. (2017, March 09). The tube-based CRT TV has officially run out of time. Retrieved from http://televisions.reviewed.com/features/the-tube-based-crt-tv-has-officially-run-out-of-time

Nielsen. (2017a, August 25). Nielsen Estimates 119.6 Million TV Homes in the U.S. for the 2017-18 TV Season. Retrieved from http://www.nielsen.com/us/en/insights/news/2017/nielsen-estimates-119-6-million-us-tv-homes-2017-2018-tv-season.html.

Nielsen. (2017b, November 16). *The Nielsen Total Audience Report: Q2 2017.* Retrieved from http://www.nielsen.com/us/en/insights/reports/2017/the-nielsen-total-audience-q2-2017.html.

Orsini, L. (2017, February 13). Crunchyroll Commemorates 1 Million Subscribers With New Anime Convention. Retrieved from https://www.forbes.com/sites/laurenorsini/2017/02/09/crunchyroll-commemorates -1-million-subscribers-with-new-anime-convention/#49a1897c3285.

Pew. (2017, January 12). Mobile Fact Sheet. Retrieved from http://www.pewinternet.org/fact-sheet/mobile/

Pierce, D. (2018, January 10). The Future of TV Is Just Screens All The Way Down. Retrieved from https://www.wired.com/story/ces-2018-the-future-of-tvs/.

Perez, S. (2017, April 10). Netflix reaches 75% of US streaming service viewers, but YouTube is catching up. Retrieved January 16, 2018, from https://techcrunch.com/2017/04/10/netflix-reaches-75-of-u-s-streaming-service-viewers-but-youtube-is-catching-up/.

Spangler, T. (2018, January 10). Comcast Wants to Turn Xfinity Into an 'Internet of Things' Smart-Home Platform. Retrieved from http://variety.com/2018/digital/news/comcast-xfinity-internet-of-things-home-automation-1202657823/.

Spangler, T. (2017, April 25). AT&T Drops 233,000 TV Subscribers as DirecTV Satellite Customers Remain Flat in Q1. Retrieved January 15, 2018, from https://variety.com/2017/digital/news/att-directv-q1-tv-subscribers-1202395243/.

Statista. (2016). LCD TV market share manufacturers worldwide 2008-2016 | Statistic. Retrieved from https://www.statista.com/statistics/267095/global-market-share-of-lcd-tv-manufacturers/

Slefo, G. (2017, April 26). Desktop and Mobile Ad Revenue Surpasses TV for the First Time. Retrieved from http://adage.com/article/digital/digital-ad-revenue-surpasses-tv-desktop-iab/308808/.

Vincent, J. (2017, February 16). 99.6 percent of new smartphones run Android or iOS. Retrieved from https://www.theverge.com/2017/2/16/14634656/android-ios-market-share-blackberry-2016/

17

Digital Imaging & Photography

Michael Scott Sheerin, M.S.*

"no time for cameras
we'll use our eyes instead"

—from the song *Cameras* by Matt and Kim

Overview

For some time now, digital photography has been by far the most popular visual medium of all communication technologies. But the way we capture images is beginning to change. We'll look back at the history of photography from its camera obscura analog beginnings, through the advent of digital photography, and into the current state of the medium.

Light-field photography, lens-less cameras and Artificial Intelligence (AI) software are some of the newer innovations that are starting to play major roles in the future direction of image capture. These new advancements all change the way a digital image is captured, processed and shared. So, exactly what is a digital image? It is a numerical representation of a two-dimensional image. The smallest element of the image, the pixel (derived from picture element), holds the values that represent the brightness of a color at any specific point on the image. Thus, the digital image is a file that contains information in the form of numbers. The way we capture this information, the software processes that can be applied to it, the way we share it, and future trends in the industry are all the focus of this chapter.

Introduction

Advancements in the field point to a time when we'll only need our eyes to take pictures, as image capture technology, to borrow from the Greek, has become "phusis," defined as a natural endowment, condition or instinct (Miller, 2012). Look around you, and I bet there is a device within three feet that you can use to capture an image; perhaps that device is even in your hand or on your wrist as you read this chapter on your mobile phone. For Millennials (and, let's be honest, older generations as well), the mobile phone is a natural appendage. And an appendage that we use a lot, as it has been estimated that we will take 1.4 trillion photos in 2020 (Lee, 2016), with nearly 85% of them captured on a mobile phone (Cakebread, 2017). These digital images are ultimately shared on social networks like Snapchat (20,000 photos shared/second) (Aslam, 2018b), Whatsapp (4.5 Billion shared/day) (BT Online, 2017), Facebook (300

---

* Associate Professor, School of Journalism and Mass Communications, Florida International University (Miami, Florida)

million photos uploaded/day) (Strategic Insights, 2018), or Instagram (95 million uploaded/day) (Aslam, 2018a), for anyone to see, making the digital image ubiquitous in our culture.

George Orwell's novel, *Nineteen Eighty-Four*, published in 1949, fictitiously predicts a dystopian society that has had its civil liberties stripped due to government surveillance, i.e. Big Brother. But as we see today, it's not Big Brother alone that is watching, as the plethora of images shared on social media platforms confirms that everyone is participating in the surveillance, including self-participation. According to Infograms' *Selfie Statistics*, one million selfies are taken globally every day, with more than 54 million images that are tagged #selfie residing on Instagram (Muller, 2018).

This opt-in social surveillance via digital self-imaging is especially popular with the Millennials. In a study conducted by SelfieCity, the average age of selfie takers was 23.7 years for females and 28 years for males, with the vast majority of the images (62.8%) taken and posted by women (SelfieCity, 2016). Thus, Orwell's imagined surveillance takes place with complete compliance by its subjects. As Fred Ritchin, professor of Photography and Imaging at New York University's Tisch School of the Arts, points out, "we are obsessed with ourselves" (Brook, 2011). Now couple this "social surveillance" trend with facial-recognition technology, and it's easy to see how someone could be "tagged" and tracked very easily. In fact, Facebook's new facial recognition algorithm, called Photo Review, will let users know about photos that they are in even if they haven't been tagged. For those who don't want to use this facial recognition algorithm, Facebook offers "a new overarching photo and video facial recognition opt out privacy setting that will delete its face template of you and deactivate the new Photo Review feature as well as the old Tag Suggestions" feature (Constine, 2017). These and other social media detection algorithms concern many that deal with privacy issues, including Ralph Gross, who conducts post-doctoral research on privacy protection at Carnegie Mellon University's CyLab. Noting that Facebook's algorithm is impressive, Gross states, "If, even when you hide your face, you can be successfully linked to your identity—that will certainly concern people.

Now is a time when it's important to discuss these questions" (Rutkin, 2015).

These findings suggest that the photograph is not just a standalone part of the digital image industry. It's fully converged with the computer and cell phone industries, among others, and has changed the way we utilize our images "post shutter-release." These digital images are not the same as the photographs of yesteryear. Those analog photos were continuous tone images, while the digital image is made up of discreet pixels, ultimately malleable and traceable, to a degree that becomes easier with each new version of photo-manipulating software, GPS, AI, and detection algorithms. And unlike the discovery of photography, which happened when no one alive today was around, this sea change in the industry has happened right in front of us; in fact, we are all participating in it—we are all "Big Brother." This chapter looks at some of the hardware and software inventions that continue to make capturing and sharing quality images easier, as well as the implications on society, as trillions of digital images enter all our media.

Background

Digital images of any sort, from family photographs to medical X-rays to geo-satellite images, can be treated as data. This ability to take, scan, manipulate, disseminate, track, or store images in a digital format has spawned major changes in the communication technology industry. From the photojournalist in the newsroom to the magazine layout artist, and from the social scientist to the vacationing tourist posting to Facebook via Instagram, digital imaging has changed media and how we view images.

The ability to manipulate digital images has grown exponentially with the addition of imaging software, and has become increasingly easier to do. Don't like your facial skin tones on that selfie you just took? Download Perfect 365 to your iOS or Android device. The app automatically detects facial points and "allows you to remove blemishes, dark circles and spots, and even lets you apply the desired look while taking photos or videos" (Upright, 2018). Want to make your images look like they were shot with a 35-mm film camera? Use Faded, an iPhone and iPad app that brings "the nostalgia and beauty of classic film to your photos" (C/Net, 2018).

Looking back, history tells us photo-manipulation dates to the film period that the Faded app attempts to capture. Images have been manipulated as far back as 1906, when a photograph taken of the San Francisco Earthquake was said to be altered as much as 30% according to forensic image analyst George Reid. A 1984 National Geographic cover photo of the Great Pyramids of Giza shows the two pyramids closer together than they are, as they were "moved" to fit the vertical layout of the magazine (Pictures that lie, 2011). In fact, repercussions stemming from the ease with which digital photographs can be manipulated caused the National Press Photographers Association (NPPA), in 1991, to update their code of ethics to encompass digital imaging factors (NPPA, 2017). Here is a brief look at how the captured, and now malleable, digital image got to this point.

The first photograph ever taken is credited to Joseph Niepce, and it turned out to be quite pedestrian in scope. Using a technique he derived from experimenting with the newly-invented lithograph process, Niepce was able to capture the view from outside his Saint-Loup-de-Varennes country house in 1826 in a camera obscura (Harry Ransom Center, University of Texas at Austin, 2018). The capture of this image involved an eight-hour exposure of sunlight onto bitumen of Judea, a type of asphalt. Niepce named this process heliography, which is Greek for sun writing (Lester, 2006). It wasn't long after that the first photographic self-portrait, now known as a selfie, was recorded in 1839. (See Figure 17.1) Using the daguerreotype process (sometimes considered the first photographic process developed by Niepce's business associate Louis Daguerre), Robert Cornelius "removed the lens cap, ran into the frame and sat stock still for five minutes before running back and replacing the lens cap" (Wild, 2018).

The next 150 years included significant innovation in photography. Outdated image capture processes kept giving way to better ones, from the daguerreotype to the calotype (William Talbot), wet-collodion (Frederick Archer), gelatin-bromide dry plate (Dr. Richard Maddox), and the now slowly disappearing continuous-tone panchromatic black-and-

white and autochromatic color negative films. Additionally, exposure time has gone from Niepce's eight-hour exposure to 1/500th of a second or less.

Figure 17.1
First 'Selfie'

Source: Library of Congress

Cameras themselves did not change that much after the early 1900s until digital photography came along. Kodak was the first to produce a prototype digital camera in 1975. The camera, invented by Steve Sasson, had a resolution of .01 megapixels and was the size of a toaster (Zhang, 2010). In 1981, Sony announced a still video camera called the MAVICA, which stands for magnetic video camera (Carter, 2015a). It was not until nine years later, in 1990, that the first digital still camera (DSC) was introduced. Called the Dycam (manufactured by a company called Dycam), it captured images in monochromatic grayscale and had a resolution that was lower than most video cameras of the time. It sold for a little less than $1,000 and had the ability to hold 32 images in its internal memory chip (Aaland, 1992).

In 1994, Apple released the Quick Take 100, the first mass-market color DSC. The Quick Take had a resolution of 640 × 480, equivalent to a NTSC TV image, and sold for $749 (Kaplan & Seagan, 2008). Complete with an internal flash and a fixed focus 50mm lens, the camera could store eight 640 × 480 color images on an internal memory chip and could transfer images to a computer via a serial cable. Other mass-market DSCs released around this time were the Kodak DC-40 in 1995 for $995 (Carter, 2015b) and the Sony Cyber-Shot DSC-F1 in 1996 for $500 (Carter, 2015c).

The DSCs, digital single lens reflex (DSLR) and mirrorless (MILC) cameras work in much the same way as a traditional still camera. The lens and the shutter allow light into the camera based on the aperture and exposure time, respectively. The difference is that the light reacts with an image sensor, usually a charge-coupled device (CCD) sensor, a complementary metal oxide semiconductor (CMOS) sensor, or the newer, live MOS sensor. When light hits the sensor, it causes an electrical charge.

The size of this sensor and the number of picture elements (pixels) found on it determine the resolution, or quality, of the captured image. The number of thousands of pixels on any given sensor is referred to as the megapixels (MP). The sensors themselves can be varied sizes. A common size for a sensor is 18 × 13.5mm (a 4:3 ratio), now referred to as the Four Thirds System (Four Thirds, 2017). In this system, the sensor area is approximately 25% of the area of exposure found in a traditional 35mm camera. Many of the sensors found in DSLRs are full frame CMOS sensors that are 35mm in size (Canon, 2018), whereas are most of the MILC sensors are either Four Thirds or Advanced Photo System Type-C (APS-C), which are slightly larger than Four Thirds sensors (Digital Camera, 2018).

The pixel, also known in digital photography as a photosite, can only record light in shades of gray, not color. To produce color images, each photosite is covered with a series of red, green, and blue filters, a technology derived from the broadcast industry. Each filter lets specific wavelengths of light pass through, according to the color of the filter, blocking the rest. Based on a process of mathematical interpolations, each pixel is then assigned a color. Because this is done for millions of pixels at one time, it requires a great deal of computer processing. The image processor in a DSC must "interpolate, preview, capture, compress, filter, store, transfer, and display the image" in a very short period (Curtin, 2011).

This image processing hardware and software is not exclusive to DSCs, MILCs, or DSLRs, as this technology has continued to improve the image capture capacities of the camera phone. Starting with the first mobile phone that could take digital images—the Sharp J-phone, released in Japan in 2000—the lens, sensor quality, and processing power have made the mobile phone the go-to-camera for most of the images captured today (Hill, 2013). An example of the improved image capture technology in mobile phones can be found on the Sony Exmor RS. Equipped with a 22.5 MP stacked sensor. The Exmor RS was the first mobile phone to use phase detection autofocus pixels that are "stacked" on the phone's imaging sensors (Horaczek, 2016). And, of course, any photograph can be digitally modified using photo editing software such as Photoshop (see Figure 17.2).

Figure 17.2
Photo Editing—Before and After

Source: M. Sheerin

Recent Developments

According to Saffo's 30-year rule, the digital imaging industry has just about reached full maturity (1990–2018). Thus, it can be expected that growth in some areas has slowed. One example of this is seen in the lack of new camera models (DSC, DSLR and MILCs) coming to market. There has been a steady decline since 2014, with only 33 new models introduced in 2017 (there were 98 new models in 2014) (Sylvester,

2018). But that doesn't imply that it is not still a dynamic industry.

New areas in light-field photography, mobile phone apps, and AI continue to push the technological boundaries. Change has always been a part of the photographic landscape. Analog improvements, such as film rolls (upgrades from expensive and cumbersome plate negatives) and Kodak's Brownie camera (sold for $1.00) brought photography to the masses at the start of the 1900's. The advent of the digital camera and the digital images it produced in the late 20th century was another major change, as the "jump to screen phenomenon" gradually pushed the film development and printing industry off to the sidelines (Reis et al., 2016). And when this new digital imaging phase converged with the telephone industry at the start of the 21st century, the industry really underwent a holistic change. The rapid rise in the use of the mobile phone as camera is the technological advancement that made the largest impact on the digital imaging and photography industry up to now, as 85% of all images taken in 2017 were captured on a smartphone (Cakebread, 2017).

But we don't stop there. Technological advancements in the field continue, and they usually happen in one of two ways. The first is by incremental improvements, and this can be illustrated with a few examples. One is digital sensor improvement, while another is lens speed, as determined by its aperture. Currently, an f/1.4 aperture is in the works, while LG's V30 is available with a f/1.6 with a glass lens (Moon, 2017). Perhaps stretching the use of the term incremental is the advent of the gigapixel camera, such as the one being developed in Chile at the Large Synoptic Survey Telescope (LSST) Project. This 3.2 gigapixel camera will be the world's largest camera and is expected to be completed in 2023. Each panoramic snapshot that the LSST captures will be of an area 40 times the size of a full moon (LSST, 2018). A dual lens camera phone is another example of incremental improvement, as is the implementation of a 2X optical zoom on a camera phone, as found on the new iPhone X (Ismail, 2018). Two other innovations that deal with camera lenses begin to pull the technological advancements from the incremental innovation column to the disruptive innovation column, the second of the two types of innovation.

The disruptive innovation that digital imaging is currently undergoing is going to have the biggest impact on the future of the industry. From advances in light-field and computational photography, to lens-less camera innovation, this is the area to watch. It's not just the ubiquitous use of mobile phones for capturing images, but it's also the new ways that these images are captured that contribute to this disruptive innovation. A leader in this field is the L16 camera by Light, a 52 MP, 16-lens camera prototype that is the size of a smartphone. (See Figure 17.3) Released in 2017, the camera has had mixed reviews. David Pierce of Wired wrote that the L16 "offers an early, decidedly imperfect look at how algorithms will dictate the future of photography. Something like the tech in the L16 will power your next smartphone, or maybe the one after that" (Pierce, 2017). To that end, the L16 follows the first camera to delve into light-field photography, the Lytro plenoptic camera, first released in 2011. However, perhaps due to its advanced technology, the L16 may have caused Lytro to stop further production of their camera, as the company has stopped all support of the still camera and have moved their operations to concentrate on light-field based 3D video.

Figure 17.3
L16 Camera

Source: Light

The way light-field cameras work is they capture data rather than images and process the data post-shutter release to produce the desired photo, including the ability to reinterpret it by focusing on any part of the image field afterwards. Light field, defined by Arun Gershun in 1936 and first used in photography by a team of Stanford researchers in 1996, is part of a broader field of photography know as computational photography, which also includes high-dynamic range (HDR) images, digital panoramic images, as well as the aforementioned light-field cameras and technology. The images produced in computational photography are "not a 1:1 record of light intensities captured on a photosensitive surface, but rather a reconstruction, based on multiple imaging sources" (Maschwitz, 2015). To clarify, optical digital image capture in general is not a 1:1 record of the light entering thru the camera lens. Only about one third of the sensor's photosites are used, as "two thirds of the digital image is interpolated by the processor in the conversion from RAW to JPG or TIF" (Mayes, 2015). The way we see with our own eyes, in fact, can be considered more like a lossy JPG than a true 1:1 recording, as "less than half of what we think of as 'seeing' is from light hitting our retinas and the balance is constructed by our brains applying knowledge models to the visual information" (Rubin, 2015).

To take it a step further, computational photography can be done with light captured by a lens-less camera. The optical phased array (OPA) camera that Ali Hajimiri of CalTech is currently working on is an example of this type of innovation. He states, "We've created a single thin layer of integrated silicon photonics that emulates the lens and sensor of a digital camera, reducing the thickness and cost of digital cameras. It can mimic a regular lens, but can switch from a fish-eye to a telephoto lens instantaneously—with just a simple adjustment in the way the array receives light" (Perkins, 2017). There have been some new developments with camera lenses as well. Researchers at the Harvard John A. Paulson School of Engineering and Applied Sciences (SEAS) have developed a flat, metalens that will allow for more optic control of light. "Traditional lenses for microscopes and cameras—including those in cell phones and laptops—require multiple curved lenses to correct chromatic aberrations, which adds weight, thickness and complexity," said Federico Capasso, Robert L. Wallace Professor of Applied Physics and Vinton Hayes Senior Research Fellow in Electrical Engineering (Burrows, 2017).

What computational photography really does is to allow the final image to be reinterpreted in ways never seen before in the industry. We are starting to

see this in the Augmented Reality (AR), or Mixed Reality (MR), apps that can be applied to camera phone images. For AR to work on digital photographs, an element of 3D space needs to be identified along with the 2D image. Though difficult to do on the limited size of the camera phone, Apple's ARKit, as well as ARCore for Android phones, are software solutions that use computational photography innovation (Nield, 2017). And some cameras can now capture images for you using AI. Google's Clips uses an AI chipset in the camera that "learns familiar faces, then favors those people (and pets!) when deciding when to take pictures. It looks for smiles and action, novel situations and other criteria" (Elgan, 2017). Using the fixed 130-degree lens, the camera takes bursts of 15 still frames, allowing the user to later pick and choose which ones to keep.

The disconnect between the optical capture of old, which itself was an interpretation of the actual object being photographed—a "willful distortion of fact," as eloquently stated in 1932 by Edward Weston, and the new data capture methods has opened a realm of possibilities (Weston et al., 1986). The old camera obscura method that has been in place for upwards of 160 years, and used in both analog and digital image capture methods, with light passing thru the lens aperture onto a light recording mechanism, has changed. And digital photography has changed with it, as it enters "a world where the digital image is almost infinitely flexible, a vessel for immeasurable volumes of information, operating in multiple dimensions and integrated into apps and technologies with purposes yet to be imagined" (Mayes, 2015).

In the past, photography represented a window to the world, and photographers were seen as voyeurs—outsiders documenting the world for all of us to see. Today, with the image capture ability that a mobile phone puts in the hands of billions, photographers are no longer them—outsiders and talented specialists. Rather, they are us—insiders that actively participate in world events and everyday living, and, thanks to social media sites such as Snapchat, Whatsapp, Facebook, and Instagram, we share this human condition in a way that was never seen before. This participation also falls under the disruptive innovation column. In *Bending the Frame*, Fred Ritchin writes, "The photograph, no longer automatically thought of as a trace of a visible reality, increasingly manifested individuals' desires for certain types of reality. And rather than a system that denies interconnectedness, the digital environment emphasizes the possibility of linkages throughout—from one image to another" (Ritchin, 2013). But Ritchin and others have questioned if the steady stream of "linked" images posted to sites like Instagram are really photographs. These images, of mostly food, dogs, and cats as documented in the SelfieCity London study, instead represent a flow of information, and are related to cinema more than they are to a static photograph (Ritchin, 2015).

This argument about what is or isn't a photograph, in part spurred on by technologic advances in the medium, has played out before, starting with George Eastman's mass production of the dry plate film in 1878 (prior to this innovation, photographers made their own wet-collodion plates). Lewis Carroll, noted early photographer and author, upon examination of a dry plate film, stated, "Here comes the rabble" (Cicala, 2014). Soon after, film became the capture method of choice in photography and the term "snapshot," defined as shooting without aiming, came into vogue at the turn of the 20th century (Rubin, 2015). "No one likes change," freelance photographer Andrew Lamberson says. "People who shot large format hated on the people who shot medium format, who in turn hated on the people who shot 35mm, who in turn hated on people who shot digital" (McHugh, 2013).

There is already a rumble stating that digital photography is dead. But this dystopian view is usually based on areas of stagnant innovation, based on metrics that once drove camera upgrades, such as megapixel counts and sensor sizes (Iqbal, 2017). Based on the exciting innovations in light-field and computational photography, the use of AR, MR, and AI, and the possibilities that lens-less cameras bring, it seems that the industry is just undergoing a mediamorphosis and is not actually dying. To drive that point home, an estimated 234,000 million images where just uploaded to Facebook while you read the last two sentences (Aslam, 2018c)!

Some post-shutter release developments that are not a function of the camera itself, but have still contributed to this innovative disruption, have occurred mainly in the mobile phone arena. Apps for image manipulation continue to evolve. Two interesting ones, among the thousands available, are Prisma and DeepArt. Both apps manipulate images captured on your camera phone using AI to "recreate the image using a style of your choice" (Motley, 2017). Google's Cardboard Camera app allows any iOS (v. 9.0 or higher) or android phone (v. 4 or higher) to capture still digital images on a continuous 360-degree plane. These images are then stitched together by the software to produce one's own interactive VR 360 experience.

Current Status

The number of digital images taken each year continues to grow at an exponential rate, increasing from 350 billion in 2010 (Heyman, 2015) to an estimated 1.4 trillion in 2020 (Lee, 2016). This means that every two minutes, humans take more photos than ever existed in total during the first 150 years of photography (Eveleth, 2015). It is estimated that 85% of the 1.4 trillion images taken in 2020 will be taken on mobile phones (Cakebread, 2017). In fact it's estimated that "more than 90% of all humans who have ever taken a picture, have only done so on a camera phone, not a stand-alone digital or film-based 'traditional' camera" (Ahonen, 2013). So, what happens to all these captured digital images? Of the estimated 1 trillion images taken in 2015, it is estimated that 657 billion of them were uploaded to social media sites (Eveleth, 2015). Thus, the vast majority of these images will not be printed, but will instead "jump to screen," as we view, transfer, manipulate, and post these images onto high-definition televisions (HDTV), 5K monitors, computer screens, tablets, and mobile phones via social media sites. Because of the Wi-Fi capabilities of our mobile phones, we send images via email, post them in collaborative virtual worlds, or view them on other mobile phones, tablets, and handheld devices, including DSCs and DSLRs. Some of these images will only "jump to screen" for a limited time, due to the increased use of photo apps such as Snapchat, making then somewhat "ephemeral" in nature.

As mobile phone are used more as camera, digital camera purchases have started to level off after half a decade of declining sales. DSLR sales still dropped 9% in the first four months of 2017 as compared to 2016, but other interchangeable lens-type cameras (DSCs and MILCs) grew 50.6 % over the same period. DSLRs still have a larger market share, as 2.4 million DSLRs were sold vs. only 1.3 million MILCs (Grigonis, 2017).

The field of digital imaging and photography has matured since that first DyCam recorded a black and white low-resolution image back in 1992. With the new innovations discussed, it seems that it has now entered a transformative period. The infinite flexibility of the digital image, the information sources that it can carry, and the possibilities for future usage in apps and yet-to-be developed technologies all are indicative of how robust and dynamic the field is. And the exciting thing is that we are all able to take part in this growth, as the only tool we need to participate is in our hands.

Factors to Watch

The End of Moore's Law?

Well, maybe, as the technology had been stuck at a 14 nm process since 2014. But chip manufacturer Intel thinks the expectations set by Moore's Law decades ago (see Chapter 11) needs to be revised as the rate of smaller-scale chips has slowed. By looking at their manufacturing techniques, Intel finally delivered a 10 nm chip in 2017, and hope to continue the tenants of Moore's Law, albeit slower, by reducing the scale of its chip designs (Pressman, 2017).

Film Makes a Comeback

The analog world continues to make a comeback. Evidenced by the rise of vinyl sales, that, according to Nielsen, have grown 260% since 2009 (Farace, 2015), film photography is now experiencing a similar bump. Fujifilm's instant film camera Instax outsold their own digital cameras four to one in 2015 and 2016 (Iqbal, 2017).

Advances in Camera Battery Technology

While Shree K. Nayar of the Columbia Vision Laboratory at Columbia University works on a camera that can be powered by the same light that produces the image (Gershgorn, 2015), researchers at the University of Central Florida are trying to improve battery technology. With the goal of charging your digital camera's battery in seconds, and have it last for day, the researchers are creating supercapacitors that can quickly transfer electrons to newly discovered, two-dimensional materials. "If they (sic) were to replace the batteries with these supercapacitors, you could charge your mobile phone in a few seconds and you wouldn't need to charge it again for over a week," says researcher Nitin Choudhary (Zhang, 2016).

Sci-Fi Comes to Life

If you've ever watched the British television series Black Mirror, you may be familiar with the episode *The Entire History of You*, where a memory implant records your life through a contact lens camera, with the ability to play it back at any time. That idea might not be as far-fetched as one might have thought in 2011 when that episode aired. The aforementioned "contact lens camera" technology coupled with ground-breaking intelligent camera technology currently being worked on by Mimi Zou, as well as the Japanese company Neurowear, might make the idea of that episode a reality. Zou's Biometric Sensing camera can be operated by your eye movements (blink twice to capture an image, for example), while Neurowear's technology can sense brainwave activity. This activity—your brain likes a certain scene that your eyes frame—triggers a camera to capture said scene (Scoblete, 2017).

Getting a Job

The job growth market for digital photographers is expected to decline 8% over the next eight years. Reasons given for the lack of growth were the decreasing cost of cameras coupled with the increased number of amateur photographers. However, opportunity for growth is seen in some sectors of the field. Drone photography and portrait photography are two areas where demand may increase, and it is expected that self-employed photographers will grow 7% over the same span. Perhaps self-employment is a survival mechanism, as the decline in the newspaper industry will lead to a projected decline of 34% for photojournalists over the same eight years (Bureau of Labor Statistics, 2017).

Bibliography

Aaland, M. (1992). *Digital photography*. Avalon Books, CA: Random House.

Ahonen, T. (2013). The Annual Mobile Industry Numbers and Stats Blog. *Communities Dominate Brands*. Retrieved January 15, 2018 from http://communities-dominate.blogs.com/brands/2013/03/the-annual-mobile-industry-numbers-and-stats-blog-yep-this-year-we-will-hit-the-mobile-moment.html.

Aslam, S. (2018a). Instagram by the Numbers: Stats, Demographics & Fun Facts. *Omnicore*. Retrieved from https://www.omnicoreagency.com/instagram-statistics/.

Aslam, S. (2018b). Snapchat by the Numbers: Stats, Demographics & Fun Facts. *Omnicore*. Retrieved from https://www.omnicoreagency.com/snapchat-statistics/.

Aslam, S. (2018c). Facebook by the Numbers: Stats, Demographics & Fun Facts. *Omnicore*. Retrieved from https://www.omnicoreagency.com/facebook-statistics/.

Brook, P. (2011). Raw Meet: Fred Ritchin Redefines Digital Photography. *Wired*. Retrieved from http://www.wired.com/rawfile/2011/09/fred-ritchin/all/1.

BT Online. (2017). WhatsApp users share 55 billion texts, 4.5 billion photos, 1 billion videos daily. *Business Insider*. Retrieved January 6, 2018 from http://www.businesstoday.in/technology/news/whatsapp-users-share-texts--photos-videos-daily/story/257230.html

Bureau of Labor Statistics. (2017). Photographers. Job Outlook. *Occupational Outlook Handbook*. Retrieved from https://www.bls.gov/ooh/media-and-communication/photographers.htm#tab-6.

Burrows, L. (2017). Flat lens opens a broad world of color. New & Events. Retrieved from https://www.seas.harvard.edu/news/2017/02/flat-lens-opens-broad-world-of-color.

C/Net. (2018). Faded Photo Editor for iPhone. *Digital Photo Tools* Retrieved from http://download.cnet.com/Faded-Photo-Editor/3000-12511_4-76282908.html.

Cakebread, C. (2017). People will take 1.2 trillion digital photos this year—thanks to smartphones. *Business Insider*. Retrieved January 6, 2018 from http://www.businessinsider.com/12-trillion-photos-to-be-taken-in-2017-thanks-to-smartphones-chart-2017-8.

Canon. (2018). Technology Used in Digital SLR Cameras. *Canon*. Retrieved from http://www.canon.com/technology/now/input/dslr.html.

Carter, R. L. (2015a). *DigiCam History Dot Com*. Retrieved from http://www.digicamhistory.com/1980_1983.html.

Carter, R. L. (2015b). *DigiCam History Dot Com*. Retrieved from http://www.digicamhistory.com/1995%20D-Z.html.

Carter, R. L. (2015c). *DigiCam History Dot Com*. Retrieved from http://www.digicamhistory.com/1996%20S-Z.html.

Cicala, R. (2014) Disruption and Innovation. *PetaPixel*. Retrieved from http://petapixel.com/2014/02/11/disruption-innovation/.

Constine, J. (2017). Facebook is also adding Facebook's facial recognition now finds photos you're untagged in. *Tech Crunch*. Retrieved from https://techcrunch.com/2017/12/19/facebook-facial-recognition-photos/.

Curtin, D. (2011). How a digital camera works. Retrieved from http://www.shortcourses.com/guide/guide1-3.html.

Deloitte. (2016). 3.5 million photos shared every minute in 2016. Press Release.

Digital Camera (2018). Mirrorless Camera Sensor Sizes. *Database*. Retrieved from https://www.digicamdb.com/mirrorless-camera-sensor-sizes/.

Elgan, M. (2017). Google's Clips camera offers a snapshot of things to come. *Mobile*. Retrieved from https://www.computerworld.com/article/3230132/mobile-wireless/google-s-clips-camera-offers-a-snapshot-of-things-to-come.html.

Eveleth, R. (2015). How Many Photographs of You Are Out There In the World? *The Atlantic*. Retrieved from http://www.theatlantic.com/technology/archive/2015/11/how-many-photographs-of-you-are-out-there-in-the-world/413389/.

Farace, J. (2015). 7 Trends That Will Change Photography Next Year: Camera & Technology Preview For 2016. *Shutterbug*. Retrieved from http://www.shutterbug.com/content/7-trends-will-change-photography-next-year-camera-technology-preview-2016.

Four Thirds. (2017). Overview. *Four Thirds*: Standard. Retrieved from http://www.four-thirds.org/en/fourthirds/whitepaper.html.

Gershgorn, D. (2015). Photography Without a Lens? Future of Images May Lie in Data. *Lens*. *N.Y. Times*. Retrieved from http://lens.blogs.nytimes.com/2015/12/23/the-future-of-computational-photography/?_r=1.

Grigonis, H. (2017). Camera sales are continuing to pick up in 2017 after years of decline. *Photography*. Retrieved from https://www.digitaltrends.com/photography/camera-industry-cipa-april-2017/.

Harry Ransom Center-The University of Texas at Austin. (2018). The First Photograph. Exhibitions. Retrieved from http://www.hrc.utexas.edu/exhibitions/permanent/.

Heyman, S. (2015). Photos, Photos Everywhere. *N.Y. Times*. Retrieved from http://www.nytimes.com/2015/07/23/arts/international/photos-photos-everywhere.html? r=1.

Hill, S. (2013). From J-Phone to Lumia 1020: A complete history of the camera phone. *Digital Trends*. Retrieved from http://www.digitaltrends.com/mobile/camera-phone-history/.

Horaczek, S. (2016). Sony's New Exmor Stacked Smartphone Camera Sensor Is The First To Use Hybrid Autofocus. *Popular Photography*. Retrieved from http://www.popphoto.com/sonys-new-exmor-stacked-smartphone-camera-sensor-is-first-to-use-hybrid-autofocus.

Iqbal, T. (2017). The Death of Digital Photography as We Know It. PetaPixel. Retrieved from https://petapixel.com/2017/04/04/death-digital-photography-know/.

Ismail, A. (2018). Best Smartphone Cameras 2018. *Tom's Guides*. Retrieved from https://www.tomsguide.com/us/best-phone-cameras,review-2272.html.

Kaplan, J. and Segan, S. (2008). 21 Great Technologies That Failed. *Features*. Retrieved from http://www.pcmag.com/article2/0,2817,2325943,00.asp.

Lee, E. (2016). How Long Does it Take to Shoot 1 Trillion Photos? *Insisghts from InfoTrends*. Retrieved from http://blog.infotrends.com/?p=21573.

Lester, P. (2006). Visual communication: Images with messages. Belmont, CA: Wadsworth.

LSST. (2018). The Large Synoptic Survey Telescope. *Public and Scientists Home*. Retrieved from http://www.lsst.org/lsst.

Maschwitz, S. (2015). The Light L16 Camera and Computational Photography. *Prolost*. Retrieved from http://prolost.com/blog/lightl16.

Mayes, S. (2015). The Next Revolution in Photography Is Coming. Time. Retrieved from http://time.com/4003527/future-of-photography/.

McHugh, M. (2013). Photographers tussle over whether 'pro Instagrammers' are visionaries or hacks. *Digital Trends*. Retrieved from http://www.digitaltrends.com/social-media/are-professional-instagrammers-photographic-visionaries-or-just-hacks/.

Miller, A.D. (2012). A Theology Study of Romans. USA: Showers of Blessing Ministries International Publishing.

Moon, M. (2017). LG V30's camera has the lowest f-stop in a smartphone. *Engadget*. Retrieved from https://www.engadget.com/2017/08/10/lg-v30-camera-f-stop-aperture/.

Motley, J. (2017). From photo manipulation to beer recipes, AI is finding its way into everything. TheNextWeb. Retrieved January 8, 2018 from https://thenextweb.com/contributors/2017/11/28/photo-manipulation-beer-recipes-ai-finding-way-everything/.

Muller. M. (2018). Selfie 'Statistics'. *Infogram*. Retrieved from https://infogram.com/selfie-statistics-1g8djp917wqo2yw.

National Press Photographers Association. (2017). NPPA Code of Ethics. Retrieved from https://nppa.org/node/5145.

Nield, D. (2017). Everything You Need to Know About Augmented Reality Now That It's Invading Your Phone. *Field Guide*. Retrieved January 8, 2018 from https://fieldguide.gizmodo.com/everything-you-need-to-know-about-augmented-reality-now-1809069515.

Perkins, R. (2017). Ultra-Thin Camera Creates Images Without Lenses. *News*. Retrieved from http://www.caltech.edu/news/ultra-thin-camera-creates-images-without-lenses-78731.

Pictures that lie. (2011). *C/NET News*. Retrieved from https://www.cnet.com/pictures/pictures-that-lie-photos/24/.

Pierce, D. (2017). REVIEW: LIGHT L16. *Gear*. Retrieved from https://www.wired.com/review/light-l16-review/.

Pressman, A. (2017). Here's How Intel Is Finally Getting Back on Track With Moore's Law. Fortune. Retrieved from http://fortune.com/2017/01/05/intel-ces-2017-moore-law/.

Reis, R. et al. (2016). Making the Best of Multimedia: Digital Photography. *Writing and Reporting for Digital Media*. Dubuque, IA: Kendall Hunt.

Ritchin, F. (2013). Bending the Frame: Photojournalism, Documentary, and the Citizen. New York, NY: Aperture Foundation.

Ritchin, F. (2015). Is Instagram Photography? Does it Matter? *International Center of Photography*. Retrieved from http://www.icp.org/perspective/is-instagram-photography-does-it-matter.

Rubin, M. (2015). The Future of Photography. *Photoshop Blog*. Retrieved from https://theblog.adobe.com/the-future-of-photography/.

Rutkin, A. (2015). Facebook can recognise you in photos even if you're not looking. *New Scientist*. Retrieved from https://www.newscientist.com/article/dn27761-facebook-can-recognise-you-in-photos-even-if-youre-not-looking/#.VYlaDBNViko.

Scoblete, G. (2017). What Does the Future of Photography Look Like? *PDNPULSE*. Retrieved from https://pdn-pulse.pdnonline.com/2017/08/predictions-for-the-future-of-photography.html.

SelfieCity. (2016). London Selfie Demograhics. Retrieved from http://selfiecity.net/london/#intro.

Strategic Insights. (2018). The Top 20 Valuable Facebook Statistics. *Zephoria Digital Marketing*. Retrieved from https://zephoria.com/top-15-valuable-facebook-statistics/.

Sylvester, C. (2018). 2017 Camera Wrap-Up – Where Have all the Cameras Gone. Retrieved from http://blog.infotrends.com/?p=23641.

Upright. (2018). 7+ Best Photo Editing Apps for Spotless Facial Retouching. *The Future Photographer*. Retrieved from http://thefuturephotographer.com/7-best-photo-editing-apps-for-facial-retouching/.

Weston, E. et al. (1986). Edward Weston: Color Photography. Tucson, AZ: Center for Creative Photography.

Wild, C. (2018). 1839. The First Selfie. *Mashable*. Retrieved from http://mashable.com/2014/11/07/first-selfie/#NMKmtmAcnsqw.

Zhang, M. (2010). The World's First Digital Camera by Kodak and Steve Sasson. PetaPixel. Retrieved from https://petapixel.com/2010/08/05/the-worlds-first-digital-camera-by-kodak-and-steve-sasson/.

Zhang, M. (2016). Future Camera Batteries Might Charge in Seconds and Last for Days. PetaPixel. Retrieved from https://petapixel.com/2016/11/23/future-camera-batteries-might-charge-seconds-last-days/.

eHealth

Heidi D. Blossom, Ph.D. & Alex Neal, M.A.*

Overview

The convergence of digital technologies in the area of healthcare is creating an age of digital health that will be highly individualized through the use of mobile communication, wireless sensors, wearable wireless devices, and super networks of big data. This convergence allows diagnostics not solely based on symptoms, but rather on up-to-the-minute individual health factors. Micro sensors built into clothing and wearable devices send health data to Internet capable devices or directly to the cloud and connect users, caregivers, and healthcare providers to vital signs that could allow for better healthcare decisions. This is "Big Health Data," and it is bringing our health information together in ways that will improve personal health and aid in advancements in artificial intelligence and robotics which have the potential of saving lives through innovative diagnostic and treatments. No matter how you look at it, eHealth is driving healthcare transformation with technological innovations that are saving lives and reducing healthcare costs.

Introduction

When was the last time you were at the doctor, the emergency room, or any other healthcare provider's office? When you were there, how many encounters with technology did you have? Did you fill out patient information online? Did the nurse check your blood pressure and temperature with a digital device? Did the doctor use a computer or tablet in the exam room? Did the prescription you received go directly to the pharmacy via electronic delivery? When you left the office, did you think about the diagnosis and treatment and seek further clarification or information online or on a mobile app? This is eHealth—the use of electronic information and communication technologies in the field of healthcare (Cashen, Dykes, & Gerber, 2004).

eHealth affects everyone and is one of the fastest growing areas of innovations in communication technology. Throughout this textbook you have read about the impacts of the digital revolution and how it has affected every aspect of our lives. The rapidly decreasing size of computer technology in the form of micro-computers, near ubiquitous Internet access and increasing bandwidth speeds, the connectedness of social media, and cloud computing have created the perfect storm for eHealth to burst on the scene and fundamentally change healthcare as we once knew it (Topol, 2013).

Essentially, eHealth is empowering consumers to play a bigger role in managing their own health. Patients now have access to their own electronic health records (EHR) and can access information about symptoms, diagnostics, and treatments about every common illness. This shifts the balance of healthcare

---

* Blossom is Associate Professor and Chair and Neal is Instructor, Dept. of Mass Communication, North Greenville University (Greenville, South Carolina).

and creates a way for individuals to take control of their own health by focusing on personalized health plans and prevention.

Definitions

- *eHealth*—the use of electronic information and communication technologies in the field of healthcare.

- *mHealth*—the use of mobile technologies such as cellphones and tablets for health communication and delivery of health information.

- *eHealth Games*—electronic games used to promote health and wellness.

- *Telehealth*—the use of telecommunications and virtual technologies to deliver health services outside of the traditional healthcare setting (WHO, 2017).

Background

Convergence of digital technologies has given health communication increased traction over the past 30 years. However, the history of medicine is rich with information-sharing from the earliest days of recorded history. The first recorded evidence of healthcare predated the use of paper, pens, or even books. They were chronicled in cave paintings in France dating back thousands of years (Hall, 2014). The cave art depicted the use of plants to treat ailments. Early evidence of surgeries and the use of anatomy in diagnosis were found in Egypt around 2250 BC (Woods & Woods, 2000). These drawings were some of the first forms of health communication because they documented health conditions and treatments for future generations of physicians.

It wasn't until the 19th century that the field of medicine advanced exponentially with developing sciences and growing knowledge of chemistry, anatomy, and physiology. The mass dissemination of information can be credited to innovations in mass communication. Medical information was communicated through the use of telegraphs, printed journals, books, photography, and telecommunications that delivered medical information to doctors and healthcare workers (Kreps et.al, 2003).

Computers were introduced in healthcare in the early 1970s and served as an administrative tool to store patient information and medical practitioner records. Developments in surgical and diagnostic instruments continued, but it wasn't until 1995 that the information floodgates were opened to provide individuals access to a pool of knowledge about every aspect of their healthcare through the Internet (Kreps et al., 2003).

Innovations in communication technology also created concerns regarding the protection of patient information on the Internet. In 1996, the Health Insurance Portability & Accountability Act (HIPAA) propelled the healthcare industry to adopt new ways of safeguarding the privacy of health records. HIPAA gave more control of information to the consumer, but because of the regulated safeguards, the healthcare industry has been slow to adopt technologies that would give patients ready access to their own medical records. The days of limited access to personal health information are long gone as innovations of communication technologies that are designed to facilitate the flow of information between the healthcare provider and the healthcare consumer have been established.

Recent Developments
mHealth

93% of physicians believe that mHealth apps can improve patient's health (Greatcall, 2017)

mHealth, or mobile health, is the use of mobile technologies such as cellphones and tablets for health communication and delivery of health information. Gone are the days when cell phones were used to simply make a call; cellphones are multifunctional tools performing as photographic and video devices, word processors, electronic organizers, and now even ECGs (electrocardiograms) or thermometers to monitor your health.

Technologies in mHealth have been featured at major technology shows worldwide for the past few years as an introduction to a rapidly growing industry. That growth can be attributed in part to ubiquitous mobile technologies which connect users to health information. More than 77% of American adults own a smartphone, and that number is expected to continue to rise (Pew, 2018). Worldwide, the number of smartphone users is expected to grow from 2.1 billion in 2016 to around 2.5 billion users by 2019 (Statista, 2017).

Apps

More than 6.1 billion users are expected to use mobile apps on their smartphones by the year 2020, and the healthcare and business sectors are quickly jumping on the opportunity to engage individuals through mHealth apps (Ericsson, 2016). There are more than 325,000 mHealth apps available for Apple and Android users with 78,000 of those apps being added in 2017 (R2G, 2018). These apps allow smartphones to carry out diverse eHealth functions from monitoring physical activity to monitoring heart rhythms with a mobile ECG (Costello, 2013).

Figure 18.1
Fitbit Health/Fitness Apps

Source: Fitbit

mHealth apps make it possible for patients to access their medical records on the go. It used to be that if patients wanted to check their medical records, they would have to logon to each medical provider's website and search for the information needed. Apple is the first tech company to offer users the ability to sync their medical records with the Health Records section within the Health app. Johns Hopkins Medicine, Cedars-Sinai, and other participating hospitals and clinics are among the first to make this feature available to their patients (Apple, 2018).

App publishers consistently identify those with chronic illness as their major target group, with people interested in health and fitness as the secondary focus (R2G, 2016). Corporations and makers of mHealth apps have found that their success comes

from consistent engagement with the use of gamification, which is the application of game playing to encourage engagement with a product or service (Weintraub, 2012). Gamification serves as a way to promote and sustain healthy behaviors by using strategies such as goal setting, feedback, reinforcement, progress updates, and social connectivity (Edwards et al., 2016). Because of past success, makers of mHealth apps continue using gamification to keep users engaged in healthy lifestyles, including diet and exercise.

Social Media

The healthcare industry has turned to social media to share information that will promote both individual and community health practices. Hospitals, physicians and healthcare providers are connecting with patients on social media providing connections with preventative healthcare, health crisis readiness, and general healthcare information on popular social media sites such as Facebook, Instagram, Twitter, YouTube, and Pinterest. With seven in ten American adults using social media, it's not surprising that the healthcare industry would use social media to connect with patients (Pew, 2018). The majority of hospitals have a Facebook page (94.4%) and Twitter account (51%) while nearly all have a Yelp page for reviews (99%) and Foursquare page for patients to check-in online (99%). While the vast majority have adopted social media, the consistent use is more prevalent among larger hospital systems in urban areas (Griffis, et al., 2014).

Research shows that there is a high level of trust with social media. Nearly 90% of respondents between the ages of 18 to 24 said they would trust medical information shared by others on their social media networks (PWC, 2015). Another 41% have used social media to make decisions about which doctors or hospitals to use while 34% said the information they find in social networks affects their decision about what medication to take (PWC, 2015).

Big Data

The next big breakthroughs in healthcare may come from advances in technology that utilize Big Data. Every time you use your cell phone, watch a video online, use an App, interact on social media, make a purchase with a credit card, or fill out a form online, that data is collected and stored for companies and governments to use to understand customer behaviors and preferences.

Big Health Data is a subset of Big Data and is derived from electronic health records (EHR), mHealth device, health sensors, social networking data, clinical notes, medical imagery, lab results, and medical research. Big Health Data is the new normal for healthcare organizations that want to understand their customers, but there are issues. Most Big Health Data is collected and stored in massive data centers where it sits, mostly unused leaving the potential for great insight and discoveries untapped. In fact, more than 88% of Big Health Data collected is uncategorized data that is ultimately unusable because there is no way to make reliable connections that would lead to usable insights (Reddy, 2017).

Data scientists are busy extracting usable health data because it is believed that the analysis of this data has life-saving implications for patients, from managing long-term care to discovering innovative cancer treatment plans. Data scientists are making big discoveries using artificial intelligence, or AI, to process and analyze both structured and unstructured data that exists to reveal patterns, trends, and insights that could lead to innovations in healthcare. It is believed that the analysis of this data has life-saving implications for patients, from managing long-term care to discovering innovative cancer treatment plans.

Artificial Intelligence

Artificial Intelligence is one of the most promising areas of technological advancement in eHealth. AI are computer systems that are designed to mimic the human brain's ability to learn, process information, and adapt to change. AI machines such as IBM's Watson are being used in medical research and have been found to be accurate in diagnosing critical ailments such as cancer and have advised on innovative life-saving treatment options that medical specialists had not discovered.

Researchers in the U.K. used AI to find predictions of heart disease in patients. The AI analyzed 10 years of patient data and discovered 22 predictors of heart disease including ethnicity, arthritis, kidney dis-

ease, age, and other factors that had not been considered. Researchers compared what the AI had found to current patients and found that the AI predicted heart disease more accurately than the standard and accepted clinical guidelines. The AI also identified the strongest predictors for heart attacks that were not included in the American Heart Association guidelines (Hutson, 2017). The advances in AI in healthcare are moving at a fast pace as scientists discover new diagnostics, treatments, and pharmaceutical discoveries that could potentially save the healthcare industry in the U.S. more than $150 billion a year by 2026 (Collier et al., 2017).

eHealth Games

Video games are mostly a sedentary activity, but there is a growing trend for companies to use video games to promote health education, physical activity, physical therapies, mental health treatments, and to generally encourage a healthy lifestyle. Major video game companies such as Nintendo, Sony, and Microsoft have developed consoles, games, and accessories that not only provide entertainment enjoyment, but also improve overall quality of life (QOL) by engaging consumers with health literacy and education to help them take control of their health (Nintendo, 2015).

eHealth gaming is being used by 70% of the top companies in the world and is expected to bring in $5.5 billion in revenue in 2018 (Lewis, 2018). The Nike+ program enables users to compare exercise results and challenge friends to earn 'fuel points.' Nike+ is a growing community of more than 11 million subscribers (Lewis, 2018). Another example involves the monitoring of chronic illnesses such as diabetes. The Bayer Didget was the first blood glucose monitor designed specifically for children with Type 1 diabetes. The unit connects to Nintendo gaming systems and rewards children for consistent testing of their blood sugar (Bayer, 2014).

The most common use of eHealth games is in health and nutrition. Users track nutrition, calorie intake, and physical activity and earn points, badges, and virtual rewards while engaging with family, friends, and other online competitors. Companies such as Fitbit have created communities of healthy competitors that use competition to help individuals achieve their own health goals while also motivating competitors to achieve more.

Another use of eHealth games is for exergaming which combines the interactivity of a video game with physical exercise. Gaming systems such as Nintendo Switch or Xbox 360 use technology that incorporates motion detecting cameras or handheld motion sensors to measure movement and physical activity of users. Popular video games such as Just Dance or Kinect Sports involves exergaming activities such as golf or bowling and pits competitors against each other in physical competitions.

Exercise equipment such as treadmills, elliptical machines, or stationary bikes commonly use exergame integration to simulate races or challenges that encourage users to compete for points or rewards. Most recently, exergames are incorporating virtual reality (VR) or augmented reality (AR) to provide a more immersive experience for the user.

Consumers can expect to see more innovations in eHealth games in the future as different organizations tap into the success of eHealth games. Major health insurance companies such as UnitedHealthcare, CIGNA, and Kaiser Permanente use eHealth games in their wellness programs to encourage group participation in healthy activities, such as diet and exercise, while also lowering costs resulting from claims. Participants are encouraged to participate to earn points, badges, monitor their progress online, and compare their results with other users. The results have been positive. Blue Shield claimed that 80% of its employees that participated in their program had a 50% drop in smoking which ultimately results in lower healthcare costs (Lewis, 2018).

Wearables

Wearables are eHealth technologies that record and communicate biometric data regarding your heart rate, blood pressure, glucose levels, and more. Wearables include clothing, wristbands, watches, earphones, sensor rings, computerized contact lenses, and dermal patches equipped with sensors that wirelessly collect and transmit health data over extended periods of time with minimal lifestyle disruptions. This is one of the fastest growing areas of eHealth with one in six Americans owning some form of wearable

technology (Piwek, et. al., 2016). The wearable market is expected to grow in sales from $10 billion in 2017 to almost $29 billion by 2022 (CCS, 2018).

Pedometers and Wristbands

The most widely used wearable device for health and fitness is the pedometer, which measures the number of steps an individual has taken. The pedometer has undergone several levels of innovation and has evolved into a more powerful digital device. Wristbands dominate the wearable fitness device market, and many companies have designed wristbands that wirelessly communicate with smartphones to measure steps, physical activity, sleep patterns, and calories burned.

The Federal Drug Administration (FDA) has issued guidelines for regulatory review of health apps intended to be used as an accessory to a regulated medical device or transform a mobile platform into a regulated medical device. The only mobile apps that are currently regulated by the FDA are those that are used as an accessory for a regulated medical device or transform the mobile platform into a regulated medical device (FDA, 2016).

Figure 18.2
Wristbands and Smartwatches

Source: Fitbit

Smartwatches

Smartwatches combine the functionality of simple timekeeping with sensor and computational power that can monitor biometrics and instantly record and report the data to your mobile device or online monitoring account. Smartwatches can perform a myriad of tasks including tracking steps taken, distance traveled, activity rates, tracking sleep patterns, and alerting the user when they have achieved health goals. Smartwatches are the fastest growing eHealth wearable product in the health and fitness technology sector. In 2017, the Apple Watch sold more than 17 million units making it one of the best-selling wearable trackers in the world (Canalys, 2018). Other health and fitness wristbands available include Fitbit Ionic, Garmin Vivosmart 3, and Samsung Gear Sport which all work similarly by recording physical data and connecting wirelessly to a smartphone loaded with the designated app.

Developers are working on advanced sensors that will take eHealth wearables to a new level in health monitoring. Smart Monitor's Smartwatch Inspyre™ is a device with mobile connectivity that alerts the user of impending convulsions, tremors, or seizures. The watch connects with a smartphone and alerts users when repetitive motion is detected and then connects with chosen emergency contacts with alerts and GPS coordinates of where the individual SmartWatch user is located. Other data collected by the SmartWatch is then compiled for review by the user and can be shared with a healthcare provider (SmartMonitor, 2018).

Smart Clothing

Smart clothing is apparel with embedded sensors and other electronics that communicate precise health metrics to a smart device through Bluetooth or Wi-Fi connectivity. Smart clothing is not a new concept but is still in its infancy. In 1984, Adidas introduced the first smart running shoe that measured distance, speed, and calories burned from a sensor located in the tongue of the shoe (Bengston, 2015). Since that time, tech companies have innovated sensor and electronic enhanced fabrics and textiles and are incorporating them into socks, shirts, yoga pants, running or biking shorts, sports bras, and other smart clothing to measure biometrics as well as kinesthetic techniques such as cadence and movement for specific sports.

Figure 18.3
Sensor Clothing

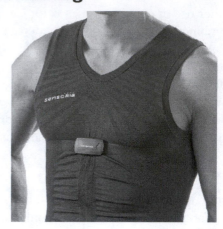

Source: Sensoria Fitness

The adoption of smart clothing is expected to grow by more than 550% by 2022 as specialized smart clothing finds its place with the consumer (Moar, 2018). Regulations and safety concerns are the biggest hindrances for this industry. Rollouts of new products take time in the research and development stages and are sometimes held back because of government regulations that are catching up with the pace of the industry.

Dermal Devices

While wearables such as wristbands have definitely moved into the forefront of eHealth hardware technology, the next innovation in wearables is found in dermal technologies that seamlessly integrate with the body allowing freedom of movement and more precise health monitoring (Ahlberg, 2011).

Harvard University researchers have developed Dermal Abyss, a biosensitive ink that allows smart tattoos to change color as changes in the body occur. The inks measure glucose, sodium and pH concentrations in the interstitial fluid surrounding cells. Dermal Abyss has a green ink that gets darker as the patient's sodium levels increase. The company is testing a pink ink that turns blue as the body's pH levels rise (Nelson Jr., 2017). The Dermal Abyss tattoo inks can be incorporated into long lasting tattoos for chronic conditions or temporary designs for short-term monitoring. Researchers are still developing and testing new inks but have already made great strides. One of the major areas of use of Dermal Abyss ink is for diabetics who could check their glucose level by looking at the color of a tattoo and avoiding having to draw blood from finger pricks several times a day (Vega et al., 2017).

Telehealth

- 71% of healthcare providers use Telehealth tools (HIMSS Analytics, 2017).

- *The Creating High Quality Results & Outcomes Necessary to Improve Chronic Care Act* (CHRONIC) was signed into law on February 9, 2018 and sets out to expand telehealth options for 19 million Americans.

Telehealth involves the use of telecommunications and virtual technology to deliver health services outside of the traditional healthcare setting (WHO, 2017). This is an umbrella term that is used to refer to the delivery of a multitude of healthcare services such as long-distance medical care, health education for patients and professionals, public health, health administration, and prevention (La Rosa, 2016). Trends in telehealth are now moving from hospitals and satellite clinics to the home and mobile devices (Dorsey & Topol, 2016).

Telemedicine

Telemedicine is a type of telehealth and involves the use of clinical services to patients from a distance through teleconference exams, shared diagnostic data, as well as phone and computer-mediated conversations (du Pré, 2017). Telemedicine is favored by young and affluent patients who are more comfortable utilizing technology (du Pré, 2017). This type of healthcare is most often used for recurrent concerns, such as getting a prescription refilled (Uscher-Pines & Mehrotra, 2014).

Telemedicine has grown exponentially since the first remotely-performed surgery in 2001 and has affected many aspects of healthcare (Collen & Ball, 2015). Telemedicine provides a convenient and consistent way to transmit medical information, imaging, and other data a healthcare provider might need for diagnostics and prescriptive assessments (Topol, 2016).

Telemedicine is not just a healthcare option for people living in rural or remote locations. It is now being used to connect healthcare practitioners with patients as a means of healthcare cost savings by meeting patients wherever they are through a mobile device or computer connectivity. Companies such as Doctors on Demand provide access to board certified doctors or psychologists at a much lower cost than an urgent care facility.

Figure 18.4

Doctors on Demand

Source: Doctors on Demand

Telemedicine could help reduce healthcare costs by more than $6 billion a year, according to analysis by Towers Watson (2013), a global financial and technological consulting firm. Towers Watson found that more than 78% of employers offer telemedicine services as an alternative to ER or doctor's' office visits for non-emergency health issues. While employers see the financial benefits of telemedicine, employees are lagging behind in actual use with less than 10% of employees utilizing this relatively new technology (Watson, 2017).

Voice Activated Personal Assistants

Another technology that has great implications for telehealth is voice activated personal assistants. One in six U.S. adults owns a voice-activated smart speaker, and this technology is set to outpace the adoption of smartphones and tablets in the next few years. Voice activated personal assistants are voice command devices integrated with a virtual assistant that can help users find information, purchase products online, set timers and alarms, control smart home devices, and listen to music and books on command. Personal assistants are increasingly being used as eHealth devices, and the healthcare industry is testing the devices to assist home bound patients with medical information and reminders while also allowing for direct connectivity with the medical providers (Edison, 2017).

Robotics

It is likely that you will encounter a healthcare robot in your near future as robotics are increasingly used for day-to-day care, health monitoring, as well as more complex health procedures such as surgeries. While we will still rely on doctors, nurses, and other clinicians in the future, advances in robotics are making strides to provide more efficient and quality care while also reducing healthcare costs (Kilgannon, 2016).

Robots provide accuracy and precision that are not always attainable by humans. Hospitals are using robots for high precision surgeries resulting in high success with smaller incisions, less pain, reduced complications, and lessened healing time (Nuzzi & Brusasco, 2018). Robots are also being used in pharmacies to help reduce prescription errors. Pharmacists who fill a prescription have to read the prescription, find the medicine, count or measure the prescribed medicine, fill the prescription, and then label the prescription based on the dose and instructions.

When a pharmacist is busy, these steps may be missed and human errors may occur. It is believed that 2.8% of prescriptions have errors. Pharmaceutical robots have reduced the number of errors in prescriptions to 0%, further reducing risks and costs (Bui, 2015).

Rural areas in the U.S. continue to experience shortages in doctors. Fortunately, new technologies in robotics have allowed doctors to meet remotely with patients. Remote robotic access in eHealth is the ability for the doctor to be off-site and project themselves to another location. They have the ability to move, see, hear, and talk as though they were actually there with the patient (Huiner, 2016). These robots can maneuver throughout the hospital making a doctor who is offsite feel as if she is on-site. Ultimately, this technology reduces costs and improves efficiency and safety at hospitals and healthcare facilities (Lee, 2013).

Robots were in operating rooms as early as 1985, with the use of the PUMA 200 industrial robot used for CT-guided brain biopsy (Kwoh et al., 1988). Medical advancements have come a long way since then, with the world's first telesurgical procedure taking place in 2001 in which a doctor located in New York used a joystick to operate a robotic arm in Strasbourg, France (Sheynin, 2016). Telesurgery is fast on its way

to becoming available in "remote areas and rural communities where doctors and formal medical care are scarce and hard to come by" (Sheynin, 2016).

Healthcare facilities use robots to help prevent the spread of disease. With flu epidemics becoming more common in recent years, healthcare facilities are concerned with protecting their patients' health and safety by maintaining germ free facilities. Hospitals across the U.S. have started using disinfecting robots that use xenon ultraviolet (UV) light to destroy bacteria, viruses and other pathogens that can cause the spread of disease.

The Xenex LightStrike Germ-Zapping robot is one of the latest robot models being adopted by hospitals in the U.S. After a normal cleaning has been conducted in a room, the LightStrike robot is brought in to kill any remaining microscopic germs or bacteria by shooting high-intensity UV rays to all surfaces (Bailey, 2018).

Figure 18.5

Xenex LightStrike Germ-Zapping Robot®

Source: Xenex

The implications for telehealth are vast and so are the issues. There is a concern that decreased human interaction could prevent a medical practitioner from identifying issues that may only be recognized in direct face-to-face interactions. The healthcare sector has shown caution in the adoption of newer telehealth technologies because of patient safety and liability concerns because medical advice given remotely without a full physical exam could lead to misdiagnosis and serious or even fatal errors for the patients. Another growing concern is in the area of patient privacy. Patient data breaches cost the U.S. healthcare industry $6.2 billion in 2016 (Ponemon Institute, 2016). The number of patient data breaches will continue to be an issue, and companies using telehealth technologies will need to invest in ways to protect the privacy of patient information.

Developing Countries

Developing countries are set to benefit the most from telehealth technologies. Millions of people die each year in developing countries due to lack of access to proper medical care. The World Health Organization reports a global shortage of 7.2 million healthcare workers in the next 20 years, but that number is expected to nearly double (GHWA & WHO, 2013). Advances in eHealth technologies have the potential of aiding in this acute shortage.

Telehealth expands healthcare access to remote areas of the globe that never dreamed of having modern medical care. Technologies such as live video conferencing, uploading of MRI or other scans to a cloud network, and remote monitoring of vital signs such as body temperature, blood pressure, and glucose levels, are all used by healthcare providers located at a distance and sometimes across the globe. All that is needed is a smartphone or computer device coupled with Internet access at both ends to connect patients with quality healthcare.

In emerging and developing nations around the globe, mHealth is being used where doctors are scarce and resources are non-existent. In developing countries, about a third of the adult population now owns a smartphone (Poushter, 2016). With a projected 5.13 billion mobile phone users worldwide by 2017, innovations in telemedicine could be a solution in these underserved areas (eMarketer, Dec 2014).

Healthcare professionals use of mobile devices around the world is also increasing. A global study of patients found that 44% have seen medical professionals using a mobile device during treatment or diagnosis. In Qatar and Saudi Arabia more than half of consumers have seen mHealth used in treatment or in diagnosis (AVG, 2015).

With a continued projected growth of smartphone users worldwide and the adoption of new applications of use, mHealth can serve as a foundation to economic growth in these developing countries by improving the health of its citizens. The major barriers to the progress of eHealth technologies in developing countries are the lack of technological resources, conflicting health system priorities, and the lack of legal frameworks.

Figure 18.6
Healthcare Worker Using a Cellphone in Ghana

Source: Nana Kofi Acquah and Novartis Foundation

Current Status

The health system in the United States is facing many challenges. The cost of healthcare in the United States is the highest in the world with the average per capita cost reaching US $10,348 (CMS, 2017, and the aggregate healthcare spending in the United States is expected to reach US $4.8 trillion by 2021 (CMS, 2011). On December 20, 2017, the Senate passed a Republican party tax bill, which in part repealed the individual mandate initiated under the *Affordable Care Act* (ACA), better known as "Obamacare." The tax bill eliminates the individual mandate penalty for not purchasing health insurance beginning in 2019.

The *Patient Protection and Affordable Care Act* (PPACA) is fueling eHealth by restricting health spending and creating new business models for healthcare that are more value-based. Historically, the healthcare industry was not rewarded for keeping patients healthy and avoiding expensive treatments; however, PPACA is forcing change to that model by propelling healthcare organizations into the adoption of new technologies to engage and empower patients to take more control of their healthcare. As this chapter was written, the PPACA was still intact, but further legislative action could be taken to repeal the *Act*.

The use of new technologies and apps that engage consumers is not enough. Full adoption of these new technologies as well as new applications of use will take time and will require a change in consumer behavior. These changes will require incentives for use and compatibility with current technologies, making it easy for users to adopt. As the proliferation of new eHealth technologies continues, it may be harder for consumers to navigate the many options available to them.

Security Issues in Digital Health

Security concerns are the biggest inhibitor of adoption of new eHealth technologies. The key security issues in eHealth include security for patient confidentiality, security that enables authentication of electronic health records, and systems security that ensures secure transmission, processing, and storage of health data (PWC, 2017). Electronic Health Records (EHR) were believed to be the solution for improved management of an individual's health, and to be the tool that would give the patient control of their own health by providing access to his or her own medical history (Waegemann, 2002). EHRs have opened the doors for patient access, but have also opened the doors to security risks. The issues of protection of EHRs against intrusion, data corruption, fraud, and theft are of highest concern for healthcare professionals and the patients they serve.

Since 2009 there have been more than 1,437 major breaches of protected health information affecting more than 154 million patient records. Of the 154 patient health records compromised, 113,208,516 occurred in 2015. In 2016, the healthcare industry saw a 320% increase in the number of providers victimized by attackers (Redspin, 2017). Over 80% of patient records breached in 2016 were a direct result of hacking attacks (Redspin, 2017). Patient information will remain a top priority for governmental agencies, healthcare, and the technology sectors.

Factors to Watch

We live in a highly personalized era where technologies have increased access to a wide range of goods and services. For example, you don't have to get in your car and drive to the store to get everything you need to survive in this world. You have access to thousands of health products online and you can have them delivered to your door, in some cases, the same day. Mobile phones have thousands of personalized accessories, case options, and more than 325,000 mHealth apps to choose from—and you don't even have to leave the comfort of your home to see a doctor when you're sick (R2G, 2018). This culture of individualized access has created a generation of consumers that expect products and services that are customized to their individual needs. Watch for eHealth options that are highly individualized as these cultural influences continue to expand into the healthcare sector.

Getting a Job

There are tremendous career opportunities in eHealth. A growing field is health informatics. According to the Bureau of Labor Statistics, employment of health information specialists is expected to grow faster than any other occupation in the U.S., 22% by 2022. The University of South Florida (n.d.) defines health informatics as "the acquiring, storing, retrieving and using of healthcare information to foster better collaboration among a patient's various healthcare providers." Careers in health informatics include nurse informaticist, chief medical information officers, director of clinical informatics, clinical data analyst, and IT consultant. Getting a Master's Degree in Health Informatics or Health Information Management is a good first step. There are also career opportunities in the creation, sales, and management of health-related technologies such as wearables, apps, and monitoring

Bibliography

AVG (2015) Global mHealth & Wearables Report 2015. Retrieved from http://www.mobileecosystemforum.com/solutions/analytics/mef-global-mhealth-and-wearables-report-2015/.

Ahlberg, L. (2011). Smart skin: Electronics that stick and stretch like a temporary tattoo. Physical Sciences News Bureau: Illinois. Retrieved from http://news.illinois.edu/news/11/0811skin_electronics_JohnRogers.html.

Apple (2018) Apple announces effortless solution bringing health records to iPhone. Press Release. Retrieved from https://www.apple.com/newsroom/2018/01/apple-announces-effortless-solution-bringing-health-records-to-iPhone/.

Bailey, W.S. (2018, Feb. 2) Xenex amps up war against potentially deadly infections. Retrieved from https://www.bizjournals.com/sanantonio/news/2018/02/02/boy-xenex-amps-up-war-against-potentially.html?utm_content=1517868915&utm_medium=social&utm_source=multiple.

Bayer. (2014). Didget Meter: Play with purpose. Bayer Diabetes. Retrieved from https://www.bayerdiabetes.ca/en/products/didget-meter.php.

Bengston, Russ (2015) A Brief History of Smart Sneakers. Complex. Retrieved from http://www.complex.com/sneakers/2012/08/a-brief-history-of-smart-sneakers/adidas1-basketball

Bui, Q. (2015). Watch Robots Transform a California Hospital. Planet Money. Retrieved from http://www.npr.org/sections/money/2015/05/27/407737439/watch-robots-transform-a-california-hospital.

CCS (2018) Wearables Market to Be Worth $25 Billion by 2019. Retrieved from https://www.ccsinsight.com/press/company-news/2332-wearables-market-to-be-worth-25-billion-by-2019-reveals-ccs-insight

Canalys (February 6, 2018) 18 million Apple Watches ship in 2017, up 54% on 2016. Canalys. Retrieved from https://www.canalys.com/newsroom/18-million-apple-watches-ship-2017-54-2016

Cashen, M. S., Dykes, P., & Gerber, B. (2004). eHealth Technology and Internet resources: barriers for vulnerable populations. Journal of Cardiovascular Nursing, 19(3), 209-214.

Centers for Medicare & Medicaid Services (2018). National Health Expenditure [Fact sheet]. Retrieved from https://www.cms.gov/research-statistics-data-and-systems/statistics-trends-and-reports/nationalhealthexpenddata/nhe-fact-sheet.html.

Center for Medicare and Medicaid Services. (2011). National Health Expenditures Projections 2011-2021. CMS.gov. Retrieved March 19, 2014 from www.cms.gov/Research-Statistics-data-and-Systems/Statistics-Trends-and-Reports/NationalHealthExpendData/Downloads/Proj2011PDF.pdf.

Collier, M., Fu, R., & Yin, L. (2017). Artificial intelligence: Healthcare's new nervous system. Accenture. Retrieved from https://www.accenture.com/us-en/insight-artificial-intelligence-healthcare.

Collen, M. F. & Ball, M. J. (2015). The History of Medical Informatics in the United States (2 ed) Springer.

Costello, S. (2013), How Many Apps are in the iPhone App Store. About.com/iPhone/iPod. Retrieved from http://ipod.about.com/od/iphonesoftwareterms/qt/apps-in-app-store.htm.

Dorsey, E. R., & Topol, E. J. (2016). State of telehealth. *The New England Journal of Medicine, 375*, 154-161. DOI: 10.1056/NEJMra1601705.

du Pré, A. (2017). Communicating about health: Current issues and perspectives. New York, NY: Oxford University Press.

Edwards, E.A., Lumsden, J., Rivas, C., Steed, L., Edwards, L. A., Thiyagarajan, A., Sohanpal, R., Caton, H., Griffiths, C. J., Munafo, M. R., Taylor, S., & Walton, R. T. (2016). Gamification for health promotion: Systematic review of behavior change techniques in smartphone apps. *BMJ Open, 6*. doi:10.1136/bmjopen-2016-012447.

eMarketer. (2014). Smartphone Users Worldwide Will Total 1.75 Billion in 2014. eMarketer.com Retrieved from http://www.emarketer.com/Article/Smartphone-Users-Worldwide-Will-Total-175-Billion-2014/1010536#hRDqhsPlE3A7dIZA.99.

Edison Research (2017). The Smart Audio Report from NPR and Edison Research, Fall-Winter 2017. (Report) Retrieved from http://nationalpublicmedia.com/wp-content/uploads/2018/01/The-Smart-Audio-Report-from-NPR-and-Edison-Research-Fall-Winter-2017.pdf

Ericsson, A. B. (2016). Ericsson mobility report: On the Pulse of the Networked Society. February 2016.

Food and Drug Administration (2016) Mobile Medical Applications Retrieved from http://www.fda.gov/medicaldevices/digitalhealth/mobilemedicalapplications.

Global Health Workforce Alliance & World Health Organization (2013) A Universal Truth: No Health Without a Workforce: Third Global Forum on Human Resources for Health Report. Retrieved from http://www.who.int/workforcealliance/knowledge/resources/hrhreport2013/en/.

Greatcall (2017) Is Mobile Healthcare the Future? [infographic] Greatcall. Retrieved from http://www.greatcall.com/greatcall/lp/is-mobile-healthcare-the-future-infographic.aspx.

Griffis, H. M., Kilaru, A. S., Werner, R. M., Asch, D. A., Hershey, J. C., Hill, S. & Merchant, R. M. (2014). Use of social media across US hospitals: descriptive analysis of adoption and utilization. *Journal of medical Internet research*, 16(11), e264.

HIMSS (2015) HIMSS Mobile Technology Survey. HIMSS Analytics. 2015. Retrieved from http://www.himss.org/2015-mobile-survey.

HIMSS Analytics. (2017). Essentials brief: 2017 inpatient telemedicine study. Retrieved from http://pages.himssanalytics.org/HA-Resource-Landing-Page.html?resource=3311&site=www.himssanalytics.org.

Hall, T. (2014). History of Medicine: All the Matters. New York, NY: McGraw-Hill.

Huiner, C. (2016) Remote Presence Robotics. InTouch Health. Retrieved from http://www.intouchhealth.com/SurgProtCollat-color_d2.pdf.

Hutson, M. (2017, April 14) Self-taught artificial intelligence beats doctors at predicting heart attacks. Retrieved from *Science*: http://www.sciencemag.org/news/2017/04/self-taught-artificial-intelligence-beats-doctors-predicting-heart-attacks.

Kilgannon, T. (2016) Hospital Adds Robots to Improve Disinfecting Routine. *TWC News*. Retrieved from http://www.twcnews.com/nys/binghamton/news/2016/03/9/robots-disinfect-hospital-rooms.html.

Krebs, P., & Duncan, D. T. (2015). Health App Use Among US Mobile Phone Owners: A National Survey. JMIR mHealth and uHealth, 3(4).

Kwoh, Y. S., Hou, J., Jonckheere, E. A., & Hayati, S. (1988). A robot with improved absolute positioning accuracy for CT guided stereotactic brain surgery. *IEEE Transactions on Biomedical Engineering*, 35(2), 153-160.

La Rosa, F. G. (2016). Telehealth and telemedicine: A new paradigm in global health. Proceedings from Colorado School of Public Health Global Health Lecture Series. Aurora, Colorado.

Lee, J. (2013) Robots get to work: More hospitals are using automated machines, but jury's still out on success. *Modern Healthcare*. Chicago, IL. Retrieved from http://www.modernhealthcare.com/article/20130525/MAGAZINE/305259957.

Lewis, C. (2018) Effective Gamification in the Workplace. Retrieved from https://www.trainingzone.co.uk/community/blogs/carolynlewis/effective-gamification-in-the-workplace.

Moar, J. (2018) Health & Fitness Wearables Retrieved from Juniper: https://www.juniperresearch.com/researchstore/smart-devices/health-fitness-wearables/vendor-strategies-trends-forecasts.

Nintendo. (2016) What is Wii Fit Plus? Retrieved from http://www.nintendo.com/wiiu/what-is-wii-fit-plus/.

Nelson Jr., K. (2017). Dermal Abyss smart tattoo may know more about your health than you do. Digital Trends. Retrieved from https://www.digitaltrends.com/cool-tech/smart-tattoos-health-wearable-news/.

Nintendo. (2015). Nintendo Top Selling Software Sales Units: Wii Fit. Nintendo. 2015-09-30. Retrieved from https://www.nintendo.co.jp/ir/en/sales/software/wii.html.

Nuzzi, R., & Brusasco, L. (2018). State of the art of robotic surgery related to vision: brain and eye applications of newly available devices. Eye and Brain, 10, 13.

Pew (2018). Social Media Fact Sheet. Pew Research Center: Internet & Technology. Retrieved from http://www.pewinternet.org/fact-sheet/social-media/.

Piwek, L., Ellis., D. A., Andrews, S., & Joinson, A. (2016). The rise of consumer health wearables: promises and barriers. *PLoS Medicine, 13*(2), e1001953.

Poushter, J. (2016) Emerging, developing countries gain ground in tech revolution. Pew Research Center. Retrieved from http://www.pewresearch.org/fact-tank/2016/02/22/key-takeaways-global-tech/.

Ponemon Institute LLC. (2016, May). Sixth Annual Benchmark Study on Privacy & Security of Healthcare Data. Retrieved from https://www.ponemon.org/local/upload/file/Sixth%20Annual%20Patient%20Privacy%20%26%20Data%20Security%20Report%20FINAL%206.pdf.

PWC (2015) Top health industry issues of 2016 Thriving in the New Health Economy. Retrieved from https://www.pwc.com/us/en/health-industries/top-health-industry-issues/assets/2016-US-HRI-TopHCissues-ChartPack.pdf.

PWC (2017). Top health industry issues of 2018: A year for resilience amid uncertainty. Retrieved from http://www.medical-buyer.co.in/images/reports/pwc-health-research-institute-top-health-industry-issues-of-2018-report.pdf.

Research2guidance (2016). mHealth app developer, economics 2016: The current status and trends of the mHealth app market. Retrieved from https://research2guidance.com/product/mhealth-app-developer-economics-2016/.

Research2guidance (2018). 325,000 mobile health apps available in 2017 – Android now the leading mHealth platform. Retrieved from https://research2guidance.com/325000-mobile-health-apps-available-in-2017/.

Reddy, Tripps (2017) 5 ways to turn data into insights and revenue with content analytics. IBM Analytics. Retrieved from https://www.ibm.com/blogs/ecm/2017/07/28/5-ways-turn-data-insights-revenue-content-analytics/.

Redspin (2017). Breach Report 2016: Protected Health Information (PHI) Carpinteria, CA: Retrieved from https://www.redspin.com/resources/download/breach-report-2016-protected-health-information-phi/.

Sheynin, D. (2016, June14) Wireless Connections Reliable Enough For Doctors To Perform Remote Surgery. *Forbes.* Retrieved from https://www.forbes.com/sites/huawei/2016/06/14/robot-surgeons-may-lack-bedside-manner-but-they-may-save-your-life/#397381af2d83.

Smart Monitor. (2018). Smart Watch Monitoring Device. Retrieved from http://www.smart-monitor.com/.

Smith, A. (2015) U.S. Smartphone Use in 2015. Pew Research Center. Research Report. Retrieved from http://www.pewinternet.org/2015/04/01/us-smartphone-use-in-2015/#.

Statista (2015). Number of apps available in leading app stores as of March 2017. Retrieved from https://www.statista.com/statistics/276623/number-of-apps-available-in-leading-app-stores/.

Statista (2017). Number of smartphone users worldwide from 2014 to 2020 (in billions). Retrieved from https://www.statista.com/statistics/201182/forecast-of-smartphone-users-in-the-us/.

Topol, E. (2013). The Creative Destruction of Medicine: How the Digital Revolution will Create Better Healthcare. 2013.

Topol, E. (2015). The patient will see you now: the future of medicine is in your hands. New York, NY: Basic Books.

Topol, E. (2015). The patient will see you now: the future of medicine is in your hands. New York, NY: Basic Books.

University of South Florida (n.d.) What is health informatics. Retrieved from http://www.usfhealthonline.com/resources/key-concepts/what-is-health-informatics/#close.

Uscher-Pines, L. & Mehrotra, A. (2014). Analysis of Teledoc use seems to indicate expanded access to care for patients without prior connection to a provider. *Health Affairs, 33*(2), 258-264.

Vega, K., Jiang, N., Liu, X., Kan, V., Barry, N., Maes, P., Yetisen, A., & Paradiso, J. (2017, September 11-15). The Dermal Abyss: Interfacing with the skin by tattooing biosensors. Proceedings of the 2017 ACM International Symposium of Wearable Computers, Maui, Hawaii. doi:10.1145/3123021.3123039

Watson, T. (2013). National Business Group on Health. Reshaping Health Care: Best Performers Leading the Way. 18th Annual Employer Survey on Purchasing Value in Health Care.

Watson, T. (2017, August 2). U.S. employers expect health care costs to rise by 5.5% in 2018, up from 4.6% in 2017. [Press Release.] Retrieved from https://www.willistowerswatson.com/en/press/2017/08/us-employers-expect-health-care-costs-to-rise-in-2018.

Waegemann, C. P., Status Report (2002). Electronic Health Records, Medical Records Institute, Retrieved from www.medrecinst.com/.

Waegemann, C. P. (2003, September). Confidentiality and Security for e-Health. Paper presented at Mobile Health Conference, Minneapolis, MN.

Weintraub, A. (2012). Gamification Hits Healthcare as Startups Vie for Cash and Partners. Retrieved from http://www.xconomy.com/new-york/2012/06/21/gamification-hits-healthcare-as-startups-vie-for-cash-and-partners/.

World Health Organization (WHO) (2017). Teleheath. Retrieved from http://www.who.int/sustainable-development/health-sector/strategies/telehealth/en/.

Woods, M., & Woods, M. B. (2000). Ancient Medicine: From Sorcery to Surgery. Minneapolis, MN: Twenty-First Century Books.

World Bank (2016, March 30) Health Expenditures Per Capita.

Xin, L., & Vega, K. (2017) DermalAbyss: Possibilities of Biosensors as a Tattooed Interface. Retrieved from MIT: https://www.media.mit.edu/projects/d-Abyss/overview/.

19

Esports

Jennifer H. Meadows, Ph.D. & Max Grubb, Ph.D.*

Overview

From its beginnings as simulations for human computer interaction research, video games have evolved in the last 60 years from simple entertainment for individuals and competition among friends to esport leagues and tournaments. Now the realm of professionals and collegiate players and teams, esports is quickly evolving into a billion-dollar industry with no end in sight to its growth. This chapter examines the development of esports, its technology and its future.

Introduction

Mat Bettinson first coined the term "eSport" during the December 1999 launch of the Online Gamers Association (The OGA, 1999). Esports has been defined several ways but at its core is professional competitive gaming. Popular esports are often team-based multiplayer games played in tournaments for large audiences both in person and online (Dwan, 2017).

Today's esports can trace their heritage to the 1950s when computer scientists created simple games and simulations for their research. Esports can be classified into two periods—the arcade and Internet eras (Lee and Schoensted, 2011). As computer technologies evolved, so did video games, leading first to competitive gaming in arcades and player experiences at home to esports today; all in conjunction with the growth of home computer technologies and the Internet.

What technology is needed for esports? Players need a high-powered gaming computer—usually a desktop computer with a powerful graphics card, CPU, and lots of memory. Peripherals needed include controllers, displays, microphones, headphones, and speakers. Players also need a broadband Internet connection and of course, the game! Esports includes a multitude of video game genres from traditional sports video games such as *Madden 18* but mostly focuses on first person shooters (FPS), real-time strategy (RTS), and multiplayer battle arena games such as *Dota 2*, *League of Legends*, and *Counter-Strike: Global Offensive*.

Background

Competitive gaming has been around as long as video games. A 1962 MIT newspaper, *Decuscope*, reported that students competed at *Spacewar*, a two-player game that was programmed on a data processor. This inspired students to tinker with the code which involved variations of the *Spacewar* game. Eventually this competition led to a small 24 student tournament, the Intergalactic Spacewar Olympics, held at Stanford University Artificial Intelligence Laboratory on October 19, 1972 (Baker, 2016).

---

* Meadows is Professor and Chair, Department of Media Arts, Design, and Technology, California State University, Chico (Chico, California). Grubb is with Youngstown State University; Youngstown, Pennsylvana

But it was the economic incentive that motivated entrepreneur Nolan Bushnell, founder of Atari, that led to the growth of video games in arcades. Atari's *Pong* is considered the first video arcade game which quickly grew in popularity. At its peak there were over 35,000 *Pong* machines in the United States, averaging a staggering $200 a week per machine (Goldberg, 2011).

Arcade gaming continued to grow in popularity and with that popularity came competitive arcade gaming. For example, a nationwide competition was launched to see who could get the highest score on the arcade game *Donkey Kong*. Scores were conveyed through word of mouth and published in trade magazines. Atari sponsored the first arcade competition, for *Space Invaders* in 1980 with more than 10,000 players (Edwards, 2013).

Arcade gaming competitions, though, pale against today's esports which really began to take off with the development of Internet enabled PC and console gaming in the 1990's. Improvements in Internet connectivity speeds contributed to the rise in multiplayer online role-playing games, the precursor to today's prolific online gaming environment (Overmar 2012).

Where entrepreneurs saw the opportunities in video games, both in arcades and at home, gamers moved from simple competition in the arcade and tournaments for the highest scores to a more robust gaming experience online. Thus, the social aspects of competition were evolving, stirring the beginnings of electronic sports (esports).

While there had been tournaments since the 1970s, the first high-stakes tournament took place in 1997 when *Quake* hosted "Red Annihilation." It attracted over 2,000 participants, and the winner was awarded the Ferrari then owned by the developer of the game, John Carmack (Edwards, 2013). That year, the Cyberathlete Professional League (CPL) was created leading to the emergence of the first eSport professionals. Later that year CPL held its first tournament, and the next year it offered $15,000 in prize money (Edwards, 2013). One professional to appear during this time was Johnathan "Fatal1ty" Wendel who reportedly won about a half-million dollars during his eSport career (Wendel, 2016). Also, it was

during this time that esports was judged to be an emerging spectator sport

During this time most games were FPS, arcade style, and sports games. Edwards (2013) observed Real-Time Strategy (RTS) games were then developed in the late 1990s. Similar to chess, RTS games require thought and long-term planning. As a RTS game, *Starcraft* was a significant factor in the growth of esports with its unlimited strategic possibilities reaching its pinnacle of popularity after 2000.

As the world entered the new millennium, the world of esports experienced seismic growth in popularity. In 2000, two major international tournaments were introduced, the World Cyber Games and the Electronic Sports World Cup. Both became annual events, serving as international platforms for esports competitors and providing the template for major esports tournaments (Edwards, 2013). These and the growing number of competitions gave gaming companies that hosted them the opportunity to feature the two principal game types, FPS and RTS. Games such as *Quake, Halo,* and the *Call of Duty* series are FPS. These and other FPS games require quick reaction and reflexes with the need for fast implementation and use of buttons.

Gamers are attracted to RTS games because they can choose from a variety of characters with different skills and abilities. In addition, RTS competitions normally involve 5-member teams competing against each other. Collaboration among team members is a must for success. RTS games attract competitors from around the world with thousands attending championship events. Games such as *Starcraft, World of Warcraft and League of Lengends (LoL)* are RTS games that contributed to the rapidly expanding esport industry.

Esports tournaments and leagues began to take off in South Korea in the early 2000s with the game *Starcraft*. When Riot Games' *League of Legends* came out in 2009, the sport exploded (Wingfield, 2014). Leagues for esports have since been established around the world. The largest is the ESL, formerly known as the Electronic Sports League, formed in 2000. Major League Gaming was formed in New York City in 2002 for professional gamers. Leagues are also formed around specific games such as *Overwatch*.

There is a great deal of money in professional esports. For example, 10 million people watched the Overwatch League in January 2018. Blizzard charged teams a total of $20 million to participate in the league, with the fee expected to grow to $60 million next year (Conditt, 2018)! Figure 19.1 shows the Florida Mayhem team entering Blizzard Arena for an Overwatch League Event.

Figure 19.1

Florida Mayhem Entering Blizzard Arena for an Overwatch League Event

Source: Blizzard Entertainment

Esports isn't just for the players, as the number of spectators for esports has grown exponentially. Millions of people watch videos of game play on YouTube and the live streaming video game service Twitch. Tournaments can fill stadiums and arenas with devoted fans.

Recent Developments

Esports continue to grow in popularity, and new developments in esports include continued growth and investment in professional esports, televised competitions on cable channels, arenas built solely for esports, collegiate esports teams, and even talk of including esports in the Olympics.

The popularity of esports hasn't escaped the owners and managers of professional sports including FIFA, NHL, MLB, NBA and NFL. The NHL entered the arena with the 2018 NHL Gaming World Championship. Using EA's *NHL* on the PS4 and Xbox One, players competed in 18 rounds leading to the

championship in Las Vegas during the NHL Awards in June 2018 (Associated Press, 2018). The NBA is going so far as to form their own esports league, NBA 2K. Formed in 2017, the league will have 17 participating NBA teams in its inaugural season in 2018 using *NBA 2K*, a game from Take Two Interactive Software (NBA, 2017). The NBA is drafting 102 professional players who will receive 6-month contracts with benefits and compensation around $35K (Harrison, 2018).

Want to watch all these gamers? Fans can watch online, on television, or go to live events. Twitch was introduced in 2011 and has become the most popular online service used to watch live gaming. Calling itself "the world's leading social video service and community for gamers," Twitch allows users to watch live gaming with audio (About Twitch, 2018). Twitch is also available as an app for a wide range of devices including smartphones, game consoles, Google's Chromecast, and Amazon FireTV. Users can just watch other gamers but the draw of Twitch is the opportunity to interact with chat and cheers. Popular gamers have huge followings on Twitch. For example, Ninja had an average of almost 80,000 viewers and DrDisRespectLIVE had about 27,000 viewers in March, 2018. Professional leagues also have large followings on Twitch. For example, the Overwatch League has an average of 60,000 viewers (The Most, 2018; About Twitch, 2018). To watch competitive esports on Twitch, users use official league channels. Other popular ways to watch online include Mixer, MLG.tv, and YouTube Gaming.

The concept of watching esports on television was scoffed at just a few years ago. Now ESPN2 carries esports and has a deal with MLG. The network has broadcast the finals of a collegiate tournament of Blizzard's *Heros of the Storm* and major events such as the International *Dota 2* Championships and BlizzCon. TBS has ELeague—a partnership between IMG and Turner Sports. ELeague has esports events for games including *Streetfighter V, Counter Strike: Global Offensive*, and *Overwatch* (ELeague, n.d.).

Esports fan argue that nothing is better than seeing an event live. 12,000 fans watched Team Newbee win $1 million at the The International *Dota 2* Championship in Seattle in 2014 (Wingfield, 2014). Since then, the numbers have grown. Over 174,000 spectators filled a stadium in Poland for the IEM (Intel Extreme Masters) World Championships in 2017 (Elder, 2017). In the U.S., regular events are held in arenas in major cities. For example, ESL One, a *Counter Strike: Global Offensive* competition, was held at Brooklyn's Barclay Center in 2017 with full capacity and over five million streams (Kline, 2017).

There are now dedicated esports arenas. The first Esports Arena was built in Orange County, CA in 2015 and an Oakland, CA facility is in development as of mid-2018. The Luxor Hotel and Casino in Las Vegas has the flagship Esports Arena Las Vegas. The arena opened on March 22, 2018 in a 30,000 square-foot former nightclub space and is part of Allied Esports International's network of esports arenas. The Las Vegas arena features a competition stage and state of the art streaming and television production studios. (See Figure 19.2) The company has plans to build 10-15 more Esports Arenas across the United States in the next few years (Esports Arena, 2018).

Are esports just another sport like basketball and track? Across the world, high schools, colleges, and universities are featuring esports teams. Often beginning as student clubs and then club sports, esports teams are now fully-funded athletics teams at many universities. Robert Morris University Illinois was the first higher education institution to offer scholarships for esports athletes in 2015. The initial team focused on *League of Legends* and recruited from the League of Legends High School Starleague (HSEL) (RMU Becomes, 2014). As of March 2018, there were at least 50 varsity esports programs at universities in the U.S.

Figure 19.2
Esports Arena Las Vegas

Esports Arena, the first of its kind to open in North America, specializes in large Esports events including the very first tournament for Blizzard Entertainment's newly released game *Overwatch*.

Source: Esports Arena

The governing body for collegiate esports is the National Association of Collegiate Esports (NACE) (Morrison, 2018). Founded in 2016, NACE members institutions are "developing the structure and tools needed to advance collegiate esports in the varsity space" (About NACE, n.d.). NACE has partnered with the High School Esports League (HSEL), the National Junior College Athletic Association (NJCAA), and the National Association of Intercollegiate Athletics (NAIA). What's missing? The NCAA of course. The NCAA announced in December 2017 that it was investigating collegiate esports. It will be interesting to see if esports becomes an NCAA sport. Support seems to be building. The 2018 Fiesta Bowl hosted the *Overwatch* Collegiate Championship at Arizona State University. Already student esport athletes are getting NCAA-like support at some universities. For example, NACE member the University of California at Irvine fully supports their esports team with dedicated practice facilities, an esports arena, and scholarships. UC, Irvine supports teams for *Overwatch* and *League of Legends*. Collegiate teams generally play for scholarship money (Conditt, 2018).

With amateur collegiate and professional esports exploding, inclusion in the Olympics could be in the future. The International Olympic Committee began to explore the possibility in 2017 for inclusion in the Paris summer games of 2024. The IOC is interested in esports for its appeal to youth but state that the sport must conform to Olympic values (Esports:International, 2017). This standard could be an issue for some games considering their violent content. Indeed, in March 2017, the IOC clearly stated that it will not consider any games that contain violence. Games will probably be related to traditional sports rather than popular esport games like *Call of Duty* and *Overwatch* (Orland, 2018).

Current Status

Newzoo released its 2018 Global Esports Market Report in February 2018. The report listed 7 key takeaways:

- Global esports revenues will reach $906 million in 2018, a year-on-year gowth of +38.2%. North America will account for $345 million of the total and China for $164 million.

- Brands will invest $694 million in the esports industry, 77% of the total market. This spending will grow to $1.4 billion by 2021, representing 84% of total esports revenues.

- The number of esports Enthusiasts worldwide will reach 165 million in 2018, a year-on-year growth of +15.2%. The total esports audience will reach 380 million this year.

- The global average revenue per esports enthusiasts will be $5.49 this year, up 20% from $4.58 in 2017.

- In 2017, there were 588 major esports events that generated an estimated $59 million in ticket revenues, up from $32 million in 2016.

- The total prize money of all esports events held in 2017 reached $112 million, breaking the $100 million mark for the first time.

- The *League of Legends* World Championship was the most watched event on Twitch in 2017 with 49.5 million viewing hours. It also generated $5.5 million in ticket revenues (NewZoo, 2018).

Nielsen (2017) reports that the average age of the esports audience member is 26 and 71% of the audience is male. Male fans are more likely to stream esports but 25% of female fans stream at least once a week. The study also found that esports fans watch comparatively little linear television but are avid streamers. For example, U.S. esports fans watch an average of just 4.4 hours of television a week.

The Nielsen study found that excluding Twitch, most fans followed both players and games via YouTube. The most popular team in the U.S. and U.K. was Cloud9 while Fnatic was the most popular in France and Germany. Top broadcasters in the U.S. for esports for the first six months of 2017 were Disney XD with 13 shows, TBS with 7, ESPN2 with 5, NFL Network with 3, and ESPNU, the CW and ESPN with 1 each. Interestingly this study also asked whether esports should be considered an actual sport (53%), a university sport (41%), and an Olympic sport (28%) (Nielsen, 2017).

According to esportsearnings.com (2018), the top earning games for 2017 were *Dota 2* with $38 million in prize money over 159 tournaments with 959 players.

In second place, *Counter-Strike: Global Offensive* awarded $19 million in prize money with 894 tournaments and 4,754 players. The classic *League of Legends* came in third with $12 million, 149 tournaments and 1,624 players. For fans, *Counter Strike: Global Offensive* is the most followed PC exclusive esports game while *Call of Duty* topped multi-platform games

Factors to Watch

Expect continued growth in the esport audience, prize money, leagues, and revenues. Television broadcasts should continue to proliferate as networks see the possibilities of esports programming and its tough-to-reach audience of young men.

Expect collegiate esports to be embraced by the NCAA which will help further standardize the sport for both institutions and players. As of this writing, this is difficult because esports teams range from NACE supported varsity teams to small student clubs. There is even a wide disparity in which university division these teams are located on a campus. Some are in athletics, while others are in student affairs or other divisions.

The big players in the game industry such as Blizzard and Valve will continue to dominate but look for smaller games to make inroads. The social aspect of how audiences consume games through sites such as Twitch and YouTube means that smaller games can easily go viral and stand out.

One hot aspect of gaming may soon become part of the esports ecosystem, virtual reality (VR). It will be important to watch the progress of the VR Challenger League, the first VR esports league supported by Oculus, Intel, and ESL. This league will have 4 events in 2018, and players will share $200,000 in prize money (VR Challenger League, n.d.). The games will be Insomniac's *Unspoken* and Ready at Dawn's *Echo Arena*.

Getting a Job

If your goal is to become a professional esports player then what are you doing reading? You should be practicing! As with any other professional athlete, esports players put in grueling hours of practice every day. Don't have the skills? Consider platforms like Twitch and YouTube where you can host your own channel. A good living can be made attracting subscribers and viewers.

If your interest is in esports events and broadcasting, consider studying multicamera live production. The same skills needed to produce a football game are needed to produce an esports tournament with everyone from play-by-play announcers to directors and camera operators needed for a successful show.

Bibliography

About NACE (n.d.). National Association of Collegiate Esports. Retrieved from https://nacesports.org/about/.

About Twitch (2018). Twitch. Retrieved from https://www.twitch.tv/p/about/.

Associated Press (2018). NHL takes esports on ice with gaming tournament. *The New York Times*. Retrieved from https://www.nytimes.com/aponline/2018/03/09/business/ap-hkn-starting-esports.html.

Baker, C. (2016) Game Tournament. *Rolling Stone*. Retrieved from https://www.rollingstone.com/culture/news/stewart-brand-recalls-first-spacewar-video-game-tournament-20160525.

Conditt, J. (2018). College esports is set to explode, starting with the Fiesta Bowl. *Engadget*. Retrieved from https://www.engadget.com/2018/02/22/college-esports-is-set-to-explode-starting-with-the-fiesta-bowl/.

Dwan, H. (2017). What are esports? A beginner's guide. *The Telegraph*. Retrieved from https://www.telegraph.co.uk/gaming/guides/esports-beginners-guide/.

Edwards, T. (2013). Esports: A Brief History. *Adani*. Retrieved from http://adanai.com/esports/.

Elder, R. (2017). The esports audience is escalating quickly. *Business Insider*. Retrieved from http://www.businessinsider.com/the-esports-audience-is-escalating-quickly-2017-3.

ELeague (n.d.) TBS. Retrieved from http://www.tbs.com/sports/eleague.

Esports Arena (2018). Esports Arena Las Vegas. Retrieved from https://esportsarena.com/lasvegas/.

Esports: International Olympic Committee considering esports for future games. BBC. Retrieved from http://www.bbc.com/sport/olympics/41790148.

Goldberg, F. (2011). The origins of the first arcade video game: Atari's Pong. *Vaniety Fair*. Retrieved from https://www.vanityfair.com/culture/2011/03/pong-excerpt-201103.

Harrison, S. (2018). The Big Breakaway. *The Outline*. Retrieved from https://theoutline.com/post/3616/the-big-breakaway-nba-esports-league-nba-2k.

Kline, M. (2017). Gaming fans are crowding into stadiums to watch esports events. Here's why. *Mashable*. Retrieved from https://mashable.com/2017/09/22/esports-events-are-filling-stadiums/#iDJLIPMP_Oq7.

Lee, D. and Schoensted, L. (2014). Comparison of eSports and traditional sports consumption motives. *Journal of Research*, v6, 2. pp 39-44.

Morrison, S. (2018). List of varsity esports programs spans America. *ESPN*. Retrieved from http://www.espn.com/esports/story/_/id/21152905/college-esports-list-varsity-esports-programs-north-america.

NBA (2017). 17 NBA Teams to Take Part in Inaugural NBA 2K esports league in 2018. Retrieved from http://www.nba.com/article/2017/05/04/nba-2k-esports-league-17-nba-teams-participate-inaugural-season.

NewZoo (2018). *Newzoo 2018 Global Esports Market Report*. Retrieved from https://resources.newzoo.com/hubfs/Reports/Newzoo_2018_Global_Esports_Market_Report_Excerpt.pdfc.

Nielsen (2017). *The Esports Playbook*. Retrieved from http://www.nielsen.com/content/dam/corporate/us/en/reports-downloads/2017-reports/nielsen-esports-playbook.

Orland, K. (2018). Violent video games not welcome for Olympic esports consideration. *Arstechnica*. Retrieved from https://arstechnica.com/gaming/2018/03/olympic-committee-open-to-esports-but-only-without-violence/.

Overmars, M. (2012). A brief history of computer games. *Stichingspel*. Retrieved from https://www.stichtingspel.org/sites/default/files/history_of_games.pdfbritis.

RMU Becomes First University to Offer Gaming Scholarships With the Addition of eSports to Varsity Lineup (2014). Robert Morris University Illinois. Retrieved from www.rmueagles.com/article/907.

The Most Popular Twitch Streamers, March 2018 (2018). TwitchMetrics. Retrieved from https://www.twitch-metrics.net/channels/popularity.

The OGA (1999). Eurogamer.net. Retrieved from http://www.eurogamer.net/articles/oga.

VR Challenger League (n.d.) Retrieved from https://vr.eslgaming.com/.

Wendel, J. (2016). The original. *The Players' Tribune*. Retrieved from https://www.theplayerstribune.com/fatal1ty-esports-the-original/.

Wingfield, N. (2014). In E-Sports, video games draw real crowds and big money. *The New York Times*. Retrieved from https://www.nytimes.com/2014/08/31/technology/esports-explosion-brings-opportunity-riches-for-video-gamers.html?ref=bits.

20

Ebooks

Steven J. Dick, Ph.D. *

Overview

Ebooks are a group of technologies that allow the distribution of traditionally printed material to computers and handheld devices. Ebooks offer the opportunity to disrupt traditional publishing through inexpensive distribution and the opportunity for new voices. Groups like Project Gutenberg have transferred the great works and documents to electronic form for inexpensive distribution. Independent authors now have the potential to circumvent the traditional big publishers to reach an audience. Interactive technology allows new opportunities from simple audio versions of the books, text-to-speech, interactive stories, and expansion into multimedia content.

This technology that was supposed to disrupt the publishing industry has, itself, been disrupted by new marketing techniques and options. Amazon, the market leader in electronic paper, now has been accused anti-competitive activity as it revamps the way it distributes ebooks to its powerful Prime services and competitors work to answer the challenge. The question is if the new marketplace will be good for independent authors and the wider audience.

Introduction

The idea of distributing books in digital form is as old the computer communication itself. Even with slower transmission speeds, book-length files could be distributed without unnecessary delay or cost. The goal was to free the publishing industry from paper while expanding the audience of books. While ebooks have become an important part of the publishing industry, adoption of ebooks has fallen far short of expectation amid competing formats, devices, and business models.

This book is available in both electronic and paper formats. Why did you choose the format you are reading? Some like the mobility of carrying several books without adding weight, the ability to acquire a book quickly, or rent a book without worrying about physically returning it. Some prefer the feel of paper in their hands, the "under the tree" experience of reading anywhere they want and avoiding having to learn the software or buy new hardware. Others want to disconnect from technology while reading or find a format easier to share. Does book content affect your choice? For example, would a Bible or another religious book feel the same in electronic form? How does price affect your choice? The relationship that people have with the technology affects the market for ebooks. That relationship may and in some ways has changed. This chapter will examine the factors that have changed ebooks over the years.

* Senior Research Scientist, Communications Department, University of Louisiana at Lafayette, Lafayette, Louisiana

Background

Ebooks require the combination of content, software, hardware, and organizational infrastructure. Also, the level of interactivity or computer-aided intelligence has varied by the system. There has been no consistent solution; individuals and organizations have promoted models for each.

Content

The content was the first necessary component. In 1971, University of Illinois student Michael Hart enjoyed access to ARPANET computers (the predecessor to the Internet). He used his access to type in and distribute the *Declaration of Independence* followed by *Bill of Rights, American Constitution*, and the Christian *Bible* (The Guardian, 2002). The effort continues today as Project Gutenberg currently offers more than 56,000 free ebooks (www.Gutenberg.org). Project Gutenberg and other non-profit organizations worked to digitize public domain text either through labor-intensive typing or scanning into computer formats. Later, some authors began to release work through creative commons or without copyright rather than market through the normal publication process.

These publicly available books became the backbone content to distribute ebooks, but soon more commercial products entered the field. Three main companies, Ereader, Bibliobytes, and Fictionworks began the sale of ebooks in the 1990's (The Telegraph, 2018). Google created its book search engine in 1994 (later Google books), and by this time, the effort to convince authors to distribute content in electronic form came in full force. Also, the public domain books were enhanced with better formatting, annotation, or illustrations to produce work sold for a low, but profitable rate. Independent authors, wishing to become professional were the first to offer content. In 2007, Amazon joined the ebook market when it introduced its Kindle ereader (discussed later) and started to sell content from major authors. Amazon was followed by Barnes & Noble, Apple, and Google (among others) and the potential seemed clear enough that major publishers would offer their books to electronic sales.

In 2011, Amazon announced that ebooks began to outsell traditional books (Miller & Bossman, 2011).

Software

Effective ebook distribution requires two types of software—file formats and display. Over the years, there have been multiple versions of each. Some people may not consider file formats software because the intention is to provide content rather than instructions to the hardware. However, most file formats contain markers for several aspects of the ebook experience. Display software can provide simple information like chapter markers, page turn locations , display of images, and interaction with dictionaries. File formats evolved from text storage (ASCII or American Standard Code for Information Interchange) and word processing formats (Adobe PDF). Many of the early abandoned formats (like LIT and MOBI) are still available in online ebook resources so you can still use them on your devices.

Display software evolved as ebook distributors needed to enforce licensing through Digital Rights Management (DRM), and users wanted a better reading experience. DRM enforces the book licenses, including approved reader accounts, time and geographic restrictions, and allowable devices. Other aspects of the display system may allow interactivity (discussed below) and help manage files. Thus, a reader may be allowed to share or lend a book, enforce time restrictions, or move a book from one company's hardware to another.

As ebooks have moved from dedicated hardware ereaders to other devices, the software has become more important and complex. The software for computer, web, smartphones, and tablets had to provide the same experience as the dedicated ereader as well as the ability to manage files across devices. Screen quality, colors, and processing power allow the image quality needed for graphic novels and comic books, Adobe PDF documents, audiobooks, and interactive books for children and multimedia.

Table 20.11:

Ereader File Formats

Format	Creator	Introduced	Notes
ASCII	American National Standards Institute	1963	Many have built on this standard to add additional capability such as "rich text."
ePub	International Digital Publishing Forum	2007	Widely used by Google, Android, and Apple. An open format but modified by some
AZW	Amazon	2007	DRM added to MOBI for Amazon Kindle. Replaced by KF8.
LIT	Microsoft	2000	Microsoft discontinued support in 2012
ODF	OASIS and OpenOffice	2005	Used in many alternatives to Microsoft Office. XML
MOBI	Mobipocket	2000	Altered original Palm format then purchased by Amazon.
PDF	Adobe	1993	Designed for page layout.

Source: S. Dick

Comic book and graphic novel ebooks demand high-resolution images. Online origins came more from HTML (web) coding than text formats. Merging the DRM demanded by publishers with the high-quality graphics, and the need to display a logical progression of images rather than text, was not easy. These graphic format books were not possible in classic ereaders and demanded software for tablets, smartphones, computer, and web interfaces (Wilson, 2015).

Before the widespread use of smaller, lighter hardware such as tablets, notebook computers, and smartphones, PDF documents required too much memory and did not format well for smaller screens. Some smartphone software such as Foxit, allowed readers to take the text out of PDFs and "reflow" it for easier reading on the small screens. Breakthroughs came with larger touch-screen PDF software allowing readers to annotate the document on the fly by writing (pen or finger) or typing directly to the document as if it were on paper. It makes PDF documents truly feel like electronic paper—a major goal of ebooks and a popular feature for businesses as people can edit and sign documents with all the formatting of a finished commercial product.

In 1932, the tests of audio recorded books included a chapter from Hellen Keller's book *Mainstream* and Edgar Allan Poe's *The Raven* (Audio Publishers Association, n.d.). These "Talking Books"

from the Library of Congress were intended for the blind. The start was slow but got a push with the introduction of the audio cassette tape in 1963 and books on tape increased popularity in the 1970s and 1980s. The company, Audible created a practical digital audio player in 1997 and joined Apple's iTunes library in 2003. In 2008, digital downloads began to surpass CD audiobooks. Audiobook software is both dependent on the intelligence of computer technology and carefully crafted to provide the same experience across platforms, so the iOS version and the Android version looks and operates similarly.

Interactivity

In 1987, two systems introduced interactivity to reading. Eastgate systems introduced *Afternoon*, a hypertext fiction by Michael Joyce (The Guardian, 2002). Apple introduced Hypercard, an interactive software package that could allow authors to create an interactive storyline with multiple branches, enabling readers to jump from point to point in a text. Hypercard could be used for presentations, reference works, or learning tools that allowed the reader to choose topic areas (Kahney, 2002). These early interactivity experiments transitioned to content written in HTML and other web formats. Now, the content often transitions to interactive storybook apps.

Display software has begun to allow readers to communicate across devices. Interactivity across devices allow a reader to move from device to device and continue reading at the same point including a seamless flow between text and audiobooks. Interactivity can extend across readers as well including shared personal notes, highlights, and social media posts. Religious books such as Bibles supplied by Olive Tree allow users to compare passages across translations, immediately find commentary, maps, dictionaries and other study aids. Dedicated apps like WebMD and cookbooks like Tasty could be considered interactive ebooks.

Children books are attracting the most aggressive software investment for interactive texts. Children's ebooks require the display capability of graphic novels, sound from audiobooks, and interactive or branching capability from HTML. One app called Novel Effects reaches across formats by using voice recognition to "hear" a parent reading a traditional book to a child and adding sound effects.

Hardware

In some cases, ebooks have been simply a companion product or feature to a technology. In others, ebooks are the killer app or primary reason to buy a product. The first handheld devices for ebooks were a flurry of personal digital assistants (PDA). As a precursor to the modern smartphone, these devices were meant to be used to store appointment, contacts, and other data. But optional apps became available. Apple entered first in 1993 with it short-lived Newton followed by Palm Pilot in 1996 and the Microsoft Reader in 2000. The handheld devices allowed the user to both read material and write on something about the size of the modern smartphone.

The most visible ebook hardware is the ereader with E-Ink. E-Ink is a form of electronic paper (Primozic, 2015) that was featured with the Kindle in 2007 (Wagner, 2011). The black and white screens on Kindles mimic paper's reflective property. Since they reflect light rather than produce it (the way a television or computer screen does), the ereader can operate with much longer battery life. E-Ink uses pixels (picture elements) made up of the clear solution with negatively charged dark particles and positively charged white particles. When a positive charge is placed on the top

of the pixel, the negatively charged dark particles are attracted to it, and the pixel goes dark. The opposite is true with the white particles. Later, grey pixels were possible by manipulating the charge (see Figure 20.1).

Consumers' desire for color images and capabilities beyond ereaders were answered by a group of tablet devices often built on Google's Android operating system and Apple's iOS. Today, the smartphone and the tablet are quickly becoming the most popular hardware readers for ebooks (Haines, 2018).

Figure 20.1:
Pixel Construction with E-Ink

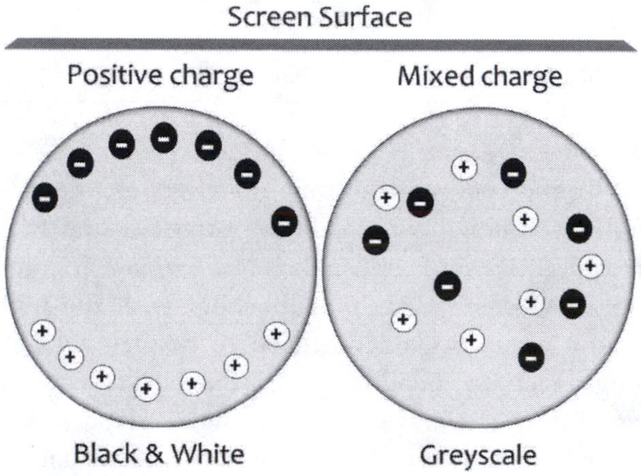

Source: S. Dick

Organizational Infrastructure

The assumed disruptive influence of ebooks on the traditional distribution model of the industry caused publishers and distributors to change their marketing plans. The problems went beyond DRM. The first choice was to license the books rather than sell them—an important change from print. Print books are sold, and the physical book then belongs to the buyer—other than the copyright. In the U.S., the buyer can do whatever they want with that one copy under what is called the "First Sale Doctrine." (Department of Justice, n.d.). That includes selling, lending, or giving away the book. The publisher only makes money on the first sale. Licensing changes that relationship. Instead of buying the book, you are buying the right to read that book. Many current licenses allow you to "lend" or "give" it to another person but they also only get a license to read it. Licensing a book rather than selling it allows publishers to profit from more sales.

Figure 20.2
Ebook Sales in Billions of US Dollars by Year

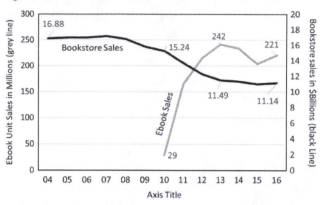

Source: Statista

The relationship between the new ebooks and traditional publishing was essential to the new marketing plans. By culling data from US Census Bureau and publishing industry reports, Statista (2017) provides enough information to understand the relationship between bookstore sales and ebook sales (see Figure 20.2) Amazon's Kindle was introduced in 2007 and bookstore sales (including items other than books) began to drop. Between 2010 and 2013, ebook sales exploded from 29 to 242 million books while bookstore sales dropped by nearly $4 billion. There had to be discussion as to how to distribute ebooks without killing bookstores.

The ereader, specifically the Amazon Kindle, was designed for impulse purchases. While the first ereaders were connected to a computer by wire, later devices quickly adopted wireless technology—at first cell and later Wi-Fi (Pierce, 2017). The goal was to create a device that would allow a user to hear about a new title and buy it immediately. The combination of a single lightweight device that would hold many books, no need to go to the store, and the excitement of the latest technology was undeniable. The expectation was that the disruptive influence of electronic distribution was going to come to publishing, and eventually most books would be replaced by ebooks.

However, more was happening than a simple technological change. Not only were ebooks replacing traditional books but also physical bookstores were being replaced by online bookstores. Pricing was also affected as Amazon pushed for a simple standard of $9.99 for most major books. Apple was already using flat rate pricing for music sales. It seemed natural to settle on the single price. The goal was to create a system where the customer would not put much thought into the order (Pierce, 2017). Major publishers, unhappy with the payment structure forced reconsideration of the flat rate price.

Libraries face a special challenge when distributing ebooks (Meadows, 2017). Publishers, fearing revenue loss to ebooks, have been far more restrictive to libraries. Realistically, the experience is very different. To borrow a paper book from the library takes two trips to the physical location, and the book may be in poor condition (The Authors Guild, 2018). Ebooks can be borrowed online and are always in original condition. The software for online libraries can automatically "return" books, so the online library and online bookstore are nearly identical experiences other than the time of ownership (access). Publishers also worry about the security of the software—possibly allowing borrowers to obtain ebooks free of digital rights management (Meadows, 2017).

Specific online interfaces manage the borrowing processes for libraries such as OverDrive that serve many libraries and provides a dedicated ereader software for local libraries. National libraries have developed such as Project Gutenberg (https://www.gutenberg.org/), and The Internet Archive's Openlibrary (https://openlibrary.org/). The OpenLibrary buys paper copies of books and acquires or creates digital copies of books. The library then loans electronic copies to match paper books it holds. This interpretation of the Fair Use Doctrine exception to Copyright Law is currently a source of dispute between the OpenLibrary, publishers, and the Authors Guild (Meadows, 2017). The outcome of this dispute will determine the rights for online libraries in the future. Google Books (not Google's Play Store) is a hybrid model allowing users to search the contents of books. If the user finds a book she wants to read, she will be linked to locations where she can buy (Play Store) or borrow books—depending on availability.

Recent Developments

Book publishers earn about $7 billion a year (Statista, 2017) not including retail markup. Market share for ebooks has begun to fall from a high of 26% in 2014 and down to 17% in 2017 according to an industry trade group (American Association of Publishers, n.d.). The one exception was a dramatic growth in audio-books with market share growing from 3% in 2014 to 5.6% in 2017.

Some point to the devices (Milliot, 2016) as the cause of sales decline. These people believe that ereaders are not attractive to consumers—especially among young people. Ereaders, tablets, and smart-phones represent work, and people are using reading as an opportunity to disconnect. Handheld electronics are expensive and in the case of ereaders, only useful for one purpose. Even the simplest ereader can cost between $50 and $200. It might be better to pay for the books you are actually going to read than pay for the ereader plus the books.

Figure 20.3
Market Share by Publishing Product

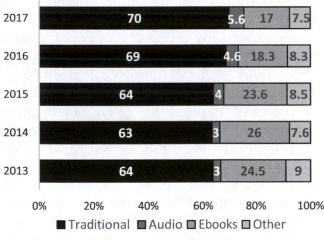

Source: S. Dick

Certain categories of books remain popular for paper publishing. The latest fad of adult coloring books, children's, and hardback books lead the categories that sell better in paper (Sweney, 2016). People seem to have a natural affinity for relaxing with paper rather than electronic ink.

Cheap and easy book rentals, professor published material, and the ease of carrying several electronic books across campus caused several universities to create initiatives to transfer from paper to electronic book formats (Redden, 2009). With textbooks, some books are sold in both formats to read one at home and carry the other to the classroom.

It is also possible that sales numbers do not accurately reflect ebook sales. Interactive books and games for children, medical texts, and religious publications may not calculate into publishing sales figures and can quickly cross a line that would cut them out of the traditional publishing sales figures. For example, Book publisher Olive Tree publishes bibles and bible study products outside normal publishing channels. Chu Chu TV (www.youtube.com/user/TheChuChutv) is a YouTube Channel and app. It rewrites traditional children's stories and nursery rhymes. Finally, the original *Grey's Anatomy* has been repackaged in the app *Visual Anatomy*. Absent interactive technology, these interactive applications might be produced by the publishing industry, and it is doubtful that most are included in the ebook sales figures.

New pricing models affect ebook sales calculations. First, as major publishers have entered the market, calculations are not as accurate for independent publishers and writers concentrating instead on the best-selling books. Second, free or promotional books are easy to distribute (Trachtenberg, 2010). Rather than selling your title outright, give away a book in exchange for reputation or corporate promotion (Anderson, 2009). Finally, in July 2014, Amazon introduced Kindle Unlimited—a substantial collection of ebooks available at no additional cost to subscribers of their Prime service. Instead of buying or renting a single book, the pricing is changed to a subscription model for a library of titles and authors are paid a share of subscriptions instead of a per-book royalty. Both Barnes & Noble and Kobo have started subscription services as well. Because all these changes, the systems that count book sales may be undercounting in a changing marketplace (Pierce, 2017).

Current Status

According to the American Association of Publishers (2018), October 2017 sales for the publishing industry were very strong resulting in a year to date increase of 1.7% over 2016. Unfortunately, ebook sales dropped by 5.5%. The one category that showed real gains was audiobooks, with sales up 27.9% compared to the previous year. The moderating effect may be in part due to ebooks becoming more expensive as large publishing houses force ebook price increases to more closely match the prices of traditional books.

Pricing inconsistency is becoming an issue. While Amazon has tried to enforce sales contracts with authors that give them the best possible price. There are two basic models for setting the price. In the wholesale model, the publisher or author suggests a price but sells the book to the distributor at a much lower wholesale price. The distributor is then free to set whatever price it wishes to put on the book. So, the price to the distributor may $6.99 with a suggested price of $12.99, but the distributor may choose to sell the book for $9.99. This makes it more difficult for the publisher to sell the book at equal prices across sales channels.

The agency price model gives more control to the publisher—so it is favored by the larger publishing houses. With agency pricing, the publisher sets the price on the book, and the distributor receives a commission on each sale. Here, the publisher sets the price for $12.99 and gives the distributor 40% on each book sold. This model does not work well with Amazon desire to standardize prices and runs sales. Overall, it also has the effect of pushing the prices higher—more equal to printed books. Agency pricing has been blamed for raising the price of ebooks and contributing to the slump in ebook sales (Miller & Bosman, 2011).

In the first quarter of 2017 (one of the lowest earners because it comes after Christmas), book publishers earned $2.33 billion—a 4.9% increase over the same quarter of 2016. For the same period, downloaded audio grew 28.8%, and hardback books grew by 8.25. Paperback book sales dropped by 4.7%, and ebooks sales dropped by 5.3%. A check of Google, Amazon, and Apple app stores revealed 4.55 million downloads of ebook reader apps. Apple's iBooks is unavailable because it is preloaded and not available in any app store. Top remaining apps were split between Kindle (26%), Google (31%), and Audible (31%). Kobo only represented 7% of all downloads and Overdrive (for libraries) had only 6%. Traditional ereaders like the Kindle and Nook are quickly losing market share (Haines, 2018). Shipment of ereaders dropped to 7.1 million units in 2016, down from 23.2 million in 2011. The drop in sales of ereaders reflects the fact that more people are reading on smartphones and newer lighter tablets.

Figure 20.3

Downloads of Ereader Apps by Percent

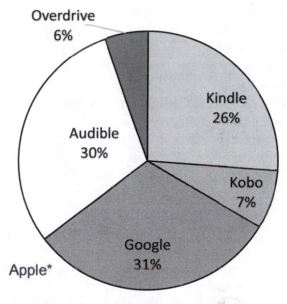

*Apple's ereader app (iBooks) comes preloaded on Apple devices and is unavailable for others.

Source: S. Dick

Factors to Watch

It is impossible to effectively evaluate the ebook industry without considering the effect of Amazon. This one company can set prices and terms, as it dominates the market with an estimated 75% of the revenue. Because of their position, Amazon can be said to enjoy "monopsony power"—a situation where there is one effective buyer to a class of goods (London, 2014). Many authors feel they must sell to Amazon if their book is to be successful.

One of the most problematic demands is Amazon's "most favored nation" (MFN) clause. The MFN requires publishers doing business with Amazon to reveal the terms of the contracts made with its competitors. Amazon would then be in a position to demand equal or better terms. By marshaling MFN power over writers, they have been able to require significant concessions that reduce their costs and move them closer to monopoly (one effective seller) status including cost sharing, promotion, and release date. The European Community agreed with Amazon to end MFN in exchange for the end of a three-year anti-trust investigation (Vincent, 2017).

The continuing problems of the biggest tech companies—especially Amazon with its success across product lines—have moved many to suggest anti-trust investigations in the United States (Ip, 2018). Even if Amazon can avoid litigation as it did in Europe, there are other possibilities. The activities of the major ebook retailers (Amazon, Barnes & Noble, Apple, and Google) or the big five publishers (Holtzbrinck Publishing Group/Macmillan, Hachette, HarperCollins, Random House, and Simon & Schuster) could be seen as an oligopolistic (undesirably few sellers). Oligopolies can lead to collusive behavior that ultimately hurts the consumer.

Getting a Job

Despite the challenges, the ebook market has grown to the point that independent publishers can market a book without excessive costs. In 2016, there were 787,000 self-published books in both print and ebook format (Statista, n.d.) You no longer have to sell your book to the major publishers, but you do need to compete with them. If you are going to publish on your own, you will have to handle some of the overhead duties (promotion, financial management, editing, design, and more) yourself. Also, you must negotiate your own deal to distribute the book. For most, that means dealing with Amazon. On Amazon, you have the choice of acting as an independent publisher or as a part of the Amazon imprint. Your choices will affect your relationship or even availability to work with other ebook distributors.

Bibliography

American Associaion of Publishers. (2018, February 28). Book Publishers Revenues up 27.6% in October 2017. American Association of Publishers: Retrieved from http://Newsroom.publishers.org/media-library.

American Association of Publishers. (n.d.). Media for Download. American Association of Publishers: Retrieved from http://newsroom.publishers.org/media-library.

Anderson, C. (2009). *FREE: The Future of a Radical Price*. New York: Hyperion.

Audio Publishers Association. (n.d.). *A History of AudioBooks*. Audio Pubishers Association:Retrieved from https://www.audiopub.org/uploads/images/backgrounds/A-history-of-audiobooks.pdf.

Department of Justice. (n.d.). *Criminal Resource Manual*. Department of Justice: Retrieved from https://www.justice.gov/usam/criminal-resource-manual-1854-copyright-infringement-first-sale-doctrine.

Haines, D. (2018, Feb 25). The Ereader Device is Dying a Rapid Death. *Just Publishing Advice*: Retrieved from https://justpublishingadvice.com/the-e-reader-device-is-dying-a-rapid-death/.

Ip, G. (2018, January 16). The Antitrust Case Against Facebook, Google, and Amazon. *Wall Street Journal*: Retrieved from https://www.wsj.com/articles/the-antitrust-case-against-facebook-google-amazon-and-apple-1516121561.

Kahney, L. (2002, Aug 14). Hypercard Forgotten But Not Gone. *Wired*: Retrieved from https://www.wired.com/2002/08/hypercard-forgotten-but-not-gone/.

London, R. (2014, October 20). Big, bad Amazon. *The Economist*: Retrieved from https://www.economist.com/blogs/freeexchange/2014/10/market-power.

Meadows, C. (2017, Dec 19). The Internet Archive's OpenLibrary Project Violates Copyright, the Authors Guild Warns. *TeleRead*: Retrieved from https://teleread.org/2017/12/19/the-internet-archives-openlibrary-project-violates-copyright-the-authors-guild-warns/.

Miller, C. C., & Bosman, J. (2011, May 19). Ebooks Outsell Print Books at Amazon. *New York Times*: Retrieved from http://www.nytimes.com/2011/05/20/technology/20amazon.html.

Milliot, J. (2016, June 17). As E-book Sales Decline, Digital Fatigue Grows. *Publishers Weekly*: Retrieved from https://www.publishersweekly.com/pw/by-topic/digital/retailing/article/70696-as-e-book-sales-decline-digital-fatigue-grows.html.

Pierce, D. (2017, December 20). The Kindle changed the Book Business. Can it change books? *Wired*: Retrieved from https://www.wired.com/story/can-amazon-change-books/.

Primozic, U. (2015, March 5). Electronic Paper Explained. *Visionect*: Retrieved from https://www.visionect.com/blog/electronic-paper-explained-what-is-it-and-how-does-it-work/.

Redden, E. (2009, January 14). Toward and All E-Book Campus. *Inside Higher Ed*: Retrieved from https://www.insidehighered.com/news/2009/01/14/ebooks.

Statista. (2017, March). Book store sales in the United States from 1992 to 2015 (in billion U.S. dollars). *Statista*: Retrieved from https://www.statista.com/statistics/197710/annual-book-store-sales-in-the-us-since-1992/.

Statista. (n.d.). Ebooks. *Statista*: Retrieved from https://www.statista.com/statistics/249036/number-of-self-published-books-in-the-us-by-format/.

Suich, A. (n.d..). From Papyrus to Pixels. *Economist Essay*: Retrieved from https://www.economist.com/news/essays/21623373-which-something-old-and-powerful-encountered-vault.

Sweney, M. (2016, May 13). Printed Book Sales Rise for the First Time in Four Years as Ebook Sales Decline. *The Guardian*: Retrieved from https://www.theguardian.com/media/2016/may/13/printed-book-sales-ebooks-decline.

The Authors Guild. (2018, January 18). An Update on Open Library. *Industry & Advocacy News*: Retrieved from https://www.authorsguild.org/industry-advocacy/update-open-library/.

The Guardian. (2002, Jan 3). Ebooks Timeline. *The Guardian*: Retrieved from https://www.theguardian.com/books/2002/jan/03/ebooks.technology.

The Telegraph. (2018, February 17). Google Editions: a history of ebooks. *The Telegraph*: Retrieved from http://www.telegraph.co.uk/technology/google/8176510/Google-Editions-a-history-of-ebooks.html.

Trachtenberg, J. A. (2010, May 21). E-Books Rewrite Bookselling. *Wall Street Journal*: Retrieved from https://www.wsj.com/articles/SB10001424052748704448304575196172206855634.

Vincent, J. (2017, May 4). Amazon Will Change Its Ebook Contracts with Publishers as EU ends Antitrust Probe. *The Verge*: Retrieved from https://www.theverge.com/2017/5/4/15541810/eu-amazon-ebooks-antitrust-investigation-ended

Wagner, K. (2011, Sept 28). The History of the Amazon Kindle So Far. *Gizmodo*: Retrieved from https://gizmodo.com/5844662/the-history-of-amazons-kindle-so-far/.

Wilson, J. L. (2015, March 25). Everything You Need to Know About Digital Comics. *PC Magazine*: Retrieved from https://www.pcmag.com/article2/0,2817,2425402,00.asp.

Section IV

Networking Technologies

21

Broadband & Home Networks

John J. Lombardi, Ph.D.[*]

Overview

Broadband is a simple way of saying "high-speed" Internet. In the U.S., transmission rates of 25 Mb/s or greater is considered "high-speed." But why is understanding what "broadband" technology is so important? More and more of our country's economy, infrastructure, and security involves broadband technology. It takes little more than to realize that Jeff Bezos, CEO of Amazon, became the nation's wealthiest person in 2018 (with an estimated net worth of more than $100 billion) to understand the prevalence of internet-based shopping. Our nation's utility grids rely on high-speed data flow to manage usage. Our nation's security, including first responders, requires that information be sent and received in a rapid, reliable, and secure manner. And this says nothing of the entertainment value of surfing the web, connecting with family and friends via social networking sites, and streaming audio and video content.

In 2010 the FCC developed a "National Broadband Plan" with the goal of vastly expanding broadband access and increasing transmission rates. Future developments will come in two primary areas: broadband accessibility (including increased transmission rates) and the number and type of broadband-connected devices. Currently, the FCC is in the process of re-claiming and reallocating unused spectrum space for broadband development. And as broadband access becomes more prevalent, more devices—phones, computers, appliances, cars, trucks, and so on—will become connected.

Introduction

> *We are all now connected by the Internet,*
> *like neurons in a giant brain.*
> —*Stephen Hawking: Theoretical Physicist*

More simply, and perhaps more commonly, Tim Berners-Lee, the man credited with developing the world wide web, has said "Anyone who has lost track of time when using a computer knows the propensity to dream, the urge to make dreams come true, and the tendency to miss lunch" (FAQ, n.d.). The world wide web is now nearly three decades old and it is deeply engrained in the daily lives of millions of people worldwide. But the world wide web is just part of the growing Internet experience. The Internet is used in virtually all aspects of our lives. In addition to surfing the web, the Internet allows users to share information with one another and to connect to a myriad of electronic devices (see Chapter 12).

The Internet can be used to send and receive photos, music, videos, phone calls, or any other type of data.

[*] Professor of Mass Communication. Frostburg State University. Frostburg Maryland

There is an estimated 7-8 zettabytes of digital information in the world (one zettabyte is a 1 followed by twenty-one 0s. And this number that is expected to double every two years (ITU, 2012a). With benefits to healthcare, energy consumption, and an improved global economy the importance of high speed Internet access around the globe cannot be overstated.

The term "broadband" is generally used to describe high speed Internet. What constitutes "high speed," however, varies a bit from country-to-country. In the U.S., the FCC defines broadband as connectivity with speeds greater than or equal to 25 Mbps download and 3 Mbps upload (FCC, 2015). The Organisation for Economic Co-Operation and Development (OECD), an organization based in France that collects and distributes international economic and social data, considers broadband any connection with speeds of at least 1 to 1.5 Mb/s (the lowest tier they measure) (OECD, 2017).

While the FCC defines broadband as connections equal to or greater than 25 Mbps upload and 3 Mbps download, actual speeds in the U.S. are generally slower, but they continue to rise. One source ranks the U.S. 16th internationally with an average download speed of 12.6 Mb/s. South Korea is at the top of the international list at 20.5 Mb/s download speeds. The global average is 5.1 Mb/s (Belson, 2015).

In recent years, the U.S. has pledged to make significant strides in national broadband coverage by reaffirming its commitment to the National Broadband Plan, which was released by the FCC in 2010. This plan includes six long-term goals. (See Figure 21.1)

Both wired and wireless broadband penetration rates continue to grow. The OECD estimates that broadband penetration rates (fixed and wireless) have reached nearly 100% within the 35 country OECD group. (OECD, 2017). This number is a bit misleading since the bulk of subscribers are using mobile broadband and countless subscribers have multiple devices (each device counting within the penetration rate data). As of 2017 wired broadband penetration rates of the 34 OECD countries average 30.1%, with Switzerland leading the list with a 50.1% wired broadband penetration rate. The U.S. is at 32.8%. (OECD, 2017).

Table 21.1
National Broadband Plan Goals

Goal 1	At least 100 million U.S. homes should have affordable access to actual download speeds of at least 100 Mb/s and actual upload speeds of at least 50 Mb/s.
Goal 2	The U.S. should lead the world in mobile innovation, with the fastest and most extensive wireless networks of any nation.
Goal 3	Every American household should have affordable access to robust broadband service, and the means and skills to subscribe if they so choose.
Goal 4	Every American community should have affordable access to at least 1 Gb/s broadband service to anchor institutions such as schools, hospitals, and government buildings.
Goal 5	To ensure the safety of the American people, every first responder should have access to a nationwide, wireless, interoperable broadband public safety network.
Goal 6	To ensure that America leads in the clean energy economy, every American should be able to use broadband to track and manage their real-time energy consumption (FCC, 2010).

Source: FCC

In terms of wireless broadband, the OECD average penetration rate has risen to nearly 100%. Again, this number is misleading since penetration rates are determined by the number of subscriptions, and individuals frequently have multiple subscriptions. Nonetheless, Japan leads this list at a 152.4% penetration rate (OECD, 2017). In 2014, a total of six countries (Australia, Denmark, Finland, Korea, Japan, and Sweden) had eclipsed the 100% penetration mark. In 2017 that list grew to 11 countries (Australia, Denmark, Estonia, Finland, Iceland, Ireland, Japan, Korea, New Zealand, Sweden, and the United States). Switzerland dropped below the 100% penetration rate. With the exponential growth of tablets and smartphones, it is becoming more and more common for people to have multiple access points to broadband services.

The increasing connection speeds associated with broadband technology allow for users to engage in such bandwidth intensive activities such as "voice-over-Internet-protocol" (VoIP) including video phone usage, "Internet protocol television" (IPTV) including increasingly complex video services such as Verizon's FiOS service, and interactive gaming. Additionally, the "always on" approach to broadband allows for consumers to easily create wireless home networks.

A wireless home network can allow for multiple computers or other devices to connect to the Internet at one time. Such configurations can allow for wireless data sharing between numerous Internet protocol (IP) devices within the home. Such setups can allow for information to easily flow from and between devices such as desktop computers, laptop computers, tablets, smartphones, audio/video devices such as stereo systems, televisions, DVD players, and even home appliances.

As an example, with a home wireless broadband network you could view videos on your television that are stored on your computer or listen to music that is stored on your computer through your home theater system. Additionally, video services such as Netflix and Hulu allow subscribers to access certain content instantly. The key device in most home networks is a residential gateway, or router. Routers are devices that link all IP devices to one another and to the home broadband connection.

This chapter briefly reviews the development of broadband and home networks, discusses the types and uses of these technologies, and examines the current status and future developments of these exciting technologies.

Background

Broadband networks can use a number of different technologies to deliver service. The most common broadband technologies include digital subscriber line (DSL), cable modem, satellite, fiber cable networks, and wireless technologies. Thanks in part to the *Telecommunication Act* of 1996, broadband providers include telephone companies, cable operators, public utilities, and private corporations.

DSL

Digital subscriber line (DSL) is a technology that supplies broadband Internet access over regular telephone lines with service being provided by various local carriers nationwide. There are several types of DSL available, but asymmetrical DSL (ADSL) is the most widely used for broadband Internet access. "Asymmetrical" refers to the fact that download speeds are faster than upload speeds. This is a common feature in most broadband Internet network technologies because the assumption is that people download more frequently than upload, and they download larger amounts of data.

With DSL, the customer has a modem that connects to a phone jack. Data moves over the telephone network to the central office. At the central office, the telephone line is connected to a DSL access multiplexer (DSLAM). The DSLAM aggregates all of the data coming in from multiple lines and connects them to a high-bandwidth Internet connection.

A DSL connection from the home (or office) to the central office is not shared. As such, individual connection speeds are not affected by other users. However, ADSL is a distance sensitive technology. This means that the farther your home is from the central office, the slower your connection speed will be. Also, this technology only works within 18,000 feet (about 3 ½ miles) of the central office (though "bridge taps" may be used to extend this range a bit).

ASDL typically offers download speeds up to 3 Mb/s. Some areas have more advanced DSL services called ADSL2, ADSL2+, and more recently, ADSL 2++ (sometimes referred to ADSL 3 or ADSL 4). These services offer higher bandwidth, up to 12 Mb/s with ADSL2, 24 Mb/s with ADSL2+, and 45 Mb/s with ADSL 2++.

FTTN

Fiber-to-the-node is a hybrid form of DSL often times referred to as VDSL (very high bit-rate DSL). This service, used for services such as Verizon's FiOS and AT&T's U-verse in most areas (some areas offer purely FTTH, discussed later). FTTN systems generally peak at 100 Mb/s downstream. This system employs a fiber optic cable that runs from the central office to a node in individual neighborhoods. The neighborhood node is a junction box that contains a VDSL gateway that converts the digital signal on the fiber optic network to a signal that is carried on ordinary copper wires to the residence.

FTTH

Fiber-to-the-home employs fiber optic networks all the way to the home. Fiber optic cables have the advantage of being extremely fast (speeds up to 1 Gb/s) and are the backbone of both cable and telecommunications networks. Extending these networks to the home is still somewhat rare due to cost constraints. However, costs are coming down, and at least one Internet company made a metropolitan area a high-speed Internet guinea pig.

In February 2012 Google announced that it would begin its test of high speed FTTH by wiring the twin cities of Kansas City, Kansas and Kansas City, Missouri with fiber optic cabling and networks cable of generating speeds up to 1 Gb/s (Kansas City is Fiber-Ready, 2012). As of early 2017 the service is also available in Huntsville, Alabama, Orange County, California, Atlanta, Georgia, Louisville, Kentucky, Charlotte, Chapel Hill, Durham and Raleigh, North Carolina, Nashville, Tennessee, Austin and San Antonio, Texas, and Provo and Salt Lake City, Utah. Google is planning to expand this service to more cities in the future (Google fiber, n.d.).

It's becoming more difficult for consumers to know whether their service is FTTN or FTTH as system-wide infrastructure upgrades are on-going. Verizon's FiOS, as an example, is FTTN in some areas and FTTH in other areas. As of 2018, Verizon's FiOS service offers speeds from 100 Mb/s for $39.99 a month up to 1 Gb/s downstream for $79.99 per month, in select areas (Verizon, n.d.).

Cable Modem

Cable television providers also offer Internet service. In their systems, a customer's Internet service can come into the home on the same cable that provides cable television service and for some, telephone service.

With the upgrade to hybrid fiber/coaxial cable networks, cable television operators began offering broadband Internet access. But how can the same cable that supplies your cable television signals also have enough bandwidth to also supply high speed Internet access? They can do this because it is possible to fit the download data (the data going from the Internet to the home) into the 6 MHz bandwidth space of a single television channel. The upload speed (the data going from the computer back to the Internet) requires only about 2 MHz of bandwidth space.

In the case of Internet through a cable service, the signal travels to the cable headend via the cable modem termination system (CMTS). The CMST acts like the DSLAM of a DSL service. From the cable headend, the signal travels to a cable node in a given neighborhood. A coaxial cable then runs from the neighborhood node to the home.

Cable modems use a standard called "data over cable service interface specifications" (DOCSIS). First generation DOCSIS 1.0, which was used with first-generation hybrid fiber/coax networks, was capable of providing bandwidth between 320 Kb/s and 10 Mb/s. DOCSIS 2.0 raised that bandwidth to up to 30 Mb/s (DOCSIS, n.d.). DOCSIS 3.0 provides bandwidth well in excess of 100 Mb/s. In fact, some modem chipsets can bond up to eight downstream channels thus creating the possibility of delivering up to 320 Mb/s (DOCSIS 3.0, 2009). More recently DOCSIS 3.1 is reportedly able to provide speeds of up to 10 Gb/s downstream and up to 1 Gb/s upstream (Featured technology, nd).

According to the Organisation for Economic Co-operation and Development (OECD), approximately 32.7% of all Internet households receive their service from cable providers (see Figure 21.1).

More cable Internet providers are moving toward FTTN systems (e.g. Verizon's FiOS and Comcast's XFINITY). As such it is becoming increasingly difficult to accurately compare costs and speeds of traditional cable modem Internet access.

Although cable Internet provides fast speeds and, arguably, reasonable rates, this service is not without problems. Unlike DSL, cable Internet users share bandwidth. This means that the useable speed of individual subscribers varies depending upon the number of simultaneous users in their neighborhood.

Figure 21.1

Fixed (Wired) Broadband Subscriptions, by Technology, December 2016

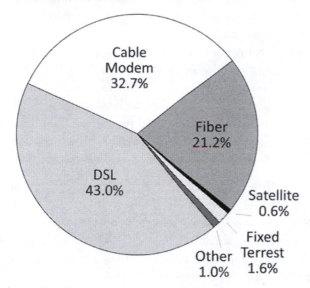

Total subscriptions: 386.8 million

Source: OECD 2016

Satellite

For those people who live out of DSL's reach and in rural areas without cable, satellite broadband Internet access is an option. With this service, a modem is connected to a small satellite dish which then communicates with the service providers' satellite. That satellite, in turn, directs the data to a provider center that has a high-capacity connection to the Internet.

Satellite Internet service cannot deliver the bandwidth of cable or DSL, but speeds are a great improvement over dial-up. For example, HughesNet offers home service with 25 Mb/s download and 3 Mb/s upload with a 10 Gb per month "soft" data limit (exceeding the data limit will result in reduced speeds) for $49.99 a month (HughesNet, n.d.). The cost is not the only disadvantage, as most satellite Internet providers impose a daily data limit too.

Wireless

There are two primary types of wireless broadband networks: mobile and fixed. Mobile broadband networks are offered by wireless telephony companies and employ 3G and 4G networks (discussed in more detail in Chapter 21). Second generation (2G) mobile broadband networks generally use the Enhanced Data GSM Environment (EDGE) protocol (some refer to this as 2.75G because it is better than traditional 2G, but not quite at the level of true 3G networks). Third generation (3G) networks generally use Evolution, Data Optimized (EVDO), or High-Speed Uplink Packet Access (HSUPA). Fourth generation (4G) networks use Long Term Evolution (LTE).

In January 2012 the International Telecommunication Union agreed on specifications for "IMT-Advanced" mobile wireless technologies (this includes what is commonly referred to as LTE-Advanced in the U.S.). It is this technological standard that is employed in 4G mobile broadband networks and is touted as being "at least 100 times faster than today's 3G smartphones" (ITU, 2012b, para. 5).

5G wireless technology is on the horizon. As discussed in Chapter 21, however, there is still debate regarding the true definition of "5G." While industry insiders continue to develop a working definition of 5G technology, cellular providers are moving forward. Recently, AT&T announced its "5G Evolution" network. Industry experts suggest that this is not true 5G technology. Instead it's an advanced, gigabit LTE network (Segan, 2017). Other providers have announced similar plans. While there are technological differences between a true 5G network and a "gigabit LTE network" the reality is that a gigabit LTE network is capable of generating a theoretical peak of 1 Gb/s data rates (Miller, 2017).

It is likely that true 5G technology will utilize networks on small cells rather than traditional cell towers because the high frequencies proposed for 5G can't easily penetrate walls. While much still needs to play out, the goal of true 5G technologies is to reach speeds up to 20 Gb/s with very little latency (Segan, 2017).

Fixed broadband wireless networks use either Wi-Fi or WiMAX. Wi-Fi uses a group of standards in the IEEE 802.11 group to provide short-range, wireless Internet access to a range of devices such as laptops, cellphones, and tablets. Wi-Fi "hotspots" can be found in many public and private locations. Some businesses and municipalities provide Wi-Fi access for free. Other places charge a fee.

WiMAX , which stands for worldwide interoperability for microwave access, is also known as IEEE 802.16. There are two versions: a fixed point-to-multipoint version and a mobile version. Unlike Wi-Fi which has a range of 100 to 300 feet, WiMAX can provide wireless access up to 30 miles for fixed stations and three to ten miles for mobile stations (What is Wi-MAX , n.d.). In 2012 Clearwire, a WiMAX service, was acquired by Sprint. The expectation, at the time, was that WiMAX coverage would rapidly grow. However, in 2016, Sprint shuttered its entire WiMAX service (Update, 2015).

BPL

Broadband over power line (BPL) was, at one time, thought to be the wave of the future. Given that power lines went into every home it was easy to understand how convenient it would be to use these cables to send broadband data into homes. The modem would actually be plugged into an electrical outlet in the subscriber's home as a means of obtaining the service. However, several factors have caused this technology to lose its appeal. BPL is quite susceptible to interference from radio frequencies, and other broadband services provide a faster and more reliable connection. Today this technology, as it applies to bringing Internet into the home, is pretty much obsolete. The use of this technology within the home, however, is still ongoing. The primary manner in which BPL is used in the home is simply to help increase the range of in-home Wi-Fi connectivity (Courtney, 2013).

Home Networks

Computer networking was, at one time, only found in large organizations such as businesses, government offices, and schools. The complexity and cost of such networking facilities was beyond the scope of most home computer owners. At one time, a computer network required the use of an Ethernet network and expensive wiring called "Cat" (category) 5 (or, more recently, "Cat 6, 7). Cat 7 cables support up to 10 Gb/s data rates. Cat 8 is still in development, but is expected to be able to deliver speeds of up to 40 Gb/s. Additionally, a server, hub, and router were needed. And all of this required someone in the household to have computer networking expertise as network maintenance was regularly needed.

Several factors changed the environment to allow home networks to take off: broadband Internet access, multiple computer households, and new, networked consumer devices and services. Because of these advances, a router (costing as little as $30) can be quickly installed. This router essentially splits the incoming Internet signal and sends it (either through a wired or wireless connection) to other equipment in the house. Computers, cellphones, tablets, televisions, DVD players, stereo receivers, and other devices can be included within the home network. As technology and connectivity evolve we see more and more home appliances like security systems, thermostats, ranges, refrigerators, washers, and dryers include Wi-Fi connectivity.

With a router, home network users can, among other things, send video files from their computer to their television; they can send audio files from their computer to their stereo receiver; they can send a print job from their smartphone to their printer; or with some additional equipment, they can use their smartphone to control home lighting and other electrical devices within the home.

There are two broad types of home networks:

- *Wired networks*—including ethernet, phone lines, and power lines

- *Wireless networks*—including Wi-Fi, Bluetooth, and Zigbee

When discussing each type of home network, it is important to consider the transmission rate, or speed, of the network. Regular file sharing and low-bandwidth applications such as home control may require a speed of 1 Mb/s or less. The MPEG-2 digital video and audio from DBS service requires a speed of 3 Mb/s; DVD-quality video requires between 3 Mb/s and 8 Mb/s, and compressed high-definition television (HDTV) requires around 20 Mb/s. As content providers move further into the 4K video world, increased network speeds will be necessary. Still, though, with compression technology content providers such as Netflix say that a steady 25 Mb/s speed is adequate. It is important to remember that not all ISPs provide these higher speeds and not all home networking technology supports the transmission of these higher speeds.

Wired Networks

Traditional networks use Ethernet, which has a data transmission rate of 10 Mb/s to 100 Mb/s. There is also Gigabit Ethernet, used mostly in business, that has transmission speeds up to 1 Gb/s. Ethernet is the kind of networking commonly found in offices and universities. As discussed earlier, traditional Ethernet has not been popular for home networking because it is expensive to install and maintain and difficult to use. To direct the data, the network must have a server, hub, and router. Each device on the network must be connected, and many computers and devices require add-on devices to enable them to work with Ethernet. Thus, despite the speed of this kind of network, its expense and complicated nature make it somewhat unpopular in the home networking market, except among those who build and maintain these networks for offices.

Many new housing developments come with "structured wiring" that includes wiring for home networks, home theatre systems, and other digital data networking services such as utility management and security. One of the popular features of structured wiring is home automation including the ability to unlock doors or adjust the temperature or lights. New homes represent a small fraction of the potential market for home networking services and equipment, so manufacturers have turned their attention to solutions for existing homes. These solutions almost always are based on "no new wires" networking solutions that use existing phone lines or power lines or are wireless.

Phone lines are ideal for home networking. This technology uses the existing random tree wiring typically found in homes and runs over regular telephone wire—there is no need for Cat 5 wiring. The technology uses frequency division multiplexing (FDM) to allow data to travel through the phone line without interfering with regular telephone calls or DSL service. There is no interference because each service is assigned a different frequency.

The Home Phone Line Networking Alliance (HomePNA) has presented several standards for phone line networking. HomePNA 1.0 boasted data transmission rates up to 1 Mb/s. It was replaced by HomePNA (HPNA) 2.0, which boasts data transmission rates up to 10 Mb/s and is backward-compatible with HPNA 1.0. HomePNA 3.1 provides data rates up to 320 Mb/s and operates over phone wires and coaxial cables, which makes it a solution to deliver video and data services (320 Mbps, n.d.).

In 2013, the HomePNA merged with the Home-Grid Forum. This body has developed a new wired home networking standard referred to as G.hn. This new standard is expected to allow for in-home networks to transfer data at speeds of up to 1 Gb/s. With the demand for more and more data to be pushed across home networks, thanks to technological evolutions such as HDTV and, increasingly, 4k video content, higher transmission speeds will be expected (Claricoats, 2016).

Wireless

The most popular type of home network is wireless. Currently, there are several types of wireless home networking technologies: Wi-Fi (otherwise known as IEEE 802.11a, 802.11b 802.11g, 802.11n, and 802.11ac), Bluetooth, and wireless mesh technologies such as ZigBee. Mesh technologies are those that do not require a central control unit.

Wi-Fi, Bluetooth, and ZigBee are based on the same premise: low-frequency radio signals from the instrumentation, science, and medical (ISM) bands of spectrum are used to transmit and receive data. The ISM bands, around 2.4 GHz, not licensed by the FCC, are used mostly for microwave ovens, cordless telephones, and home networking. Two standards, 802.11a and 802.11ac operate at the higher 5 GHz frequency.

Wireless networks utilize a transceiver (combination transmitter and receiver) that is connected to a wired network or gateway (generally a router) at a fixed location. Much like cellular telephones, wireless networks use microcells to extend the connectivity range by overlapping to allow the user to roam without losing the connection (Wi-Fi Alliance, n.d.).

Wi-Fi is the most common type of wireless networking. It uses a series of similarly labeled transmission protocols (802.11a, 802.11b, 802.11g, 802.11n, and 802.11ac). Wi-Fi was originally the consumer-friendly label attached to IEEE 802.11b, the specification for wireless Ethernet. 802.11b was created in July 1999. It can transfer data up to 11 Mb/s and is supported by the Wi-Fi Alliance. A couple years later 802.11a was introduced, providing bandwidth up to 54 Mb/s. This was soon followed by the release of 802.11g, which combines the best of 802.11a and 802.11b, providing bandwidth up to 54 Mb/s. The 802.11n standard was released in 2007 and amended in 2009 and provides bandwidth over 100 Mb/s (Mitchell, n.d.). The 802.11ac standard, released in 2012 allows for speeds up to 1.3 Gb/s. However, because this standard will operate only in the 5 GHz frequencies, the transmission range of this Wi-Fi standard could be smaller than that of 802.11n Wi-Fi (Vaughan-Nichols, 2012).

There are two other emerging standards as well. Currently, the 802.11ad standard operates in the 60 GHz frequencies with throughput speeds up to 7 Gb/s (Poole, n.d.a). Another standard is also in the developmental stages. The 802.11af or "White-Fi" standard would utilize low power systems working within "white space" (the unused frequency spectrum space between television signals). There are two primary benefits to this approach. Using frequencies in this portion of the spectrum would allow for greater coverage areas. Additionally, this standard could accommodate greater bandwidth. While this standard is still being developed, it does look promising (Poole, n.d.b).

Because wireless networks use so much of their available bandwidth for coordination among the devices on the network, it is difficult to compare the rated speeds of these networks with the rated speeds of wired networks. For example, 802.11b is rated at 11 Mb/s, but the actual throughput (the amount of data that can be effectively transmitted) is only about 6 Mb/s. Similarly, 802.11g's rated speed of 54 Mb/s yields a data throughput of only about 25 Mb/s. Tests of 802.11n have confirmed speeds from 100 Mb/s to 140 Mb/s. Actual speed of the 802.11ac protocol are expected to top out at about 800 Mb/s (Marshall, 2012).

Security is an issue with any network. Wi-Fi uses two types of encryption: WEP (Wired Equivalent Privacy) and WPA (Wi-Fi Protected Access). WEP has security flaws and is easily hacked. WPA fixes those flaws in WEP and uses a 128-bit encryption. There are

two versions: WPA-Personal that uses a password and WPA-Enterprise that uses a server to verify network users (Wi-Fi Alliance, n.d.). WPA2 is an upgrade to WPA and is now required of all Wi-Fi Alliance certified products (WPA2, n.d.).

While the most popular version of Wi-Fi can transmit data up to 140 Mb/s for up to 150 feet (depending upon which protocol), Bluetooth was developed for short-range communication at a data rate of up to 3 Mb/s and is geared primarily toward voice and data applications. Bluetooth technologies are good for transmitting data up to 10 meters. Bluetooth technology is built into devices such as laptop computers, music players (including car stereo systems), and cellphones.

Bluetooth-enhanced devices can communicate with each other and create an ad hoc network. The technology works with and enhances other networking technologies. Bluetooth 4.0 is the current standard. The main advantage of Bluetooth 4.0 is that it requires less power to run thus making it useable in more and more (and smaller and smaller) devices (Lee, 2011).

Versions of Bluetooth include 4.1 and 4.2. Version 4.1 had limited noticeable changes aside from better security encryption. Version 4.2, however, saw an increase in data transfer rates up to 2.5 times faster than v. 4.0 (Bluetooth 4.2, n.d.).

More recently Bluetooth 5.0 has emerged. The primary difference between Bluetooth 4.2 and Bluetooth 5.0 is not in transmission speeds, but in transmission range. Bluetooth 5.0 has an estimated range of approximately 200 meters (Bluetooth 5, 2017).

ZigBee, also known as IEEE 802.15.4, is classified, along with Bluetooth, as a technology for wireless personal area networks (WPANs). Like Bluetooth 4.0, ZigBee's transmission standard uses little power. It uses the 2.4 GHz radio frequency to deliver data in numerous home and commercial devices (ZigBee, n.d.).

Usually, a home network will involve not just one of the technologies discussed above, but several. It is not unusual for a home network to be configured for HPNA, Wi-Fi, and even traditional Ethernet. Table 21.2 compares each of the home networking technologies discussed in this section.

Table 21.2
Comparison of Home Networking Technologies

Protocol	How it Works	Standard(s)	Specifications
Ethernet	Uses Cat 5, 5e, 6, 6a, or 7 wiring with a server and hub to direct traffic. Cat 8 is in development	IEEE 802.3xx IEEE 802.3.1	10 Mb/s (Cat 5) to 10 Gb/s (Cat 7) 40 Gb/s (Cat 8)
HomePNA/ HomeGrid Forum	Uses existing phone lines and OFDM	HPNA 1.0 HPNA 2.0 HPNA 3.0 HPNA 3.1 G.hn	1.0, up to 1 Mb/s 2.0, 10 Mb/s 3.0, 128 Mb/s 3.1, 320 Mb/s 1 Gb/s
IEEE 802.11a Wi-Fi	Wireless. Uses electro-magnetic radio signals to transmit between access point and users	IEEE 802.11a 5 GHz	Up to 54 Mb/s
IEEE 802.11b Wi-Fi	Wireless. Uses electro-magnetic radio signals to transmit between access point and users	IEEE 802.11b 2.4 GHz	Up to 11 Mb/s
IEEE 802.11g Wi-Fi	Wireless. Uses electro-magnetic radio signals to transmit between access point and users	IEEE 802.11g 2.4 GHz	Up to 54 Mb/s
IEEE 802.11n Wi-Fi	Wireless. Uses electro-magnetic radio signals to transmit between access point and users	IEEE 802.11n 2.4 GHz	Up to 140 Mb/s
IEEE 802.11ac Wi-Fi	Wireless. Uses electro-magnetic radio signals to transmit between access point and users	IEEE 802.11ac 5 GHz	Up to 1.3 Gb/s
IEEE 802.11ad Wi-Fi	Wireless. Uses electro-magnetic radio signals to transmit between access point and users	IEEE 802.11ad 60 GHz	Up to 7 Gb/s
Bluetooth	Wireless	v.1.0 (2.4 GHz) v.2.0 + (EDR) v.3.0 (802.11) v.4.0 (802.11) v.4.1 (802.11) v.4.2 (802.11) v.5.0 (802.11n)	v.1.0 (1 Mb/s) v.2.0 (3 Mb/s) v.3. 0 (24 Mb/s) v.4.0 (24 Mb/s + lower power) v.4.1 (added security + lower power) v.4.2 (approx. 50 Mb/s + lower power + added security) v.5.0 (lower power + added security + in-creased range)
Powerline	Uses existing power lines in home	HomePlug v1.0 HomePlug AV HPCC HomePlug AV2	v. 1 (Up to 14 Mb/s) AV (Up to 200 Mb/s) AV2 (Up to 500 Mb/s)
ZigBee	Wireless. Uses Electro-magnet radio signals to transmit between access point and users	IEEE 802.15.4	250 Kb/s
Z-wave	Uses 908 MHz 2-way RF	Proprietary	100 Kbps

Source: J. Lombardi and J. Meadows

Residential Gateways

The residential gateway, also known as the broadband router, is what makes the home network infinitely more useful. This is the device that allows users on a home network to share access to their broadband connection. As broadband connections become more common, the one "pipe" coming into the home will most likely carry numerous services such as the Internet, phone, and entertainment. A residential gateway seamlessly connects the home network to a broadband network so all network devices in the home can be used at the same time.

Working Together—The Home Network and Residential Gateway

A home network controlled by a residential gateway or central router allows multiple users to access a broadband connection at the same time. Household members do not have to compete for access to the Internet, printer, television content, music files, or movies. The home network allows for shared access of all controllable devices.

Technological innovations have made it possible to access computer devices through a home network in the same way as you would access the Internet. Televisions and Blu-ray DVD players regularly are configured to access streamed audio and video content without having to funnel it through a computer.

Cellphone and tablet technology is more regularly being used to access home networks remotely. With this technology it is now possible to set your DVR to record a show or to turn lights on and off without being in the home. The residential gateway or router also allows multiple computers to access the Internet at the same time. This is accomplished by creating a "virtual" IP address for each computer. The residential gateway routes different signals to appropriate devices in the home.

Home networks and residential gateways are key to what industry pundits are calling the "smart home." Although having our washing machine tell us when our clothes are done may not be a top priority for many of us, utility management, security, and enhanced telephone services are just a few of the useful applications for this technology. Before these applica-

tions can be implemented, however, two developments are necessary. First, appropriate devices for each application (appliance controls, security cameras, telephones, etc.) have to be configured to connect to one or more of the different home networking topologies (wireless, HPNA,). Next, software, including user interfaces, control modules, etc., needs to be created and installed. It is easy to conceive of being able to go to a web page for your home to adjust the air conditioner, turn on the lights, or monitor the security system, but these types of services will not be widely available until consumers have proven that they are willing to pay for them.

Broadband technology can be used to improve home networks to allow for the control of home appliances, heating/cooling systems, sprinkler systems, and more. This allows for homeowners to continually and expeditiously monitor resource consumption. According to the FCC's National Broadband Plan, consumers who can easily monitor their own consumption are more likely to modify their usage thus eliminating or at least reducing waste.

Net Neutrality

The issue of "net neutrality" continues to evolve. Prior to the Obama Administration, it was possible for Internet Service Providers (ISPs) to block or prioritize access to web content. What this means is that ISPs could, if they chose to do so, prevent or restrict subscriber access to certain websites or web content.

The Internet had, for many years, operated as a global symbiosis. That is, each ISP would allow content to freely flow, even when that content would come from a direct competitor. (Keep in mind, ISPs provide Internet access, but they often times provide content as well). In turn, the competing ISP would also allow the free flow of content. So, while Company A may be competing with Company B, Company B would allow Company A's content to flow uninterrupted…and vice versa.

However, since 2010 there has been growing concern that this symbiosis would change. The concern was that one ISP would restrict or block access to content provided by another ISP or vice versa. This could have a significant impact on the consumer. As an example, let's say you get your Internet from Company

A, but you also subscribe to video content from Company B. Company A could slow down your access to Company B's content, prevent you from accessing Company B's content altogether, or charge you extra to access Company's B's content.

The main fear—that ISPs will deliberately alter pass-through rates of various content providers—is being realized. Netflix, one of the largest online video content providers in the U.S., is increasingly finding itself battling some of the nation's largest ISPs. Because of the increased popularity of Netflix and similar content providers, the amount of data being passed through is quickly and exponentially increasing. Internet providers, such as Comcast, that also provide video content, dislike the idea of allocating so much of their bandwidth to competing program providers. This has led to behind-the-scenes negotiations taking place in order to reach what are being made called "peering" agreements. In exchange for some level of compensation, an ISP will increase the pass-through speed for certain content providers (Gustin, 2014). This is the type of deal Netflix made with Comcast in spring 2014.

The Obama administration and the FCC, however, wanted to prevent this from happening. The proponents of net neutrality believe this is a free speech issue. They suggest that ISPs who block or prioritize access to certain Web content can easily direct users to certain sites and away from others or increase the prices for access to certain content. Regulators are concerned that large broadband players such as Verizon and Comcast could have an unfair advantage. The fear is that allowing ISPs to give preferential treatment to some content companies could stifle innovation (Wyatt, 2014).

Net neutrality opponents believe such regulations would serve only to minimize investment (Bradley, 2009). On April 6, 2010 the court ruled that the FCC has only limited power over Web traffic under current law. As such, the FCC cannot tell ISPs to provide equal access to all Web content (Wyatt, 2010).

Despite this setback, the FCC remained committed to a free and open Internet and took another crack at the net neutrality issue. In early 2014 regulators reaffirmed this commitment when they unveiled a new plan. The new plan was an attempt to accomplish virtually the same thing as the plan the U.S. Court of Appeals disliked, but with some technical differences the Commission hoped would pass muster.

In February 2015 the FCC voted, 3-2 to implement net neutrality. However, industry groups and some in Congress are not pleased and believe the FCC overstepped its authority. Oral arguments in one suit were heard in late 2015. Oral arguments in another suit were heard in early 2016 (Brodkin, 2016). Net neutrality was implemented…for a while.

With President Obama out of office in 2017, a new FCC was installed. Trump's FCC voted to repeal net neutrality despite overwhelming public support for the regulations. But proposals to reinstate net neutrality have been considered by the U.S. Congress.

In the meantime, the FCC is attempting to reclassify cellphone data as "broadband Internet". As mentioned earlier, the U.S. is attempting to implement the National Broadband Plan. Goal 1 of the plan suggests that "at least 100 million U.S. homes will have affordable access to actual download speeds of at least 100 Mg/s…" (FCC, 2010). Since more and more Americans have cellular data plans, reclassifying cellphone data as "broadband Internet" would, on paper, help achieve this goal. However, some believe that this will actually hurt long-term, widespread improvements. "This would not only camouflage many of the communities in the U.S. with no access to the internet, but could prevent them from getting necessary funding to build that access" (Rogers, 2017, para 3).

Recent Developments

In order for the FCC to make significant inroads regarding the National Broadband Plan, significant funds will need to be raised and more spectrum space will need to be allocated for wireless networking (or, more likely, reallocated).

In his 2014 State of the Union address President Obama reaffirmed his commitment to the National Broadband Plan when he called for the connection of 99% of that nation's schools to high-speed broadband service. This announcement met with quick support from the Fiber to the Home Council. The FTTH Council president, Heather Burnett Gold said "As we noted in our petition recommending this action to the Commission: these experiments will enable local creativity to identify the best options for the future, spurring innovation and job creation, AND empowering and connecting communities in areas of the country often left behind" (Brunner, 2014, para. 6).

Additionally, the FCC is reforming the universal service and inter-carrier compensation systems. The "universal service fund" is basically a surcharge placed on all phone and Internet services. Inter-carrier compensation is basically a fee that one carrier pays another to originate, transport, or terminate various telephony related signals. Jointly these fees generate about $4.5 billion annually. These fees will now go into a new "Connect America Fund" which is designed to expand high-speed Internet and voice service to approximately 7 million Americans in rural areas in a six-year period. Additionally, about 500,000 jobs and $50 billion in economic growth is expected (FCC, 2011a). The implementation of this is ongoing.

In terms of accessing additional spectrum space, there are two ways this can be done. The first way is to tap into the unused space that is located between allocated frequencies. This is referred to as "white space." In December 2011 the FCC announced that the Office of Engineering and Technology (OET) approved the use of a "white spaces database system." This is thought to be the first step toward using this available spectrum space. Expanding wireless services would be a primary use for this space (FCC, 2011b). In 2013 Google launched its "white space database." Through the Internet giant, people can quickly search for unused white space (Fitchard, 2013). Since Google pioneered this business, several other companies have launched white space databases.

There are several primary advantages to claiming and using white space. First, it is currently unused. The primary advantage, however, comes in its coverage area. Compared to, for example, a wireless Wi-Fi router (which can transmit a signal approximately 300 feet under optimal conditions) white space can be used to transmit a signal within an approximate six-mile radius (TV Whitespace, n.d.). If significant progress is going to be made in implementing the goals

of the National Broadband Plan, considerably more coverage area will be needed.

The second way, reclaiming space currently allocated to broadcasters, is a bit more challenging. Since the television transition from analog to digital transmissions in June 2009, there has been some unused spectrum space. Because digital television signals take up less spectrum space than analog signals, television broadcasters are generally using only part of the allocated spectrum space. However, not all broadcasters are willing to give up their currently unused spectrum space, even if they would receive some financial compensation. Nonetheless, in February, 2012, Congress passed a bill that gives the FCC the ability to reclaim and auction spectrum space. Additionally, the legislation creates a second digital television (DTV) transition that will allow for current signals to be "repacked" (Eggerton, 2012). After years of delays, the auction finally commenced in Spring 2016.

These spectrum auctions are "reverse auctions," where the government determines an initial value of spectrum space and presents that to an interested broadcaster. If the broadcaster accepted that amount the FCC then attempted to find a "buyer". If the FCC didn't find a buyer at that price, the broadcaster was presented with a new, lower dollar amount. The broadcaster could choose to continue the process or to drop out. This process continued until either the broadcaster dropped out or a mutually agreeable dollar amount was determined. This process concluded in 2017, with wireless companies bidding a total of about $20 billion to buy channels from TV stations.

Current Status

The statistics on broadband penetration vary widely. The Organisation for Economic Co-operation and Development (OECD) keeps track of worldwide broadband penetration. According to the International Telecommunications Union (ITU), broadband subscribers in the U.S. have gradually increased over the last few years. As Table 21.3 illustrates, an estimated 32.8% of Americans had access to broadband service by late-2016. Despite this increase, the U.S.' world standing for broadband access is considered low. The OECD has the U.S. ranked 16th in the world in broadband penetration (see Table 21.4).

Table 21.3
U.S. Fixed Broadband Penetration History

2006	2007	2008	2009	2010	2011	2012	2013	2014	2015	2016
18.2	21.7	23.9	26.5	26.1	27.3	28.8	29.9	30.9	32.1	32.8

Source: OECD (2016)

Table 21.4
Fixed (Wired) Broadband Subscribers (per 100 Inhabitants, by Technology, June 2015)

Rank	Country	DSL	Cable	Fiber	Satellite	Other	Total	Total sub-
1	Switzerland	25.0	14.8	9.6	0.2	0.6	50.1	4 198 150
2	Denmark	18.3	12.8	11.0	0.0	0.0	42.4	2 430 002
3	Netherlands	16.5	19.4	6.0	0.0	0.0	41.9	7 135 000
4	France	32.5	5.2	3.3	0.0	0.0	41.4	27 683 000
5	Norway	12.0	12.2	15.5	0.1	0.0	40.5	2 120 360
6	Korea	2.3	8.1	30.0	0.0	0.0	40.4	20 555 683
7	Germany	29.1	8.7	0.7	0.0	0.0	38.6	31 867 148
8	United Kingdom	30.9	7.5	0.0	0.0	0.0	38.5	25 250 011
9	Iceland	25.0	0.0	12.6	0.0	0.0	38.1	128 023
10	Belgium	18.3	19.4	0.1	0.0	0.1	37.8	4 265 026
11	Sweden	9.7	6.8	20.4	0.0	0.1	37.1	3 679 768
12	Canada	12.2	19.4	3.7	0.0	0.0	36.8	13 347 882
13	Luxembourg	23.9	3.7	7.0	0.0	0.2	34.8	203 100
14	Greece	33.4	0.0	0.1	0.1	0.0	33.5	3 616 705
15	New Zealand	23.0	1.4	7.0	0.1	0.0	32.9	1 554 206
16	United States	8.2	19.7	3.7	0.6	0.2	32.8	106 327
17	Portugal	8.9	10.8	10.5	0.0	0.0	32.7	3 372 571
18	Finland	13.7	7.5	9.6	0.0	0.3	31.2	1 712 000
19	Japan	2.3	5.4	22.9	0.0	0.0	30.6	38 743 212
20	Spain	13.6	5.8	10.7	0.1	0.0	30.5	14 163 442
21	Australia	19.3	4.3	5.8	0.3	0.0	30.1	7 374 000
22	Estonia	10.5	7.1	10.5	0.0	0.5	29.2	384 787
23	Ireland	19.9	7.8	0.2	0.1	0.0	29.0	1 360 309
24	Czech Republic	8.6	5.3	4.9	0.0	0.0	28.8	3 038 394
25	Austria	18.5	9.5	0.5	0.0	0.0	28.7	2 510 500
26	Hungary	8.0	14.0	5.3	0.0	0.0	28.7	2 814 523
27	Slovenia	11.3	8.5	8.0	0.0	0.1	28.3	583 540
28	Israel	18.2	8.2	0.0	0.0	0.0	26.5	2 263 051
29	Latvia	6.6	0.9	16.6	0.0	1.7	26.5	519 154
30	Italy	20.1	0.0	0.8	0.0	3.4	25.7	15 563 279
31	Slovak Republic	9.0	3.0	7.0	0.0	0.1	24.6	1 336 541
32	Poland	6.5	7.0	1.5	0.0	1.2	18.3	7 042 470
33	Chile	5.1	8.6	1.1	0.0	1.1	15.9	2 904 580
34	Turkey	9.9	0.9	2.5	0.0	0.1	13.4	10 499 692
36	Mexico	6.3	4.6	2.1	0.0	0.1	13.3	16 277 627
Av-	OECD	12.9	9.8	6.4	0.2	0.3	28.8	386 824

Source: OECD (2016)

The United States currently ranks 18th in the world for broadband penetration with 32.8 subscribers per 100 inhabitants (up from just 18.2 in 2006). The rankings are presented in Table 21.4. The United States has the largest number of broadband subscribers with over 102 million. Fiber connections were most numerous in Korea (26.8) and Japan (21.3). The U.S. is above the OECD average penetration rates for overall broadband usage and for the usage of cable modems. However, the U.S. is below the OECD average for DSL and fiber penetration.

Approximately 365 million people worldwide subscribe to some type of broadband service (up from approximately 321 million just two years ago). Worldwide DSL subscribers are the most abundant, with nearly 561 million subscribers (among the 34 OECD countries). Approximately 263 million people have cable Internet and over 170 million have fiber (see Figure 21.2). As mentioned above, Japan and Korea lead the way in terms of fiber usage. Figure 21.2 shows in Japan, over 75% of all broadband subscribers utilize fiber networks. In Korea that number is about 70%. In the U.S., however, the number is about 10%. The OECD global average for fiber usage is nearly 18%.

Home broadband penetration in the United States has plateaued according to Pew Internet. As of 2015, 67% of U.S. household had broadband service. This is down from a high of 70% in 2013. However, more Americans got their broadband through their smartphone in 2015 with 13% up from 8% in 2013. This trend aligns with the rise of cord cutters, those who discontinue use of cable and satellite television and landline phone services. The study found that smartphone-reliant users were at a disadvantage because of data caps and were more likely to stop service because of financial issues. Cost was the number one reason people said they didn't have home broadband service. (Horrigan & Duggan, 2015).

Factors to Watch

The development of home networking applications will continue to escalate if the overall market share continues to expand. With the increased penetration of tablets and other "smart devices" the home networking device market is expected to grow to more than $45 billion by 2018 (Home Networking, n.d.).

Figure 21.2

Percentage of Fiber Connections in Total Broadband Subscriptions, December 2016

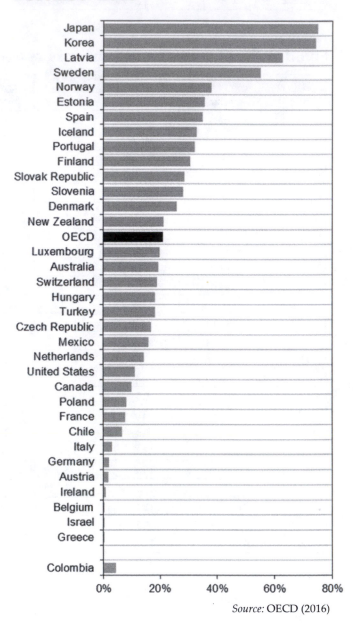

Source: OECD (2016)

Mobile broadband speeds will continue to increase. Currently, 4G LTE networks are becoming more and more prevalent. In theory, 4G can reach speeds of 100 Mb/s down and 50 Mb/s up. More realistically, however, consumers can expect speeds of 5 to 12 Mb/s downstream and 2 to 5 Mb/s upstream (Smith, 2012). Regardless, this is a significant increase over 3G speeds (generally less than 1 Mb/s down).

As mentioned earlier, 5G wireless technology is still in the development phase. What will likely come sooner is advanced 4G LTE service to include significant increases in transmission rates. Some operators, such as AT&T may even use a "5G" moniker, but it won't soon consist of true 5G technology.

Aside from continued market growth and increased speeds, there continues to be several topics that must be monitored. They include:

Usage-Based Pricing

Tied to the net neutrality issue, at least indirectly, is the issue of "usage-based pricing." Usage-based pricing—where users pay based on how much data they send and receive—is common with cellphone companies, however, it may become a reality with ISPs as well. Currently in the U.S., about 80% of broadband service is provided by cable companies. The increased demand for high bandwidth throughput for video services such as Netflix, along with the content competition such services provide, had caused some cable companies to experience revenue drops. If net neutrality becomes a reality and ISPs are prohibited from tweaking the throughput speeds of some content providers, they may be even more inclined to explore usage-based pricing (Engebretson, 2013).

Getting a Job

There are two primary areas for employment within the broadband industry. The first area involves support and development of the broadband technologies. It is clear that broadband technologies will continue to permeate our work and home settings. With the increase in technology reliance there will be an increase in the need for people to develop, repair, and maintain the hardware and software that run broadband networks.

Another major area of employment is within the content creation industry. With more and more access to devices that can display media content, there will be an increased demand for quality content. This content can come in the form of interactive games and instructional materials, audio content, and video content. As we become a more connected, society, a greater number of professional organizations will utilize various forms of content to connect to audiences, customers, clients, and employees.

Projecting the Future

The future of broadband technology in the U.S. will be directly impacted by the future outcomes regarding net neutrality and by the success/failure of spectrum auctions. Regardless, we can expect a continued increase in coverage areas and overall data transmission rates. In 2000 we were experiencing transmission rates below 1 Mb/s. Today, in home data transmission rates can eclipsed 500 Mb/s. In the next 15 years can we expect speeds to increase again by a factor of 500? There's no way to be certain, but it is safe to assume the speeds will be much faster than they are today.

Another major shift that can be expected is the number of devices that can and will be connected to wired and wireless networks. Appliances already exist that allow us to control them remotely through broadband networks. We can access our DVRs, our thermostats, our security systems, and our home computers from remote locations. Portable devices and appliances... yes... appliances are increasingly used to access audio and video content. Some cars can already interface with a home computer to allow trip itinerary to be transferred from the computer to the car's navigation system. There have been numerous successful experiments with driverless cars. Self-driving delivery vehicles, including tractor-trailers, are also on the horizon.

Tech companies like Microsoft have developed "smart homes." These homes feature numerous types of interactivity and are all built on a broadband backbone. Smart homes have a voice-activated whole-house automation system. Through verbal commands the homeowner can have the automation system look up a recipe, check voicemail, turn on the TV, and adjust the lighting. Some smart homes have multi-wall projectors that can project different pictures or themes on various walls. They have cameras that can read things like prescription bottles to help the homeowner be assured of taking the correct medicine. There is even a refrigerator concept that would allow the refrigerator to sense when food is about to go bad and then look up recipes using those ingredients.

It is truly difficult to predict the future, but clearly there is an on-going trend to remotely connect with more devices through broadband networks.

Bibliography

320 Mbps home networking specification released (n.d.). Retrieved from HomePna website: http://www.homepna.org/products/specifications/.

Belson, D (September, 2015). Akamai's [state of the internet: Q3 2015 report]. Retrieved from https://www.akamai.com/us/en/multimedia/documents/report/q3-2015-soti-connectivity-final.pdf

Bluetooth 4.2 is here – but what does it mean (n.d.). Retrieved from http://blog.csr.com/2014/12/bluetooth-4-2-is-here-but-what-does-it-mean/.

Bluetooth 5 versus Bluetooth 4.2, what's the difference? (2017, February 8). Retrieved from https://www.semiconductor-store.com/blog/2017/Bluetooth-5-versus-Bluetooth-4-2-whats-the-difference/2080.

Bradley, T. (2009, September 22). Battle lines drawn in FCC net neutrality fight. *PC World*. Retrieved from http://www.pcworld.com/businesscenter/article/172391/battle_lines_drawn_in_fcc_net_neutrality_fight.html.

Brodkin, J. (2016, February 26). One year later, net neutrality still faces attacks in court and congress. *Ars Technica*. Retrieved from http://arstechnica.com/business/2016/02/one-year-later-net-neutrality-still-faces-attacks-in-court-and-congress/.

Brunner, C. (2014, January 24). FTTH council hails fcc order to boost deployment of fiber broadband networks in rural areas. *FTTH Council Americas*. Retrieved from http://www.ftthcouncil.org/p/bl/et/blogid=3&blogaid=254.

Claricoats, A. (2016, January 8). Homegrid forum powerline technology boosting capabilities for the connected home at CES 2016. Retrieved from http://www.homegridforum.org/content/pages.php?pg=news_press_releases_item&id=342.

Courtney, M. (2013, October 15). Whatever happened to broadband over power line? *Engineering and Technology Magazine*. Retrieved from http://eandt.theiet.org/magazine/2013/10/broadband-over-power-line.cfm.

DOCSIS (n.d.): Data over cable service interface specifications. (n.d.). Retrieved from Javvin website: http://www.javvin.com/protocolDOCSIS.html.

Docsis 3.0. (2009, March 13). Retrieved from Light Reading website: http://www.lightreading.com/document.asp?doc_id=173525.

Eggerton, J. (2012, February 17). Congress approves spectrum incentive auction. *Broadcasting and Cable*. Retrieved from http://www.broadcastingcable.com/article/480721-Congress_Approves_Spectrum_Incentive_Auctions.php.

Engebretson, J. (2013, June 7). Moffett: Cablecos should impose usage-based pricing; but can they? *Telecompetitor*. Retrieved from http://www.telecompetitor.com/moffett-cablecos-should-impose-usage-based-pricing-but-can-they/.

FAQ (n.d.) w3.org. Retrieved from http://www.w3.org/People/Berners-Lee/FAQ.html.

Featured Technology (n.d.): DOCSIS 3.1- A new generation of cable technology. Retrieved from http://www.cablelabs.com/innovations/featured-technology/.

Federal Communications Commission. (2010). *The National Broadband Plan*. Retrieved from http://download.broadband.gov/plan/national-broadband-plan.pdf.

Federal Communications Commission (2011a, November 18). FCC releases "connect America fund" order to help expand broadband, create jobs, benefit consumers. [Press Release]. Retrieved from http://www.fcc.gov/document/press-release-fcc-releases-connect-america-fund-order.

Federal Communications Commission (2011b, December 22). FCC chairman Genachowski announces approval of first television white space database and device. [Press Release]. Retrieved from http://www.fcc.gov/document/chairman-announces-approval-white-spaces-database-spectrum-bridge

Federal Communications Commission (2015, January 29). 2015 Broadband Progress Report and Notice of Inquiry on Immediate Action to Calculate Deployment, FCC 15-10. Retrieved from https://apps.fcc.gov/edocs_public/attachmatch/FCC-15-10A1_Rcd.pdf

Fitchard, K. (2013, November 14). *White spaces anyone? Google opens its spectrum database to developers*. Retrieved from http://gigaom.com/2013/11/14/white-spaces-anyone-google-opens-its-spectrum-database-to-developers/.

Google fiber expansion plans (n.d.) Retrieved from https://fiber.google.com/newcities/.

Gustin, S. (2014, February 19). Here's why your netflix is slowing down. *Technology and Media*. Retrieved from http://business.time.com/2014/02/19/netflix-verizon-peering/.

Home Networking Device Market – Global Industry Size, Share, Trends, Analysis and Forecasts 2012-2018 (n.d.). Retrieved from http://www.transparencymarketresearch.com/home-networking-device-market.html.

Horrigan, J. and Duggan, M. (2015). Home Broadband 2015. Pew Internet. Retrieved from http://www.pewinternet.org/2015/12/21/home-broadband-2015/

HughesNet, (n.d.). Retrieved from https://www.hughesnet.com/get-started.

International Telecommunication Union (2012a). *Building our networked future based on broadband*. Retrieved from http://www.itu.int/en/broadband/Pages/overview.aspx.

International Telecommunication Union (2012b, January 18). IMT-Advanced standards announced for next-generation mobile technology. [Press Release]. Retrieved from http://www.itu.int/net/pressoffice/press_releases/2012/02.aspx.

Kansas City is Fiber-Ready. (2012, February 6). Retrieved from http://googlefiberblog.blogspot.com/2012/02/weve-measured-utility-poles-weve.html.

Lee, N. (2011, October 5). Bluetooth 4.0: What is it, and does it matter? Retrieved from http://reviews.cnet.com/8301-19512_7-20116316-233/bluetooth-4.0-what-is-it-and-does-it-matter/.

Marshall, G. (2012, February 1). 802.11ac: what you need to know. Retrieved from http://www.techradar.com/news/networking/wi-fi/802-11ac-what-you-need-to-know-1059194.

Miller, M. (2017, March, 15). Is gigabit lte in your future?. *PC Magazine*. Retrieved from https://www.pcmag.com/article/352395/is-gigabit-lte-in-your-future.

Mitchell, B. (n.d.). Wireless standards- 802.11b 802.11a 802.11g and 802.11n: The 802.11 family explained. About.com. Retrieved from http://compnetworking.about.com/cs/wireless80211/a/aa80211standard.htm.

OECD Broadband (2017). Retrieved from the Organisation for Economic Co-operation and Development website: http://www.oecd.org/sti/broadband/broadband-statistics/

Poole, I. (n.d.a). IEEE 802.11ad microwave wi-fi/wigig tutorial. *Radio-Electronics.com*. Retrieved from http://www.radio-electronics.com/info/wireless/wi-fi/ieee-802-11ad-microwave.php.

Poole, I. (n.d.b). IEEE 802.11af white-fi. *Radio-Electronics.com*. Retrieved from http://www.radio-electronics.com/info/wireless/wi-fi/ieee-802-11af-white-fi-tv-space.php.

Rogers, K. (2017, December 20). The fcc's next stunt: Reclassifying cell phone service as 'broadband internet'. *Motherboard*. Retrieved from https://motherboard.vice.com/en_us/article/mbpezp/fcc-smartphone-data-reclassification?utm_campaign=sharebutton.

Segan, S, (2017, May 1). What is 5G?. *PC Magazine*. Retrieved from https://www.pcmag.com/article/345387/what-is-5g.

Smith, J. (2012, January 1). Ridiculous 4G LTE speeds hit Indianapolis for super bowl 46. *Gotta Be Mobile*. Retrieved from http://www.gottabemobile.com/2012/01/19/ridiculous-4g-lte-speeds-hit-indianapolis-for-super-bowl-46/.

TV Whitespace – Breakthrough Technology (n.d.). Retrieved from http://www.carlsonwireless.com/tv-white-space/.

Update (2015): Sprint plans to shut down WiMAX network tomorrow, says only a small percentage remain on network (November 5, 2015). Retrieved from http://www.fiercewireless.com/story/sprint-plans-shut-down-WiMAX -network-tomorrow-number-customers-left-unclear/2015-11-05.

Vaughan-Nichols, S.J. (2012, January 9). 802.11ac: Gigabit wi-fi devices will be shipping in 2012. ZDNet. Retrieved from http://www.zdnet.com/blog/networking/80211ac-gigabit-wi-fi-devices-will-be-shipping-in-2012/1867.

Verizon. (n.d.). *FiOS Internet*. https://fios.verizon.com/fios-plans.html.

What is WiMAX (n.d.). Retrieved from http://www.WiMAX .com/general/what-is-WiMAX .

Wi-Fi Alliance. (n.d.). *Wi-Fi overview*. Retrieved from http://www.wi-fi.org/OpenSection/ why_Wi-Fi.asp?TID=2.

WPA2. (n.d.). Retrieved from http://www.wi-fi.org/knowledge_center/wpa2/.

Wyatt, E. (2010, April 6). U.S. court of curbs F.C.C. authority on web traffic. *The New York Times*. Retrieved from http://www.nytimes.com/2010/04/07/technology/07net.html.

Wyatt, E. (2014, February 19). FCC seeks a new path on 'net neutrality' rules. *The New York Times*. Retrieved from http://www.nytimes.com/2014/02/20/business/fcc-to-propose-new-rules-on-open-internet.html?_r=0.

ZigBee. (n.d.). *ZigBee Alliance*. Retrieved from http://www.zigbee.org/About/AboutTechnology/ZigBeeTechnology.aspx.

22

Telephony

William R. Davie, Ph.D. *

Overview

Telephony is a catch-all term for electronic communication linked to the invention and development of the telephone. In modern terms, the mobile handset offers verbal exchanges of telephony and is complemented by the broadband delivery of text, graphics, audio, and video content. Today, the smartphone is the primary instrument of telephony, serving billions of people worldwide and contributing to the global economy. The captivating experiences provided by smartphone apps have put mobile telephony at the center of the telecommunications revolution where machine-to-machine communications (M2M) and the Internet of Things (IoT) drive innovation.

Introduction

Telephony transmits streams of voice and data between distant points by wired and wireless means. By either a telephone cable or wireless electromagnetic energy, telephone instruments are used to generate commerce and communication. Mobile telephony passed a milestone in 2016 when a U.S. survey showed the majority of American homes (50.8%) had moved to wireless-only phones (Blumberg & Luke, 2017). So the thought of living without a mobile phone has become something that is, well, unthinkable for millions of Americans. Telephone communications are at the center of the global-mobile age, and how we reached this point in technology over 150 years of trial and experimentation is worthy of reflection.

Background

The telephone owes much of its invention to the keen competition between brilliant inventors and their associates. Alexander Graham Bell's vision in the 19th century overcame weaker alternatives like the acoustic phone vibrating a taut wire between two people speaking into cans. It was actually Bell's drive to help serve the deaf more than his zeal for competition that produced the world's first telephone. In pursuit of this dream, this Scottish inventor migrated from Canada to Boston, where he worked feverishly on his "talking telegraph," to assist hearing impaired Americans like his wife, Mabel Hubbard Bell, who was rendered deaf by scarlet fever as a child of five. The legend of Bell's race to the U.S. patent office arriving just ahead of his rival inventor Elisha Gray in 1876 added the spirit of competition to the story (AT&T, 2010a).

The company Bell and his associates founded, American Telephone and Telegraph, sprang to life in 1885 (AT&T, 2010b). Inventing telephony was Bell's passion; it was Theodore Vail who actually invented the business of telephony. Vail crisscrossed the United States with AT&T company wires, and it was his entrepreneurial savvy as company president that built AT&T into a monolithic enterprise overtaking smaller

---

* BORSF Regents Chair of Communication, University of Louisiana, Lafayette

phone companies and expanding to transcontinental cables in 1915 (John, 1999). By the time this corporate mogul retired in 1919, AT&T could boast with slight exaggeration it owned nearly every telephone instrument, every switch, and pole in the nation. AT&T's preeminence came with government support in 1934 when the Federal Communications Commission classified this enterprise as a protected monopoly (Thierer, 1994). It is hard to overstate AT&T's hold on early telephony and the U.S. wired networks it built referred to as POTS.

Plain Old Telephone Service (POTS) transmitted voice calls via copper wires linked to an analog exchange center where multiple calls rapidly switched back and forth between trunk and branch lines to reach their destination. The formal term describing how network cables transmitted calls was the Public Switched Telephone Network (PSTN) (Livengood, Lin, & Vaishnav, 2006). (See Figure 22.1) The mechanical PSTN would be replaced one day by a computer-driven PSTN that expanded both phone capacity and the speed of transmission.

Figure 22.1

Traditional Telephone Local Loop Network Star Architecture (POTS Network)

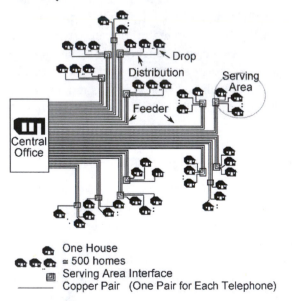

One House
≅ 500 homes
Serving Area Interface
—— Copper Pair   (One Pair for Each Telephone)

Source: Technology Futures, Inc

Over the next half century, telephony evolved through the Storage Program Control (SPC), where home phone numbers were computed and stored. The SPC was quite an improvement for switching phone calls digitally through a centralized control system. Microprocessors routed vast numbers of calls over "blocks" or series of exchanges (Viswanathan, 1992). Fiber optic glass strands would replace copper wires, and computer modems translated telephone calls into binary signals to be converted into verbal sounds.

Birth of Mobile Phones

The move from wired to wireless telephony owed its inspiration to the automobile boom in the 20th century. Telephone-equipped cars were needed to summon emergency calls for fire engines, ambulances, and police cruisers along with taxicabs. At first radio transmission was used between base stations and vehicles, and so linking to the PSTN by mobile handset was not a solution. However, in St. Louis, AT&T formed a project with Southwestern Bell to operate a Mobile Telephone Service (MTS). Motorola built the radios and the Bell System installed them. MTS was modeled after conventional dispatch radio (Farley, T., 2005), and public interest grew in MTS for private uses.

Texas ranchers and oil company executives looked for a portable means to communicate remotely with workers in the field. Thomas Carter of Dallas introduced in 1955 his two-way radiophone attachment, and the homemade Carterfone sold more than 3,500 units connecting mobile calls by AT&T wires and poles. AT&T hoped to put a stop to Carter's use of their facilities, so the company sued and took their case before the U.S. Supreme Court. The high court's decision allowed "any lawful device" to connect with those phone company cables (Lasar, 2008). This precedent-setting decision set the stage for the cellular boom to follow years later.

Today's mobile phone was born of the genius of an electrical engineer working on car phones for Motorola's research and development lab. Martin Cooper was charmed by the "communicator" he had watched on the TV show Star Trek (Time, 2007). In 1973, he took to the streets to show off his handheld communicator and puckishly placed a call to Joel S. Engel, a rival engineer at Bell Labs. Cooper's phone was bulkier than Captain Kirk's model, and weighed almost two pounds, so it was dubbed "the brick" (Economist, 2009). Building a handheld wireless phone was one accomplishment, but overcoming the challenge of networking hubs of antennas to switch calls while mobile phones are in transit was quite another.

Early land-mobile telephony found a band of frequencies for transmitting over a broad area, yet the lack of capacity limited their reach, and made mobile pricing prohibitive. For more conventional mobile phones, AT&T needed a network of carefully spaced hubs. Bell engineers Engel and Richard Frenkiel drew up a cellular map with a network of antenna links (Lemels, 2000). Their map caught the attention of FCC engineers in Washington, D.C., but there were technical issues to resolve. Telecom engineers turned toward an alternative solution (Oehmk, 2000). Instead of using broad coverage antennas, cellular communications linked to small reception areas so the same frequencies could be used in multiple, nonadjacent "cells." Figure 22.2 shows cellular network architecture.

Figure 22.2
Cellular Telephone Network Architecture

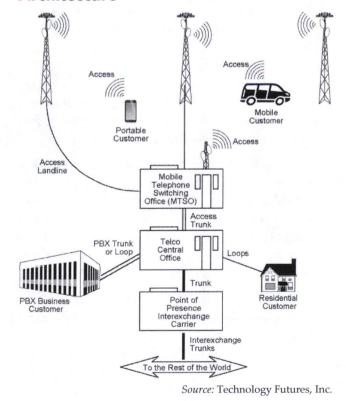

Source: Technology Futures, Inc.

Smartphone Evolution

The transition to smartphone technology began with the idea of integrating personal computer technology with a mobile operating system linking voice, text, and Internet data to a cellular network. Personal Digital Assistants or PDAs were an early stop along the way. IBM's Simon Personal Communicator introduced in 1994 was based on a design by electrical engineer Frank J. Canova. It used a touch screen to open an address book, a calculator, and a sketchpad. Simon was not a market success but a fleet harbinger signaling the mobile phone future (Sager, 2012). By converging computer and cellular technology, mobile phones continued to evolve. In 1996, Nokia moved forward with its 9000 model that integrated portable phone features with Hewlett-Packard's Personal Digital Assistant (PDA). Motorola's engineers pushed the envelope with the launch of the MPx200 in 2003. This joint venture with Microsoft used a Windows-based operating system and a full package of AT&T wireless services including email and instant messaging complementing its use of "apps."

Tiers of Smartphone Technology

At the basic level of its architecture, the smartphone is viewed as a stack of four tiers of technology (Grimmelman, 2011). First, the base tier offers users the applications or apps that can be anything from a video game to a calculator, a social media site, financial data, fitness monitor, or even a machine-to-machine (M2M) link that turns up the thermostat or starts the coffee brewing in the kitchen. These apps are easily pressed into service through image icons, but the apps rely on an operating system, which is the second tier of the smartphone architecture. It might be Apple's iOS, Android, Windows, or another operating system, but they all act as intermediaries between the applications and hardware. The third tier is the handset itself, and familiar models like the Google Pixel, Samsun Galaxy, and Apple's iPhone owe much to earlier pioneers in mobile telephony, like the Blackberry. The fourth tier of technology connects mobile devices to the rest of the world through networks known by their abbreviations, CDMA, GSM, EDGE, EVDO, and LTE, and also numbered by generational technology (1G, 2G, 3G, and 4G).

From Blackberry to Android

A Canadian firm, Research in Motion (RIM) had the cellular world in the palm of its hands when it launched the Blackberry smartphone in 2002. Blackberry's compact features for wireless email, mobile faxing, and its tiny keyboard with raised letters, numbers, and symbols attracted a popular following (Connors, 2012). It reached its popularity peak in 2008, when urban professionals transfixed by their mobile handsets were called "Blackberry Jam," and the term "Crackberry" underscored its addictive quality. Apple and Android models would enter the fray in 2007 and 2008 and render their users just as dependent. Competition grew fast in the mobile market and Blackberry fell behind, dropping to quarterly lows that approached a billion dollars in losses (Austen, 2013). RIM and Blackberry fought back with innovative rivals like the Z10, but Apple's iPhone and the up-and-coming Androids took over the race.

Apple's iPhone was an exemplar of inventive engineering when it debuted in 2007, and despite its spectacular price tag of $499, customers lined up to pay Apple's piper, Steve Jobs. Apple had built brand loyalty through its line of laptops and iPods (Vogelstein, 2008). iPhone's innovations sold the market on this new way of sliding and pressing icons under glass rather than pushing tiny keys or poking at a mobile phone with a stylus. Its operating system (iOS) with a web browser convinced millions they needed an upgrade. The engraved case of aluminum and glass with shining Apple logo became the market leader, but would soon face stiff competition.

Google's open-source Linux-based Android operating system debuted in 2008 with two manufacturers offering the same basic model with different names: HTC Dream and TMobile G-1. These Android phones used both a slide-out keyboard and touch screen, but posed no threat to the iPhone's dominance. Android phones would follow with capacitive touch screens only to push ahead of the iPhone. Samsung's smartphone gained an edge, but Apple alleged it infringed on the iPhone's innovations, and proved its case in court (Lee, 2011). Late in 2017, the U.S. Supreme Court let stand Apple's $120 million judgment against Samsung for the slide-to-unlock feature, while another Apple judgment against Samsung stood pending (Kastrenakes, 2017).

Recent Developments

In the progress of mobile telephony, leading manufacturers have enhanced innovations ranging from Augmented Reality (AR) to 360-degree cameras (Gibbs, 2017). Several smartphones changed the means of access including facial recognition and fingerprint scanning to open the full-screen, swipe across the surface, and touch desired apps. Vanishing are the home screen buttons and password keys. Apple's ID fingerprint gave way to facial recognition for the iPhone X, but then reports circulated this Apple smartphone would be cancelled due to soft sales (Kelly, 2018). At a practical level, longer battery life and wireless charging blocks gave customers the additional convenience they desired for energy maintenance.

The photographic experience on the smartphone improved through widespread adoption of a dual camera system combining one-wide angle view with a "telephoto" zoom capacity. Image resolution became sharper through Ultra High-Definition (4K) recording capacity, and the smartphone lenses improved to give a better view. While the improved 4K screens and zooming capacity enhanced some smartphone models, other innovations like screens that would bend around the wrist met with mixed reviews (Yan, S., 2017). Thermal imaging, optical zooms, and 360-degree cameras are other smartphone features designed to attract consumer interest.

Augmented Reality (AR), discussed in Chapter 15, synthesizes digital information for smartphone users in real time. This innovation became a successful gaming application for Pokémon Go, and just for fun offered the overlay of animal features on friends' faces with Snapchat filters. AR is also used to affix global positioning systems and map out routes for drivers, plus give compasses direction bearings.

The smartphone screen's refresh speeds and aspect ratios expanded the display in larger models from a horizontal to vertical ratio of 16:9 to around 18:5. Pioneering this new format was the LG G6 model, which was followed by the Samsung Galaxy S8 and S8 plus. Video and graphics produced in the

16:9 format for large screen television and movie theaters remain the standard, but the wider format on mobile phones is attracting wider video and graphic content. Smartphone screens also quickened refresh rates from 60 to 120 Hz. The faster an image refreshes on a screen the clearer the resolution appears to the viewer (Callaham, 2017). Inside the circuitry panels is the mechanism needed to make phone calls. Traditional SIM cards were replaced by eSIM technology by some manufacturers. Google Pixel 2 and Google Pixel 2XL adopted eSIM technology and gave the boot to embedded SIM cards that could be removed.

Not everyone liked the removal of familiar features like the familiar headphone jack that Apple removed in 2016 to make way for Bluetooth Earpods (Savov, 2017). But Androids followed suit by removing headphone portals and using their own Bluetooth technology. The removal of the home button (Gibbs, 2017) became part of the new look as manufacturers sought to give users a seamless experience. Even the smartphone design embraced the seamless feel by removing the *bezel*, the grooved ring holding the glass face in place. Leading manufacturers like Samsung, Huawei, Apple, Vivo, and Xiaomi began marketing "bezel-free" smartphone designs.

Current Status

The accelerating growth of m-commerce has overcome the challenge of using small smartphone screens and grew to become a $230 billion enterprise in the United States. What was commonplace in U.S. airports, sporting events, and concerts, is quite accepted worldwide. The transfer of money by mobile phone known as the digital wallet serves as accepted currency for consumers in daily retail transactions. Apple introduced its Apple Pay service in 2014 and activated one million credit cards within the first week of its launch (Rubin, 2015). Samsung Pay was in competition soon thereafter. Android Pay uses NFC (near field communication) technology, and the joint venture of AT&T, Verizon, and T-Mobile moved m-commerce forward in 2018 when Android Pay and Google Wallet combined forces to create a single system called Google Pay. The worldwide digital transformation of m-commerce through contactless payments using mobile handsets is expected to continue to grow (Juniper Research, 2016).

Wearable Technology

Cartoon detective Dick Tracy inspired American baby boomers to imagine one day they would make phone calls from their wristwatches. Fast-forward to new millennial consumers, unfamiliar with that square-jawed police hero, and the phone wristwatch became a tough sell. Apple and Samsung opened up the battlefront for the market in 2014. Their smartwatches would tell time and make phone calls through touch screens responding to voice commands, but also required companion smartphones (Shanklin, 2014). Industry observers liked the smartwatch for fitness monitoring, day planning, and online access but warned against making calls due to excessive drain on the battery and added billing costs (Hartmans, 2017b).

Popular Apps

When it comes to the most popular smartphone applications, Google rose to dominance in 2017 with Google Play, Google Search, GMail, Google Maps, and YouTube (Hartmans, 2017a), but social networking apps like Facebook, Snapchat, and Instagram remain popular according to comScore metrics. Pandora's music app also attracts smartphone users. It is noteworthy that both voice minutes and traditional text messaging are declining uses for mobile phones. What is taking their place are communicating through apps like Skype, FaceTime, Facebook messenger, and WhatsApp (FCC, 2017).

Global Frontiers

Worldwide, more people are fascinated by their mobile phones than ever before and that has led nations around the world to review the evolutionary development of cellular networks. At the end of 2016, 65% of the world's population subscribed to mobile telephony (GSMA, 2017). While nearly two-thirds of the world's population had entered the mobile-global age, Group Speciale Mobile Association (GSMA) predicted by 2020, there would be 5.7 billion mobile subscribers on the planet ranging from 87% in Europe to 50% of Sub-Saharan Africa. Smartphone marketing shifted geographically to the developing countries of India and Pakistan, where analysts predict 310 million unique subscribers by 2020. Huawei, the mobile giant of China continued its surge in sales in 2017 thanks to the booming Asian Pacific market. More than half of the mobile subscribers in the world live in Asia, where two of the most populous countries, India and China, add millions of mobile subscribers each year.

The worldwide sale of smartphones overall grew seven percent in the fourth quarter of 2016, and there was good news for Apple, according to Gartner research. "It has taken eight quarters for Apple to regain the number one global smartphone vendor ranking, but the positions of the two leaders has never been so close, with only 256,000 units difference" (Gartner, 2017). The manufacturers of smartphones from 2016 to 2017 grew at a double-digit rate for four of the top five smartphone makers in 2017 thanks to the growing demand for Samsung, Huawei, Xiaomi, and Vivo models. These lower-cost models outpaced the units sold by Apple in the emerging Asia Pacific markets, where Apple found itself in third place behind Korean manufacturer Samsung, and China's Huawei.

Factors to Watch

Both the number of wireless subscribers and the download of data by mobile carriers are rising in the United States, according to the CTIA figures. There were 396 million mobile subscribers by the end of 2016, reflecting an uptick of about 5% from the previous year. At the same time, the exchange of wireless data rose 42% to 13.7 trillion megabytes (MB) (FCC, 2017). The vast majority of Americans subscribe to one of four national service providers—Verizon, AT&T, TMobile, or Sprint, with 92% of the American population served by 3G networks. The problem is in rural areas, where only 55% of the population can access all four of the top competitors in broadband delivery. These four service providers account for 411 million connections, but the resellers and Mobile Virtual Network Operators (MVNOs) like TracFone Wireless and Google Fi reach certain segments of the market, where low-income consumers or Americans with lower data needs are able to buy a cellular plan.

Two-year contracts were once routinely accepted for U.S. wireless customers to buy smartphones. Those customer agreements meant they would pick a phone and pay a subsidized price. In return, a two-year agreement with the carrier came with a monthly service charge. Then in 2013 the two-year contract began to disappear, and unlimited minutes took its place. TMobile disrupted the two-year format with its "Un-Carrier" campaign in 2013 where customers bought smartphones in installments or at a single purchase price.

The choices for cellular agreements are defined as either pre-paid or post-paid. The prepaid option offers customers a fixed number of minutes, data, or texts for a particular period, while postpaid billing charges customers at the end of each month and tacks on extra charge for usage above a set figure. The move to unlimited data plans was viewed by some as misleading since for mobile carriers it meant that you would have to pay more or deal with the "throttling" of data downloads that makes it a slower process. The price ranges for mobile plan is $30-100, fluctuating by company and plan.

American Frontiers

The Federal Communications Commission evaluates commercial mobile wireless services in the U.S. with an emphasis on competitive market conditions. None of the four largest providers, AT&T, Verizon, TMobile and Sprint, are viewed as dominant in their share of the mobile market, but all are growing to meet the demand for wireless access by all Americans. The Cellular Telecommunications and Internet Association (CTIA) collect data for use by public and private firms, and the CTIA report shows steady growth in mobile subscribers along with a spectacular surge in data usage.

The CTIA shows the smartphone penetration rate nearly doubled over a five-year period from 42–81% while wireless data usage grew in one year by 42% (2015–2016). U.S. data costs decreased, however, by about 7% over the same period (FCC, 2017). Naturally the rise in broadband use requires additional spectrum, and as a result the FCC freed up a range of frequencies. The television channel reverse auction freed up around 220 megahertz of spectrum for wireless use in 2016-17.

The wireless service companies are now focused on fifth generation (5G) cellular networking, and in anticipation of 5G networks, the FCC made available 3250 MHz of Extremely High Frequency (EHF) bandwidth that transmits millimeter waves (mmW), which are subject to atmospheric attenuation and limited by distance. What the Next Generation Mobile Networks Alliance predicts is faster data speeds with digital access to the Internet of Things (IoT) through the use of M2M technology. The challenge in technology comes in assigning higher frequencies and wider bandwidth to meet the needs of 5G networks.

Fifth Generation (5G)

The CTIA predicts the number of cell sites will significantly increase as the wireless industry prepares for 5G and has to make its networks denser. In July 2016, the FCC adopted new licensing, service, and technical rules for deploying three spectrum bands above 24 GHz and sought comments for mobile services, in the mmW bands. The Commission's Spectrum Frontiers rulemaking proceeding also took steps towards enabling the next generation of 5G wireless technologies.

While precise networking standards for 5G have yet to be finalized as of early 2018, the major players have started 5G trials. AT&T's 5G trial in Austin used millimeter wave spectrum to deliver an "ultra-fast" internet connection expected to hit speeds of up to 1 Gbps, that could be used for Internet access, VPN, unified communications, and 4K video streams. The excitement over 5G is not just because it is the next big thing after 4G, it is also because M2M is becoming more pertinent as the demand grows for linking sensors, monitors, and machines to cellular networks.

Getting a Job

The telephone industry offers a variety of careers from creative positions in content and entertainment to technical positions in electronic telecommunications and computer science. There are even unskilled positions available in sales and customer service or at calling centers. Major firms such as AT&T, Verizon, T-Mobile, and Sprint welcome job inquiries and applications online. Experience and education are important though for technical and creative carriers with these companies, although they do offer training for positions in retail outlets and call centers. College degrees are often specified but not always required, and internships are available in many of their offices. Some technical positions must meet federal regulations and professional standards, but study guides are available for preparation and certification. AT&T in particular, offers study guides online for positions in retail sales, customer service, and technical engineering.

Bibliography

AT&T. (2010a). Inventing the telephone. AT&T Corporate History. Retrieved from http://www.corp.att.com/history/inventing.html.

AT&T. (2010b). Milestones in AT&T history. AT&T Corporate History. Retrieved from http://www.corp.att.com/history/milestones.html.

Austen, I. (2013, Sept. 27). BlackBerry's Future in doubt, keyboard lovers bemoan their own. *The New York Times*. Retrieved from http://222.nytimes.com/2013/09/28/technology/blackberry-loses-nearly-1-billion-in -quarter.html?action= click&module=Search&region=searchResults%230&ve.

Blumberg, S.J., & Luke, J.V. (2017, May). Wireless Substitution: Early Release of Estimates from the National Health Interview Survey, July-December 2016. *National Center for Health Statistics*. Retrieved from https://www.cdc.gov/nchs/data/nhis/earlyrelease/wireless201705.pdf.

Callaham, J. (2017, Dec. 21). The top three smartphone innovations of 2017. Anroidauthority.com. Retrieved from https://www.androidauthority.com/smartphone-innovation-2017-824131/.

Connors, W. (2012, March 29). Can CEO revive blackberry? *Wall Street Journal* Marketplace, B1, B4.

Economist. (2009). Brain scan—father of the cell phone. *The Economist* website. Retrieved from http://www.economist.com/node/13725793?story_id=13725793.

Farley, T. (2005, March 4). Mobile telephone history. *Telektronnik*. Retrieved from http://www.privateline.com/wp-content/uploads/2016/01/TelenorPage_022-034.pdf

FCC. (2017, Sept. 27). Twentieth Annual Report and Analysis of Competitive Market Conditions with Respect to Mobile Wireless. Retrieved at https//apps.fcc.gov/docs_public/attachmatch/DA-15-1487A1.pdf.

Gartner. (2017, Feb. 15). Gartner says worldwide smartphone sales grew 7 percent in fourth quarter of 2016. Retrieved from https://www.gartner.com/newsroom/id/3609817.

Gibbs, S. (2017, Aug. 20). Augmented Reality: Apple and Google's next battleground. *The Guardian.com*. Retrieved from https://www.theguardian.com/technology/2017/aug/30/ar-augmented-reality-apple-google-smartphone-ikea-pokemon-go.

Grimmelmann, J. (2011). Owning the stack: The legal war to control the smartphone platform. Ars technical. Retrieved from http://arstechnica.com/tech-policy/news/2011/09/owning-the-stack-the-legal-war-for-control-of-the-smartphone-platform.ars.

GSMA. (2017). The Mobile Economy 2017. Retrieved from https://www.gsma.com/mobileeconomy/.

Hartmans, A. (2017a, Aug. 29). These are the 10 most used smartphone apps. *Business Insider*. Retrieved from http://www.businessinsider.com/most-used-smartphone-apps-2017-8/#10-pandora-1.

Hartmans, A. (2017b, Dec. 23). I tried a lot of smartwatches in 2017 – here are my top picks. *Business Insider*. Retrieved from http://www.businessinsider.com/best-smartwatches-gift-guide-2017-12/#the-lg-watch-style-is-a-clean-minimalist-smartwatch-1.

John, R. (1999). Theodore N. Vail and the civic origins of universal service. *Business and Economic History*, 28:2, 71-81.

Juniper Research (2016, March 1). Apple & Samsung drive NFC mobile payment users to nearly $150M globally this year. JuniperResearch.com. Retrieved from https://www.juniperresearch.com/press/press-releases/apple-samsung-drive-nfc-mobile-payment-users

Kastrenakes, J. (2017, Nov. 6). Apple has finally won $120 million from Samsung in slide-to-unlock patent battle. Theverge.com. Retrieved from https://www.theverge.com/2017/11/6/16614038/apple-samsung-slide-to-unlock-supreme-court-120-million.

Kelly, G. (2018, Jan. 21). Apple leak reveals sudden iPhone X cancellation. Forbes.com. Retrieved from https://www.forbes.com/sites/gordonkelly/2018/01/21/apple-iphone-x-buy-sales-specs-new-iphone-release-date/.

Lasar, M. (2008). Any lawful device: 40 years after the Carterfone decision. Ars technical. Retrieved from http://arstechnica.com/tech-policy/news/2008/06/carterfone-40-years.ars.

Lee, T.B. (2011). Yes, Google stole from Apple, and that's a good thing. Forbes.com. Retrieved from https://www.forbes.com/sites/timothylee/2011/10/25/yes-google-stole-from-apple-and-thats-a-good-thing/

Lemels. (2000). LEMELS N-MIT Inventor of the Week Archive—Cellular Technology. Retrieved from http://web.mit.edu/invent/iow/freneng.html.

Livengood, D., Lin, J., & Vaishnav, C. (2006, May 16). Public Switched Telephone Networks: An Analysis of Emerging Networks. Engineering Systems Division, Massachusetts Institute of Technology. Retrieved from http://ocw.mit.edu/courses/engineering-systems-division/esd-342-advanced-system-architecture-spring-2006/projects/report_pstn.pdf.

Oehmk, T. (2000). Cell phones ruin the Opera? Meet the culprit. *New York Times* - Technology. Retrieved from http://www.nytimes.com/2000/01/06/technology/cell-phones-ruin-the-opera-meet-the-culprit.html.

Rubin, B.F. (2015, May 28). Google's Android pay to duke it out with Apple Pay. CNET. Retrieved from http://www.cnet.com/news/google-refreshes-mobile-payments-effort-with-android-pay/.

Sager, I. (2012, June 29). Before iPhone and Android came Simon, the First Smartphone. Bloomberg.com. Retrieved from https://www.bloomberg.com/news/articles/2012-06-29/before-iphone-and-android-came-simon-the-first-smartphone.

Savov, V. (2017, Oct. 5). The Pixel's missing headphone jack proves Apple was right. Theverge.com. Retrieved from https://www.theverge.com/2017/10/5/16428570/google-pixel-2-no-headphone-jack-apple-wireless-future.

Shanklin, W. (2014, Sept. 24). Apple Watch vs. Samsung Gear S. *Gizmat*. Retrieved from http://www.gizmag.com/apple-watch-vs-samsung-gear-s/33960/.

Thierer, A.D. (1994). Unnatural Monopoly: Critical Moments in the Development of the Bell System Monopoly. *The Cato Journal* 14:2.

Time. (2007). Best Inventions of 2007. Best Inventors—Martin Cooper - 1926. *Time Specials*. Retrieved from http://www.time.com/time/specials/2007/article/0,28804,1677329_1677708_1677825,00.html

Viswanathan, T. (1992). Telecommunication Switching Systems and Networks. New Delhi: Prentice Hall of India. Retrieved from http://www.certified-easy.com/aa.php?isbn=ISBN:8120307135&name=Telecommunication_Switching_Systems_and_Networks.

Vogelstein, F. (2008). The untold story: how the iPhone blew up the wireless industry. *Wired Magazine* 16:02—Gadgets—Wireless. Retrieved at http://www.wired.com/gadgets/wireless/magazine/16-02/ff_iphone?currentPage=1.

Yan, S. (2017, May 25). This bendable smartphone comes with a catch. Money.CNN.com. Retrieved from http://money.cnn.com/2016/05/25/technology/bendable-smartphones-china/index.html.

23

The Internet

Stephanie Bor, Ph.D. and Leila Chelbi, M.A.[*]

Overview

From a military communication network to a worldwide online space, the Internet is constantly evolving to encompass new functions that connect people and institutions from different fields, and with various purposes. Due to its versatility, the Internet is now omnipresent in commerce, education, banking, dating, and much more. What seemed to belong to the realm of science fiction decades ago, is now part of our daily life. The infrastructure extended considerably such that it is now connecting sophisticated devices, to form what we call the Internet of Things (IoT).

Introduction

There is no denying the significance of the Internet on human culture, as it has infiltrated almost every aspect of society. The pervasiveness of this technology is illustrated by statistics, that reveal that nearly 3.86 billion people use the Internet throughout the globe (Internet Live Stats, 2018). Since its humble beginnings as a military project during the late 1960s, the Internet has emerged as a crucial part of everyday life as people have come to rely on this technology for work, education, relationships, and entertainment.

So, what exactly is this technology that has taken over our lives? According to Martin Irvine of Georgetown University, the Internet can best be understood in three components. It is "a worldwide computing system using a common means of linking hardware and transmitting digital information, a community of people using a common communication technology, and a globally-distributed system of information" (DeFleur and Dennis, 2002, p. 219). It is important to note, however, that the Internet does not act alone in providing us with seemingly endless information-seeking and communication opportunities. An integral part of this technology is the world wide web. While the Internet is a network of computers, the World Wide Web allows users to access that network in a user-friendly way. It provides an audio-visual format and a graphical interface that is easier to use than remembering lines of computer code, allowing people the ability to browse, search, and share information among vast networks.

The impact of the Internet on its users' lives is widespread and diverse, as it influences the ways in which people understand salient issues in their lives, such as their health, government, and communities. It has disrupted traditional social conventions by changing the style and scope of communication performed by people in their interactions with friends and family, as well as with strangers. Further, it is becoming more apparent that the network structure of the Internet enhances individuals' personal autonomy by allowing people to function more effectively on their own; it is no longer necessary for people to rely on physical institutions such as banks and post offices to perform daily

[*] Bor is Adjunct Professor at the University of Denver (Denver, Colorado). Chelbi is a graduate student at the University of South Carolina, Columbia, SC

tasks such as paying bills and sending messages (Rainie & Wellman, 2012). The scene at your local coffee shop in the middle of the day exemplifies this point, as patrons are seen hunched behind their laptops using public wireless Internet to perform job functions that were once conducted in traditional office settings.

This chapter examines Internet technology by beginning with a review of its origins and rise to popularity. Next, recent developments in online marketing, social interaction, and politics will be discussed in relation to their impact on the current state of the Internet. We will conclude by briefly highlighting several issues related to the Internet that are anticipated to receive attention in future debate and research.

Background

Though it is now accessible to virtually anyone who has a compatible device, the Internet began as a military project. During the Cold War, the United States government wanted to maintain a communication system that would still function if the country was attacked by missiles, and existing radio transmitters and telephone poles were disabled. The solution was to transmit information in small bits so that it could travel faster and be sent again more easily if its path was disrupted. This concept is known as packet-switching.

Many sources consider the birth of the Internet to have occurred in 1968 when the Advanced Research Projects Agency Network (ARPANET) was founded. Several universities, including UCLA and Stanford, were collaborating on military projects and needed a fast, easy way to send and receive information about those projects. Thus, ARPANET became the first collection of networked computers to transfer information to and from remote locations using packet switching.

ARPANET users discovered that, in addition to sending information to each other for collaboration and research, they were also using the computer network for personal communication, so individual electronic mail (email), accounts were established. Email accounts allow users to have a personally identifiable user name, followed by the @ sign, followed by the name of the host computer system.

USENET was developed in 1976 to serve as a way for students at The University of North Carolina and Duke University to communicate through computer networks. It served as an electronic bulletin board that allowed users on the network to post thoughts on different topics through email. USENET then expanded to include other computers that were not allowed to use ARPANET.

In 1986, ARPANET was replaced by NSFNET (sponsored by the National Science Foundation) which featured upgraded high-speed, fiber-optic technology. This upgrade allowed for more bandwidth and faster network connections because the network was connected to supercomputers throughout the country. This technology is what we now refer to as the modern-day Internet. The general public could now access the Internet through Internet service providers (ISPs) such as America Online, CompuServe, and Prodigy. Every computer and server on the Internet was assigned a unique IP (Internet Protocol) address that consisted of a series of numbers (for example, 290.152.74.113).

Rogers' (2003) Diffusion of Innovation theory points out that low levels of complexity in an innovation aid adoption; in other words, innovations that are easy to use are more successful. That qualification presented a problem for the early versions of the Internet—much of it was still being run on "text-based" commands. Even though the public could now access the Internet, they needed a more user-friendly way to receive the information it contained and send information to others that didn't involve learning text-based commands.

In 1989, Tim Berners-Lee created a graphical interface for accessing the Internet and named his innovation the "World Wide Web." One of the key features of the world wide web was the concept of hyperlinks and common-language web addresses known as uniform resource locators (URLs). This innovation allows a user to simply click on a certain word or picture and automatically retrieve the information that is tied to that link. The hyperlink sends a request to a special server known as a "domain name server," the server locates the IP address of the information, and sends that back to the original computer, which then sends a request for information to that IP address. The user's computer is then able to display text, video, images, and audio that has been requested.

Today we know this as simple "point-and-click" access to information, but in 1989 it was revolutionary. Users were no longer forced to memorize codes or commands to get from one place to the next on the Internet—they could simply point to the content they wanted and click to access it.

It is worthwhile at this point to explain the IP address and domain name system in more detail. The domain name (e.g, google.com) is how we navigate the World Wide Web, but on the back end (which we don't see), the IP address—a set of numbers—are the actual addresses. The Internet Corporation for Assigned Names and Numbers, or ICANN, is responsible for assigning domain names and numbers to specific websites and servers. With 1.6 billion users on the Internet, that can be quite a task (ICANN, 2010). To try and keep things simple, ICANN maintains two different sets of "top level domain" names: generic TLD names (gTLD) such as .edu, .com, and .org, and country codes (ccTLD) such as .br for Brazil, .ca for Canada, and .ru for Russia.

IP addresses used to consist of a set of four numbers (e.g., 209.152.74.113), in a system known as IPv4. With 256 values for each number, more than four billion addresses could be designated. The problem is that these addresses have been allocated, requiring a new address system. IPv6 is the designation for these new addresses, offering 340,282,366,920,938,000,000,000,000,000,000,000,000 separate addresses (Parr, 2011). Without getting too detailed, it is doubtful that these addresses will be used up any time soon.

So, what made the Internet so popular in the first place? During the late 1990s and the early 2000s, the Internet became one of the most rapidly adopted mass consumer technologies in history (see Figure 23.1). By comparison, radio took 38 years to attract 50 million Americans, while the Internet took only four years to attract a comparable size audience (Rainie and Wellman, 2012). Advancements in hardware and software exist as primary factors that stimulated the widespread adoption of the Internet. Additionally, enthusiasm displayed by the U.S. federal government,

which imposed minimal legal regulations on this technology, also contributed to early penetration of this technology.

Figure 23.1
Percent of U.S. Adults Who Use the Internet (2000-2018)

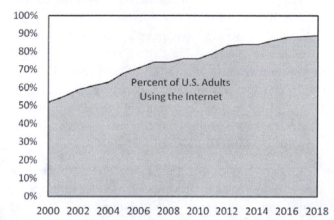

Source: Pew Research, 2018

Internet growth in the early 2000s can also be attributed to consumers' attraction to certain web features and applications. For example, online gaming, radio, instant messaging, health-focused websites, and pornography were all highly instrumental in enticing new Internet users (Rainie & Wellman, 2012). Retail shopping was another activity that attracted new users to the Internet, as businesses quickly capitalized on markets of consumers that preferred buying products online. It is worth emphasizing that the impact of the Internet on business and commerce has been significant. A new concept—e-commerce—was created to describe any transaction completed over a computer-mediated network that involves the transfer of ownership or rights to use goods or services. For example, if you purchase a song from iTunes, you are engaging in e-commerce.

E-commerce is not to be confused with e-business, which is a different term that encompasses procedures for business that are conducted over a computer-mediated network such as ordering new materials to aid in the production of goods, as well as marketing to customers and processing their orders (Mesenbourg, 1999). E-business and e-commerce continue to constitute prominent activities that engage Internet users, as both have emerged as established fields in business school curriculum and the workforce.

Recent Developments
Computer-Mediated Relationships

The Internet continues to be increasingly important for social interactions. Especially with the rise in social networking sites, people use the Internet to cultivate new relationships and to maintain existing bonds with friends and family. Computer-mediated interpersonal communications constitute a major area field of research, which ultimately reflects conflicting evidence as to whether the Internet enhances or is harmful to relationships.

Research concludes that several Internet activities such as game playing and social media can improve online and face-to-face peer relationships (Lai and Gwung, 2013). It is also suggested that online communication can have a positive influence on adolescents' sense of identity and the quality of their friendships (Davis, 2013). In regard to the parent-child social dynamic, there is some evidence that certain activities such as watching videos online can enhance relationships (Lai and Gwung, 2013). However, the Internet has also been identified as a source of tension between parents and children, and growing concerns over safety have challenged parents to negotiate rules for monitoring and controlling their children's Internet use.

In terms of romantic relationships, research shows that Americans are increasingly looking for love online (Geiger & Livingston, 2018). In the mid-2000s, studies showed that online dating carried a negative stigma, as people who used it tended to view it as a subpar way of meeting people. However, a more recent study revealed a major shift in attitudes, as the majority of Americans now say online dating is a good way to meet people (Smith & Anderson, 2016). Particularly among young adults, the number of 18–24 year olds who use online dating has nearly tripled from 10% in 2018 to 27% in 2015. Mobile dating apps such as Tinder and Bumble are largely responsible for the increase. Perhaps most interestingly, online dating use among 55–64 year olds also increased substantially, as 12% of people in this age range report using an online dating site or mobile app (Smith & Anderson, 2016).

Perhaps the most significant evidence of the influence of the Internet on romantic relationships can be recognized in the $2.7 billion in revenue reported in 2016 (Gillies, 2017). Match Group was the largest company with a 34.4 percent market share and 45 brands that include March, Tinder, and OKCupid. According to journalist Dan Slater (2013), online dating has had a profound change on society by modifying our perceptions of commitment, as well as the potential for romantic chemistry to be determined by mathematical algorithms. Research reveals that attitudes towards online dating continue to become more positive over time, suggesting their increased prominence in the future of Internet use.

Social Networking Sites and Politics

Social networking sites have become an essential tool for political influence. Websites such as Facebook and Twitter have provided the technological infrastructure for people to organize and activate massive political movements that have influenced significant events, such as elections and government upheavals.

In the context of electoral politics, social networking sites have become a standard tool in politicians' campaign toolboxes. These sites provide political candidates with an inexpensive means to spread messages and generate voter support in the months leading up to Election Day. Since their emergence during the 2008 United States election, the use of social networking sites has expanded significantly as every serious candidate in the 2016 election maintains active accounts on an array of sites such as Facebook, Twitter, Tumbler, and Instagram. Politicians use these platforms to collect information about voters such as email addresses, geographic locations, opinions, and personal interests, which could subsequently be used to customize messages sent to individual voters and to inform policymaking (Bor, 2013; Hong & Kim, 2016). Evidence of political candidates' success in generating attention on social networking sites continues to increase. We can cite the examples of Donald Trump and Hillary Clinton during the 2016 U.S. presidential election campaign who used a combination of multiple social media platforms to connect with their partisans. According to Craig (2016), politicians are expected to connect with voters in theatrical ways that often relies on social media. This connection is crucial given the fact that their followers now often rely on the content posted by candidates as their main source of information (Enli, 2017). The growth of social media use is especially pronounced in the 30–49 year old age demographic. This increasing percentage has been attributed to the increase in cellphone usage, as research reports that the proportion of Americans who use their cellphones to track political news or campaign coverage has doubled since 2010.

Beyond democratic elections, social networking sites have provided coordinating tools for political movements throughout the world. According to González-Bailón & Wang (2016), "digital technologies have accelerated the speed of communication and amplified its reach; they have also made it easier to analyze connections and improve our understanding of how networks mediate the emergence of collective action... Online technologies allegedly allow anyone with an internet connection to become an information broker and be in a position to trigger diffusion reactions."

But despite the positive impact of social media on participatory democracy and civic engagement, there are also negative effects. One significant field of research focuses on the ways in which social media exacerbates the polarization of political discourse. This research shows that social media users often gravitate toward platforms that reinforce their existing beliefs (Hong & Kim, 2016).

Another consequence of the vulnerability of online networks was the Russian meddling in the 2016 U.S. election. Besides spear-fishing attacks to manipulate votes, Russians deployed a series of misinformation campaigns (Berghel, 2017). The proliferation of trolling and fake news further polarized audiences and decreased trust in legacy media. According to Gottfried and Shearer (2016), 62% of American adults use social media as their news source. Consequently, most fake news stories circulated on social media, which led many users to believe them (Silverman and Singer-Vine 2016).

One study revealed the highly partisan nature of political discussion on Twitter, concluding that clusters of political talk are characterized by homogeneous views, which confines users in echo chambers (Hong & Kim, 2016). Echo chambers are "characterized by how

people in online debates selectively avoid opposing arguments, and therefore face little resistance" (Karlsen, Steen-Johnsen, Wollebæk, & Enjolras, 2017). Researchers suggest that social media polarization is harmful to communities, societies, and democracies (Matakos, Terzi, & Tsaparas, 2017). Social networking sites such as Twitter and Facebook have proven to be an especially effective tool for citizens during political protests and revolutions because they can provide an efficient means for conveying warnings and updates about dramatically shifting ground events such as violent conflict and home evacuations. The Arab Spring is an example of the powerful role of social networking sites in generating collective actions (Freelon, 2018). Even though 59% of social media users find political content stressful and frustrating, 49% of them post content related to politics on Facebook, and 40% on Twitter (Duggan & Smith, 2016). For more on social media see Chapter 24.

Advancements in Online Marketing

With the ongoing changes in the online experience, the strategies for Internet marketers are in constant flux. The sheer amount of traditional advertising messages on the Web has led consumers to develop an immunity against ads, causing the conversion rates for traditional advertisements to decline. Additionally, the growing focus on social connections that has been stimulated by the popularity of social networking sites has made it evident that marketing now requires a true two-way dialogue with consumers. It is not enough to write press releases or even to post advertisements to social networking sites. Marketers now have to identify conversations about their companies, products and competitors—and then actively engage in them. This "social listening" is part of the increased efforts to monitor and analyze the outcome of online marketing campaigns, which allows marketers to gain more information about potential customers and target their efforts to specific user groups.

A key factor in engaging with potential customers is content marketing. While the concept has been around for years, creating brand-specific content that actually has a value for potential customers has become even more valuable in times of online social media. Personalization also plays a key role in enhancing the effectiveness of email marketing. In fact, adding information about the email's recipient (e.g. name) increases the probability of opening the email by 20%, which increases sales leads by 31% and decreases unsubscribing by 17% (Sahri, Wheeler, & Chintagunta, 2018).

The Internet of Things

The term Internet of Things (IoT) was introduced by the technology pioneer Kevin Ashton in 1998. Even though it has been around for 20 years, it has become more relevant today, given that 49% of the world's population is connected online (Rainie & Anderson, 2017). While most users typically think about connected cars, personal assistants, and smartphones, the reality is that the IoT includes more devices such as road sensors, health-monitoring and sporting goods, to name a few. There are endless possibilities as to the nature of "things" that can be connected to exchange data. This expanding network, discussed in more detail in Chapter 12, undeniably brings convenience and connection, but will make it nearly impossible for users to unplug by 2026 (Rainie & Anderson, 2017).

Current Status

A 2017 report revealed that 49% of the world's population is connected online (Rainie & Anderson, 2017). Findings from the Pew Research Internet and American Life Project illustrate the remarkable growth in Internet adoption since the turn of the 21st century (Pew, 2018). In 2018, approximately 90% of American adults use the Internet, which is up from approximately 50% in 2000. Further, 49% of the entire world's population is connected online (Rainie & Anderson, 2017).

While there was an increase in Internet use observed across all demographic groups, it is interesting to point out distinctions among certain user populations. For example, when comparing different age groups, it is evident that younger adults are considerably more likely to use the Internet. The percentage of people between ages 18–29 who use the Internet reaches near-saturation at 98%; 97% of people ages 30–49 use the Internet; 87% of people ages 50–64 use the Internet; and 66% of people over 65 years of age use the Internet. (Pew, 2018).

Education level also appears to be a factor in predicting Internet usage (Pew, 2018). More than 93% of people with at least some college education use the Internet, while only 65% of non-high school graduates use the Internet. When it comes to economic income, nearly all households (98%) with a reported income of $75,000 or more are Internet users. This percentage steadily declines as household income decreases, but still 81% of households that make less than $30,000 per year reported using the Internet.

Other demographic categories such as gender, race, and ethnicity reflect minimal variation between groups. However, it is interesting to note that a gap still remains when comparing different ethnic groups' broadband connections at home. A survey completed in 2016 revealed that while 78% of white, non-Hispanic Internet users have high-speed Internet at their home, only 65% of black, non-Hispanic, and 58% of Hispanic adults use high-speed Internet at home (Pew, 2018). In addition to tracking user penetration statistics, it seems equally important to understand what people are actually *doing* online. According to *The Digital Future Report* (2017) that conducts an annual survey of Internet trends and issues, Internet users go online to engage in four main activities:

- Communication Services (i.e. checking email, instant messaging, posting on message boards)
- Fact-finding, Information Sources, and Education (i.e. looking up a definition, distance learning)
- Posting Information and Uploads (i.e. posting photos, uploading music videos)
- Information Gathering (seeking news, looking for health information)

Additionally, since 2010 there has been a significant increase in the percentage of Internet users making online purchases. In 2016, 83% of Internet users bought something online, with clothes and travel being the most popular items purchased (see Table 22.1).

Although Internet users generally agree that this technology has positive implications for individuals and society as a whole, a study of non-users reveals that 13% of American adults still choose not to use the Internet or email (Anderson & Perrin, 2016). Among this percentage of non-users, irrelevance and difficulty in using the technology were reported as the top two reasons for Internet avoidance.

Table 22.1

10 Most Popular Online Purchases in 2016

Item(s) Purchased Online	% of Internet users who have purchased item online
Clothes	67%
Gifts	64%
Electronics	54%
Books	51%
Travel	51%
Hobbies	43%
Software/ Computer Games	37%
Videos/ DVDs	37%
Computers	35%
Children's Goods	34%
Sporting Goods	29%
CDs	23%

Source: The Digital Future Report, USC Annenberg School Center for the Digital Future

Factors to Watch

Privacy and Personal Data

With the increasing sociability and personalization of the Internet, protecting privacy online has become an important topic. The vast majority of Americans agree that safe practices on the Internet are central not only to their own, but also the Nation's, safety (National Cyber Security Alliance, 2012). And as more and more routine tasks (e.g. banking, social security administration, healthcare, bill payments) move online, the importance of safely handling personal data in an online environment will only increase in years to come.

The rise of social media has made personal information (such as photos, birth dates, addresses, and phone numbers) available to third parties. Oftentimes, this information is given away willingly by the individuals who control the information or is collected by third parties without expressed consent. And while publicly posting vacation photos on Facebook might not seem like a serious privacy and security threat, the consequences can be severe. For example, researchers

have been able to successfully predict individuals' social security numbers using publicly available data (such as Facebook profiles) and other over-the-counter software (Acquisti & Gross, 2009).

More than 85% of Internet users have taken steps to reduce the amount of data they make available online by setting stricter privacy settings in social networking sites, changing their browsing behavior, or installing specific security software (Rainie et al., 2013). Still, 91% of Internet users think that they don't have control of how their personal data is collected and used by other entities (Rainie, 2016). Depending on the context, users find it acceptable to share their health information with their doctor (52%), and grocery stores to obtain loyalty cards (47%), whereas only 37% of users accept to share their driving habits with their insurance company (Rainie & Duggan, 2016).

Privacy is closely related to anonymity. When users have the possibility to hide their identity, they can engage in antisocial behavior such as trolling. Online trolling is "the practice of behaving in a deceptive, destructive, or disruptive manner in a social setting on the Internet with no apparent instrumental purpose" (Buckels, Trapnell, & Paulhus, 2014). Technical and human solutions are likely to be implemented in order to ensure a safe online social climate (Rainie, Anderson, & Albright, 2017). According to Anderson (2017), 41% of Americans have experienced some form of online harassment, and 79% of them said that online platforms should regulate interactions.

The Rise in Mobile Connectivity

The widespread adoption of smartphones has dramatically changed the way the Internet is used. According to a 2018 Pew Research Center fact sheet, 95% of American adults now own a smartphone, which constitutes a 77% increase since 2011. This shift from the stationary use of the Internet, which has been the standard for most of the Internet's history, is having a great impact on the way content is presented and consumed. Due to smaller screen sizes and different usage patterns (shorter, but re-occurring usage), the question becomes whether the information presented to mobile Internet users should replicate the regular Internet content, or if it should be an extension? The trend currently points towards a converged model, in which both worlds are closely related.

A study of smartphone use conducted in 2015 explored the concept of being smartphone-dependent, which is a term that refers to people who rely on their smartphone for Internet access because they either do not have traditional broadband service at home or have few options for online access other than their cellphones. (Smith, 2015). This research revealed that dependency upon smartphones is especially high among minorities and economically disadvantaged populations who often do not own a personal computer and only use cellphones when they go online. Further, lower-income smartphone-dependent users are much more likely to access their phones for career opportunities such as conducting job searches and applying for jobs.

To conclude, the Internet is clearly a constantly evolving technology that will continue to be used by humans in new and creative ways. A survey revealed that the importance of the Internet for its users continues to increase over time, as more than half of Internet users in 2013 claimed that the Internet would be, at a minimum, "very hard" to give up (Fox & Rainie, 2014).

While its capacity to make life easier remains debatable, the Internet unarguably makes information and communication more accessible. It will be important to continue monitoring the unanticipated outcomes of Internet use, and to analyze the influence of these behaviors on society.

Getting A Job

With smart devices tracking user behavior on multiple levels and the ability to store and compute vast amounts of information, one of the biggest areas of development surrounding the Internet is Big Data and analytics (discussed in Chapter 24). Both terms have become buzzwords not only in advertising, public relations, and journalism, but also in many other industries attempting to leverage Internet-facilitated information to their benefit. According to Markow et al. (2017), demand for data scientists will increase by 28% through 2020. As a result, a theoretical understanding of applications and implications of data analytics—as well as the practical skills to carry out statistical computations—can be great assets in the job market.

Projecting the Future

What will the Internet look like in 2033? Will it be a smart, personal assistant that reduces information overload and never lets us forget a birthday and the ideal present for that person? Or will it be a Big Brother-type surveillance tool tracking our behavior and allowing others to exploit it? It is difficult to predict the future of a technology that has changed so rapidly in recent years. Before the iPhone was introduced in 2007, for most people, the thought of having a powerful computer in their pocket that allowed them to connect to everybody and everything was science fiction. Today, it is commonplace.

So, what will we consider normal in 2033? Although a definite answer is almost impossible to give, some general trends emerge. For example, information sharing will be completely interwoven into daily life—so much so that it becomes invisible and effortless. Cars, buildings, cities, and especially people (think: wearable technology, implants) will have even more sensors and software that track resources, respond to crime, and take constant vital signs. These smart objects, using artificial intelligence and big data to generate predictions about any aspect of life, have formed the Internet of Things. People will be more aware of the world around them and how their own behavior in this interconnected world affects themselves and others. It will help individuals to better manage their workloads, and to monitor daily life, especially in regard to personal health. Augmented reality and wearable devices will allow real-time feedback and suggestions in a variety of life situations. Imagine sitting in a history lecture and being transported in time and space to experience the civil war in virtual, enhanced, augmented reality. Or standing in a boutique and immediately knowing not only if your size is in stock but also if it would be cheaper to buy at a different store just by looking at the price tag.

Sounds great, doesn't it? But this less cluttered, more streamlined online world—where unwanted information is automatically filtered out and life is greatly assisted by artificial intelligence—also poses big challenges to personal safety and privacy. Experts agree that it will be impossible to prevent abuse, but that it will be a constant race between hackers and providers of protective solutions detecting fraudulent behavior. The rise of data-driven services and growing connectivity leads some to worry of a dystopian, Minority Report-style future in which our course in life is mapped out for us, eroding our ability to make free choices. Imagine your health insurance being automatically canceled because you had a burger last *night? And weren't you supposed to work out? Some universities have already started tracking students' physical activity levels with outcomes affecting their grades. If technology predicts our life, big corporations, and data savvy criminals might be the ones most benefitting from all that information. Online fraud, identity theft—but also stalking, bullying, and other offenses—might become even more prevalent when our complete behavioral profiles are stored online.*

So, what will the Internet look like in 2033? We don't know. But it is most likely to fall somewhere in between these extremes.

Bibliography

Acquisti, A., & Gross, R. (2009). Predicting social security numbers from public data. *Proceedings of the National Academy of Sciences, 106* (27), 10975-10980.

Anderson, M. (2017). Key trends shaping technology in 2017 (Fact Tank). Pew Research Center, Washington, D.C. Retrieved from http://www.pewresearch.org/fact-tank/2017/12/28/key-trends-shaping-technology-in-2017/

Anderson, M, & Perrin, A. (2016). 13% of Americans don't use the internet. Who are they? (Fact Tank). Pew Research Center, Washington, D.C. Retrieved from http://www.pewresearch.org/fact-tank/2016/09/07/some-americans-dont-use-the-internet-who-are-they

Berghel, H. (2017). Oh, What a Tangled Web: Russian Hacking, Fake News, and the 2016 US Presidential Election. *Computer, 50*(9), 87-91.

Bor, S. (2013). Using social network sites to improve communication between political campaigns and citizens in the 2012 election. *American Behavioral Scientist,* doi:10.1177/0002764213490698.

Buckels, E. E., Trapnell, P. D., & Paulhus, D. L. (2014). Trolls just want to have fun. *Personality and individual Differences, 67,* 97-102.

Cannarella, J., & Spechler, J. A. (2014). Epidemiological modeling of online social network dynamics. Unpublished manuscript. Department of Mechanical and Aerospace Engineering, Princeton University. Retrieved from http://arxiv.org/pdf/1401.4208v1.pdf.

Craig, G. (2016). *Performing politics: Media interviews, debates and press conferences.* Austin, Texas: John Wiley & Sons.

Dahlberg, L. (2001). Computer-mediated communication and the public sphere: A critical analysis. *Journal of Computer-Mediated Communication, 7*(1), 0-0.

Davis, K. (2013). Young people's digital lives: The impact of interpersonal relationships and digital media use on adolescents' sense of identity. *Computers in Human Behavior, 29,* 2281-2293.

DeFleur, M. L. & Dennis, E. E. (2002) *Understanding mass communication: A liberal arts perspective.* Boston: Houghton-Mifflin.

Duggan, M.. & Smith, A. (2016). The Political Environment on Social Media. *Pew Internet and American Life Project* (Report). Pew Research Center, Washington, D.C. Retrieved from http://www.pewinternet.org/2016/10/25/the-political-environment-on-social-media/

Eloqua, & Kapost. (2012). *Content Marketing ROI: Why content marketing can become your most productive channel.* [eBook] Retrieved from http://marketeer.kapost.com/.

Enli, G. (2017). Twitter as arena for the authentic outsider: exploring the social media campaigns of Trump and Clinton in the 2016 US presidential election. *European Journal of Communication, 32*(1), 50-61.

Fox, S. & Rainie, L. (2014). The web at 25 in the U.S. *Pew Internet and American Life Project* (Report). Pew Research Center, Washington, D.C. Retrieved from http://www.pewinternet.org/2014/02/27/the-web-at-25-in-the-u-s.

Freelon, Deen. "Deen Freelon: Watching From Afar: Media Consumption Patterns Around the Arab Spring." *Policy* 2016 (2018): 2014.

Geiger, A., & Livingston, G. (2018). 8 facts about love and marriage in America. Report. Pew Research Center, Washington, D.C. Retrieved from http://www.pewresearch.org/fact-tank/2018/02/13/8-facts-about-love-and-marriage/

Gillies, T. (2017). In 'swipe left' era of mobile dating, eHarmony tried to avoid getting 'frozen in time.' CNBC. Retrieved from https://www.cnbc.com/2017/02/11/in-swipe-left-era-of-mobile-dating-eharmony-tries-to-avoid-getting-frozen-in-time.html.

González-Bailón, S., & Wang, N., (2016). Networked Discontent: The Anatomy of Protest Campaigns in Social Media. Retrieved from https://ssrn.com/abstract=2268165.

Gottfried, J., & Shearer, E. 2016. "News Use across Social Media Platforms 2016." Pew Research Center, May 26. http://www.journalism.org/2016/05/26/news-use-across-social-media-platforms-2016.

Hong, S., & Kim, S. H. (2016). Political polarization on twitter: Implications for the use of social media in digital governments. *Government Information Quarterly, 33*(4), 777-782.

ICANN Internet Corporation for Assignment Names and Numbers (2010). Accessed from http://www.icann.org/.

Internet Live Stats (2018). Retrieved from http://www.internetlivestats.com/internet-users/

Karlsen, R., Steen-Johnsen, K., Wollebæk, D., & Enjolras, B. (2017). Echo chamber and trench warfare dynamics in online debates. *European journal of communication, 32*(3), 257-273.

Lai, C., & Gwung, H. (2013). The effect of gender and Internet usage on physical and cyber interpersonal relationships. *Computers & Education, 69,* 303-309.

Lenhart, A., Anderson, M., & Smith, A. (2015). Teens, Technology and Romantic Relationships. *Pew Research Center* (Report). Pew Research Center, Washington, D.C. Retrieved from http://www.pewinternet.org/2015/10/01/teens-technology-and-romantic-relationships.

Lenhart, A. & Duggan, M. (2014). Couples, the Internet, and social media. *Pew Internet and American Life Project* (Report). Pew Research Center, Washington, D.C. Retrieved from www.perinternet.org/2014/02/11/couples-the-internet-and-social-media.

Markow, W., Braganza, S., Taska, B., Miller, S. M., & Hughes, D. (2017). *The Quant Crunch: How the Demand for Data Science Skills is Disrupting the Job Market.* Retrieved from https://public.dhe.ibm.com/common/ssi/ecm/im/en/iml14576usen/analytics-analytics-platform-im-analyst-paper-or-report-iml14576usen-20171229.pdf

Matakos, A., Terzi, E., & Tsaparas, P. (2017). Measuring and moderating opinion polarization in social networks. *Data Mining and Knowledge Discovery, 31*(5), 1480-1505.

Mesenbourg, T. L. (1999). *Measuring electronic business: Definitions, underlying concepts, and measurement plans.* Retrieved from http://www.census.gov/epdc/www/ebusiness.hum.

National Cyber Security Alliance, 2012. (2012). NCSA / McAfee Online Safety Survey. Retrieved from http://staysafeonline.org/stay-safe-online/resources/ [Accessed October 12 2013].

Parr, B. (2011). IPv4 and IPv6: A short guide. Retrieved from http://mashable.com/2011/02/03/ipv4-ipv6-guide/.

Perrin, A. (2015). One-fifth of Americans report going online almost constantly. Report. Pew Research Center, Washington, D.C. Retrieved from http://www.pewresearch.org/fact-tank/2015/12/08/one-fifth-of-americans-report-going-online-almost-constantly.

Perrin, A., & Duggan, M. (2015). Americans' internet access: 2000-2015. Report. Pew Research Center, Washington, D.C. Retrieved from http://www.pewinternet.org/2015/06/26/americans-internet-access-2000-2015.

Pew Reseach Center. (2018). Internet/Broadband Fact Sheet. Report. Pew Research Center, Washington, D.C. Retrieved from http://www.pewinternet.org/fact-sheet/internet-broadband.

Rainie, L., Kiesler, S., Kang, R., Madden, M., Duggan, M., Brown, S., & Dabbish, L. (2013). Anonymity, Privacy, and Security Online. *Pew Internet and American Life Project* (Report). Pew Research Center, Washington, D.C. Retrieved from http://www.pewinternet.org/2013/09/05/anonymity-privacy-and-security-online/.

Rainie, L. (2016). The state of privacy in post-Snowden America.(Fact Tank). Pew Research Center, Washington, D.C. Retrieved from http://www.pewresearch.org/fact-tank/2016/09/21/the-state-of-privacy-in-america/

Rainie, L. & Anderson, J. (2017). The Internet of Things Conenctivity Binge: What Are the Implications? *Pew Internet and American Life Project* (Report). Pew Research Center, Washington, D.C. Retrieved from http://www.pewinternet.org/2017/06/06/the-internet-of-things-connectivity-binge-what-are-the-implications/

Rainie, L., Anderson, J., & Albright, J. (2017). The Future of Free Speech, Trolls, Anonymity and Fake News Online. *Pew Internet and American Life Project* (Report). Pew Research Center, Washington, D.C. Retrieved from http://www.pewinternet.org/2017/03/29/the-future-of-free-speech-trolls-anonymity-and-fake-news-online/

Rainie, L., & Duhhan, M. (2016). Privacy and Information Sharing. *Pew Internet and American Life Project* (Report). Pew Research Center, Washington, D.C. Retrieved from http://www.pewinternet.org/2016/01/14/privacy-and-information-sharing/

Rainie, L., Smith, A., Schlozman, K. L., Brady, H. & Verba, S. (2012). Social media and political engagement. *Pew Internet and American Life Project* (Report). Pew Research Center, Washington, D.C. Retrieved from http://pewinternet.org/Reports/2012/Political-Engagement.aspx.

Rainie, L. & Wellman, B. (2012). *Networked: The new social operating system.* Cambridge, MA: MIT Press.Rogers, E. M. (2003). *Diffusion of Innovations.* Simon and Schuster.

Rogers, E. (1983). *Diffusion of Innovations,* 3rd Edition. New York, NY: Free Press

Sahni, N. S., Wheeler, S. C., & Chintagunta, P. (2018). Personalization in Email Marketing: The Role of Noninformative Advertising Content. *Marketing Science.*

Slater, D. (2013). *Love in the Time of Algorithms: What Technology Does to Meeting and Mating*. New York: Penguin Group.

Shirky, C. (2010, December 20). The political power of social media. *Foreign Affairs*. Retrieved from http://www. foreignaffairs. com/articles/shirky/the-political-power-of-social-media.

Silverman, C. & Singer-Vine, J. 2016. "Most Americans Who See Fake News Believe It, New Survey Says." *BuzzFeed News*, December 6.

Smith, A. (2014). Cell phones, social media and campaign 2014. Report. Pew Research Center, Washington, D.C. Retrieved from http://www.pewinternet.org/2014/11/03/cell-phones-social-media-and-campaign-2014.

Smith, A. (2015) U.S. Smartphone Use in 2015. Report. Pew Research Center, Washington, D.C. Retrieved from http://www.pewinternet.org/2015/04/01/us-smartphone-use-in-2015/

Smith, A. (2016). Online dating usage by demographic group. Report. Pew Research Center, Washington, D.C. Retrieved from http://www.pewinternet.org/2016/02/11/online-dating-demographic-tables.

Smith, A., & Anderson, M. (2016). 5 faces about online dating. Report. Pew Research Center, Washington, D.C. Retrieved from http://www.pewresearch.org/fact-tank/2016/02/29/5-facts-about-online-dating.

The Digital Future Report. (2017). The 2017 digital future report. *USC Annenberg School Center for the Digital Future* (Report). Center for the Digital Future. Los Angeles, CA. Retrieved from http://www.digitalcenter.org/wp-content/up-loads/2013/10/2017-Digital-Future-Report.pdf Wu, M., Lu, T. J., Ling, F. Y., Sun, J., & Du, H. Y. (2010, August). Research on the architecture of Internet of things. In *Advanced Computer Theory and Engineering (ICACTE), 2010 3rd International Conference on* (Vol. 5, pp. V5-484). IEEE.

24

Social Media

Rachel Stuart, M.A.*

Overview

Social media has come to permeate almost every aspect of both our online and real-world experiences. The first listservs like CompuServe enabled the evolution of social media to multi-modal platforms like Facebook, Twitter, and Snapchat. Social media provide increased documentation and augmentation of our realities, while introducing issues surrounding privacy and identity. The maturation of social media has led to changes in how we interact with the leaders of the word and how we gather and disseminate news.

> *"Despite the constant negative press covfefe"*
>
> —Donald J. Trump, 45th President of the United States, May 31st, 2017, via Twitter

Introduction

It's a Tuesday night in late May. If you are on the semester system, you have most likely sold your books for pennies on the dollar you paid for them and have abandoned the halls of higher learning for the summer. If you are on the quarter system, you are ignoring the pile of work you have in front of you and that will be in front of you for at least the next three weeks. As a distraction, you casually glance at your Twitter feed, and you find that the President of the United States has made at minimum a puzzling tweet, if not on the other end of the spectrum, an alarming tweet that makes you question his overall health. In an instant, the 45th President has created an instant meme that causes the Internet and social media to grind to a halt as all of us online openly question: what's a covfefe?

For the next several news cycles, "covfefe-gate" dominated the discussions online and around the proverbial water-cooler, pushing aside actual news stories like a large explosion in Kabul and the increasing focus on the Russian tampering with the 2016 American election. People defended the term tongue-in-cheek, saying that it was coined during the Bowling Green Massacre—a fictional terrorist attack that was made up by Trump surrogate, Kellyanne Conway in January 2017 (Flegenheimer, 2017). Others used it as fodder to call into question the mental and cognitive stability of the President and demand his removal from office. No matter whether it was funny, concerning, absurd, or everything in-between, one thing was certainly demonstrated by this event: social media is now a driving force behind how we connect, get news, and disseminate information. Social media has come to permeate almost every aspect of both our online- and real-world experiences. Even with so much

---

* Faculty member in the Communication Studies department at Highline College, Des Moines, WA.

exposure, there is still some confusion as to what constitutes a social networking site (SNS), and which of the literally millions of web pages on the Internet can be considered SNSs.

What is an SNS? According to boyd and Ellison (2008), there are three criteria that a website must meet to be considered an SNS. A website must allow users to "(1) construct a public or semi-public profile within a bounded system, (2) articulate a list of other users with whom they share a connection, and (3) view and traverse their list of connections and those made by others within the system" (boyd & Ellison, 2008, p. 211). These guidelines may seem to restrict what can be considered an SNS; however, there are still literally hundreds of vastly diverse websites that are functioning as such. Social networking sites have seemingly permanently cemented their place in the landscape of the Internet, and as we become comfortable explorers of the online social world, the nuances between social networking and social media become clearer and more defined. For over a decade, the terms social media and social networking sites have been used almost interchangeably. However, as the sites and apps dedicated to creating connections to people become more nuanced, there does seem to be a differentiation between the two.

According to *Social Media Today*, social media are forms of "...electronic communication (as web sites for social networking and microblogging) through which users create online communities to share information, ideas, personal messages, and other content (as videos)" (Schauer, 2015, par. 3), whereas, social networking sites are dedicated to "...the creation and maintenance of personal and business relationships...online" (Schauer, 2015, par. 4). One way to think about the difference between social networking sites and social media is to think about how the Internet and the world wide web are delineated: If you are on the world wide web, the you are on the Internet, but just because you are on the Internet does not necessarily mean you are on the web. The same is true for social media versus social networking sites; if you are on a social networking site, you are on a form of social media, but just because you are on social media, does not mean you are on a social networking site. As the social media and social networking sites become more nuanced and comprehensive, the distinction between

them will become ever more blurred. Rather than looking toward the future at this point, let's take a look at the background and history of social media and SNSs.

Background

Social media sites have taken on many forms during their evolution. Social networking on the Internet can trace its roots back to listservs such as CompuServe, BBS, and AOL where people converged to share computer files and ideas (Nickson, 2009). CompuServe was started in 1969 by Jeff Wilkins, who wanted to help streamline his father-in-law's insurance business (Banks, 2007).

During the 1960s, computers were still prohibitively expensive; so many small, private businesses could not afford a computer of their own. During that time, it was common practice to "timeshare" computers with other companies (Banks, 2007). Timesharing, in this sense, meant that there was one central computer that allowed several different companies to share access in order to remotely use it for general computing purposes. Wilkins saw the potential in this market, and with the help of two college friends, talked the board of directors at his father-in-law's insurance company into buying a computer for timesharing purposes. With this first computer, Wilkins and his two partners, Alexander Trevor and John Goltz, started up CompuServe Networks, Inc. By taking the basic concept of timesharing already in place and improving upon it, Wilkins, Trevor, and Goltz created the first centralized site for computer networking and sharing. In 1977, as home computers started to become popular, Wilkins started designing an application that would connect those home computers to the centralized CompuServe computer. The home computer owner could use the central computer for access, for storage and—most importantly—for "person-to-person communications—both public and private" (Banks, 2007).

Another two decades would go by before the first identifiable SNS would appear on the Internet. Throughout the 1980s and early 1990s, there were several different bulletin board systems (BBSs) and sites including America Online (AOL) that provided convergence points for people to meet and share online.

In 1996, the first "identifiable" SNS was created—SixDegrees.com (boyd & Ellison, 2008) SixDegrees was originally based upon the concept that no two people are separated by more than six degrees of separation. The concept of the website was fairly simple—sign up, provide some personal background, and supply the email addresses of ten friends, family, or colleagues. Each person had his or her own profile, could search for friends, and for the friends of friends (Caslon Analytics, 2006). It was completely free and relatively easy to use. SixDegrees shut down in 2001 after the dot com bubble popped. What was left in its wake, however, was the beginning of SNSs as they are known today. There have been literally hundreds of different SNSs that have sprung from the footprints of SixDegrees. In the decade following the demise of SixDegrees, SNSs such as Instagram, Friendster, MySpace, LinkedIn, Facebook, and Twitter have become Internet zeitgeists.

Friendster was created in 2002 by a former Netscape engineer, Jonathan Abrams (Milian, 2009). The website was designed for people to create profiles that included personal information—everything from gender to birth date to favorite foods—and the ability to connect with friends that they might not otherwise be able to connect to easily. The original design of Friendster was fresh and innovative, and personal privacy was an important consideration. In order to add someone as a Friendster contact, the friend requester needed to know either the last name or the email address of the requested. It was Abrams' original intention to have a website that hosted pages for close friends and family to be able to connect, not as a virtual popularity contest to see who could get the most "Friendsters" (Milian, 2009).

Shortly after the debut of Friendster, a new SNS hit the Internet, MySpace. From its inception by Tom Anderson and Chris DeWolfe in 2003, MySpace was markedly different from Friendster. While Friendster focused on making and maintaining connections with people who already knew each other, MySpace was busy turning the online social networking phenomenon into a multimedia experience. It was the first SNS to allow members to customize their profiles using HyperText Markup Language (HTML). So, instead of having "cookie-cutter" profiles like Friendster offered, MySpace users could completely adapt their profiles to their own tastes, right down to the font of the page and music playing in the background. As Nickson (2009) stated, "it looked and felt hipper than the major competitor Friendster right from the start, and it conducted a campaign of sorts in the early days to show alienated Friendster users just what they were missing" (par. 15). This competition signaled trouble for Friendster, which was slow to adapt to this new form of social networking. A stroke of good fortune for MySpace also came in the form of rumors being spread that Friendster was going to start charging fees for its services.

In 2005, with 22 million users, MySpace was sold to News Corp. for $580 million (BusinessWeek, 2005); News Corp. later sold it for only $35 million. MySpace, amazingly, is still a functional social media site, with approximately 20 to 50 million unique views per month (Arbel, 2016). The site, which was at one point the pinnacle of social networking sites is now focused on providing a platform for individuals to feature their music videos and songs (Arbel, 2016).

From its humble roots as a way for Harvard students to stay connected to one another, Facebook has come a long way. Facebook was created in 2004 by Mark Zuckerberg with the help of Dustin Moskovitz, Chris Hughes, and Eduardo Saverin (Newsroom, 2018). Originally, Facebook was only open to Harvard students, however, by the end of the year, it had expanded to Yale University, Columbia University, and Stanford University, with new headquarters for the company in Palo Alto, CA. In 2005, the company started providing social networking services to anyone who had a valid e-mail address ending in .edu. By 2006, Facebook was offering its website to anyone over the age of 13 who had a valid email address (Newsroom, 2018). What made Facebook unique, at the time, was that it was the first SNS to offer the "news feed" on a user's home page.

In all other social media platforms before Facebook, in order to see what friends were doing, the user would have to click to that friend's page. Facebook, instead, put a live feed of all changes users were posting—everything from relationship changes, to job changes, to updates of their status. In essence, Facebook made microblogging popular. This was a huge shift from MySpace, which had placed a tremendous amount of emphasis on traditional blogging, where people could type as much as they wanted to. Interestingly, up until July 2011, Facebook users were limited to 420-character status updates. Since November 2011, however, Facebook users have a staggering 63,206 characters to say what's on their mind (Protalinski, 2011). The length of the post on Facebook has become more irrelevant as the platform tries to highlight its usefulness in providing a multimedia experience, including a new host of emotions to communicate to our friends, allowing for the more appropriate frowny-face emoji than the generic "thumbs-up" like button when your best friend's dog died. In addition, Facebook has introduced more interactive features, like 360 panorama photos, live vlogging, and increased control over photographs with filters and augmented reality enhancements. Facebook, unlike Twitter, is no longer focused on the length of the post, but the quality of content you include with it.

Twitter was formed in 2006 by three employees of podcasting company Odeo, Inc.: Jack Dorsey, Evan Williams, and Biz Stone (Beaumont, 2008). It was created out of a desire to stay in touch with friends easier than allowed by Facebook, MySpace, and LinkedIn. Originally, Twitter took the concept of the 160-character limit the first iteration of text messaging imposed on its users and shortened the message length down to 140 (to allow the extra 20 characters to be used for a user name; McArthur, 2017). This made Twitter one of the first truly mobile social media platforms because you could tweet via a texting app. The 140-character limit had been the identifying hallmark of the platform, even as it added video and photo functionality seamlessly to its mobile and desktop interfaces. On November 7, 2017, Twitter officially expanded its tweet length to 280, effectively doubling the message length you can tweet (Perez, 2017). The response to this development has been surprising. On one hand, the expansion was celebrated because it would allow people to express themselves more clearly than 140 characters would allow. On the other, the overall sentiment toward the change has been negative, with analysis from 2.7 million tweets on the matter showing that 63% were critical of the expanded length of a tweet (Perez, 2017).

Social media sites such as Twitter and Facebook clearly helped cement social media within the landscape of the Internet. The last 15 years of social media tells a story of more dynamic, specialized social media sites entering the scene and sharing the spotlight with the above-mentioned giants. Sites like LinkedIn, Snapchat, Instagram, and Reddit point toward the increasingly diversified and specialized trajectory social media is taking. LinkedIn was created in 2003 by Reid Hoffman, Allen Blue, Jean-Luc Vaillant, Eric Ly, and Konstantin Guericke (About Us, 2018). LinkedIn returned the concept of SNSs to its old CompuServe roots. According to Stross (2012), LinkedIn is unique because "among online networking sites, LinkedIn stands out as the specialized one—it's for professional connections only" (par. 1). Instead of helping the user find a long-lost friend from high school, LinkedIn helps build professional connections, which in turn could lead to better job opportunities and more productivity. In June 2016, Microsoft bought the social media platform for $26.2 billion in cash (Darrow, 2016). Growth overall for LinkedIn has been slow. In October 2015, the business-centered social media site had over 400 million members worldwide with 100 million active users (Weber, 2015). In April 2017, the site reported to have 500 million users worldwide, and is no longer providing information about daily or monthly active users, just unique profiles on the site (Darrow, 2017).

Snapchat, which was originally released under the name Picaboo, was created by two Stanford University students, Evan Spiegel and Robert Murphy, in 2011 (Colao, 2012). The original premise of Snapchat was simple: users send friends "snaps," photographs and videos that last anywhere from one to ten seconds, and when the time expires the photos or videos disappear. In addition to the fleeting nature of the snap, if the recipient of a snap screen captures it, Snapchat will let the sender know that the person they sent it to saved it. By May 2017, Snapchat started to allow users to send snaps of any length and allowed the recipients of the snaps to keep them indefinitely (Newton, 2017). To say that Snapchat has become popular would be an understatement. By October 2012, one billion snaps had been sent, and the app averaged more than 20 million snaps a day (Gannes, 2012). By the first quarter of 2017, Snap, Inc. Snapchat's parent company reported 166 million daily active users (Constine, 2017a). In November 2013, Snapchat turned down a $3 billion offer in cash from Facebook, (Fiegerman, 2015) and as of early 2017, the company was valued at $40 billion (LaMonica, 2017).

Instagram was originally developed by Kevin Systrom and Mike Krieger in 2010 as a photo-taking application where users could apply various filters to their photographs and then upload them to the Internet. In 2011, Instagram (or "the 'gram" as some youngsters like to call it) added the ability to hashtag the photographs uploaded as a way to find both users and photographs (Introducing hashtags, 2011). Facebook purchased Instagram in 2012 from Systrom and Krieger for $1 billion in cash and stock (Langer, 2013). By the time Facebook acquired the filtered photo app giant, Instagram had over 30 million users (Upbin, 2012). While many in the tech industry saw Facebook's extravagant price paid for Instagram as a bad business decision, the move has proved to be lucrative as Instagram now has 500 million daily active users and 800 million monthly active users (Etherington, 2017). In addition, it has become one of the most popular social media sites on the planet, second only to Facebook itself in percent of unique users (Social Media Fact Sheet, 2018), and as of late 2016, the company was valued at $50 billion (Chaykowski, 2016).

Trying to keep up with the shift to visually-based, mobile social media platforms, Twitter bought the video-sharing start up Vine for $30 million in 2012 (Vine, 2018). Vine was created by Dominik Hoffman, Rus Yusupov, and Colin Kroll the year before and operated as a private, invite-only application (Dave, 2013). After its acquisition by Twitter, the number of active users of Vine shot up to over 200 million (Levy, 2015). Vine did not find the same traction that sites like Instagram or Snapchat did and in late 2016, Twitter discontinued the video platform (Foxx, 2016). One of Vine's original developers, Dominik Hoffman, is reportedly working on V2, a follow-up to his first video sharing social platform, which will have no affiliation with Twitter (Blumenthal, 2017).

Perhaps the ultimate, if not most misunderstood and underrated, example of the increasingly specialization of social media is the website Reddit. The self-proclaimed "front page of the Internet," Reddit is the epitome of the user generated content that is the hallmark of Web 2.0. Reddit was developed in 2005 by Alexis Ohanian and Steve Huffman when the pair wanted to create a website that aggregated the most popular and shared links on the Internet. In addition to moderators posting links on the site, Reddit also allowed users to post their favorite links to the site. In its first few years, the website was competing with sites like Del.icio.us and Digg, however, in 2008, Reddit allowed the ability for users to create pages for specialized content, also known as subreddits (Fiegerman, 2014).

The creation of subreddits proved to be the major turning point for Reddit, letting users to both post their own content and to create pages within the website to share content on a specific thing. As of January 2018, there were over 1.2 million subreddits hosted on the site, up from just over 1 million subreddits available at the same time the year before (New Subreddits, n.d.). To demonstrate how fast subreddits are created on the site, in one 24-hour period, between January 27th and January 28th, 2018, there were 606 new subreddits added to the site (New Subreddits, n.d.) There are subreddits for just about everything imaginable, and while this liberty has created issues surrounding questionable to illegal content (Fiegerman, 2014), the self-generated and self-policing on the site has created a massive community of Redditors, participating in a global social media experiment where you are just as liable to find a plethora of cat pictures as you are a group of Redditors saving a fellow user from the temptation of suicide. Reddit is still lagging behind the other social media giants in keeping up with the evolving demands of mobile technologies, but the site is a hallmark of the social media trend of today: giving users what they want, when they want it.

Recent Developments

Social media, like other innovations of Web 2.0, practice constant and consistent innovation. Each of the aforementioned sites has continued to grow and adapt the ever-changing wants and desires of their users and the Internet society. Social media is a mirror of the Internet as a whole—an organic, seemingly living breathing thing that is at times unpredictable and volatile. One thing is certain, though: social media are here to stay. Sites such as Facebook, Instagram, Twitter, Snapchat, and Reddit are mainstays in many of our lives, and as we grow accustomed to them, there are new ones vying for our attention. As we make room for new applications, businesses across the globe use well-established social media sites as viable platforms to vie for our attention as well. In this section, we will examine new trends on the scene for social media, including the increased documentation and augmentation of our realities, issues surrounding privacy and social media, and finally this section will end with an examination of the maturation of social media and how that has affected how we interact with the leaders of the word, and how we gather and disseminate news.

Augmenting Reality: The Evolution of User-Generated Content

Since Facebook bought the virtual reality start-up, Oculus Rift, in 2014 for $2 billion (Orland, 2014), the writing has been on the wall that social media platforms across the board were shifting toward a more immersive experience. The complete integration of virtual reality into social media platforms still feels like it might be a few years off, but the increase usage of augmented reality (AR) technologies within social media has arrived and will continue to increase. According to Reality Technologies, a company that specializes in reality-manipulating technologies including virtual reality, augmented reality, and mixed reality, augmented reality can be defined as "…An enhanced version of reality where live direct or indirect views of physical real-world environments are augmented with superimposed computer-generated images over a user's view of the real-world, thus enhancing one's current perception of reality" (Augmented Reality, 2016, para. 3).

Augmented reality has been used in a wide variety of applications over the last several years. In the area of navigation, AR programs are being paired with special headsets or helmets to "show" a cyclist the correct route. In addition, car makers have started to use AR technology to project information about the car and GPS information onto a portion of the windshield (Fisher, 2017). There have been certain mobile apps that utilized the camera on your phone to create enhancements or overlays. These include the astronomy-based Sky View, that created a constellation view overlay as you looked at the night sky through your camera, or companies like IKEA who have integrated AR into their mobile sites so that you can "see" what a piece of furniture would look like in your house (Fisher, 2017). In addition, Apple introduced the iPhone 8 and iPhone X in the fall of 2017. Both of these phones feature the new A11 bionic chip and the iOS 11 operating systems, which were specifically designed to support AR apps and features like their animojis (Costello, 2017)

Augmented reality came front and center to the collective consciousness of the Internet and mobile technology users with the introduction of the popular game *Pokémon Go*, which paired AR with tricking a whole generation of people to go outside and get exercise by challenging users to hunt around their neighborhoods and the areas they were in for Pokémon. As users came across Pokémon in the game, when they held up their phone, it would be a camera view of their surroundings with an AR Pokémon hanging out waiting to be caught. During the *Pokémon Go* craze of 2016-2017, other platforms, like the social media giant, Snapchat started heavily integrating AR technologies onto their site.

Snapchat has by far had the most success with the early adoption and integration of AR technologies into their social media platform. The AR features on Snapchat, according to Peckham (2017) "…allow you to distort your face, look stupid in front of your friends, show off your location and even transform your face for all of your followers" (para. 2). These AR features include filters and lenses. Features allow you to change the color of your picture or add an increasingly popular geofilter to your snap. Geofilters allow individuals to add location specific filters to their snaps that allow their followers to know exactly where they are. Lenses are truly AR filters. They will turn your face into a dog's face or give you crazy hair. In addition, Snapchat started allowing users to add AR emojis and objects to your videos. AR emojis are 3D objects, like cartoon characters that you can add to your video and move around (Ingraham, 2017). In late 2017, Snapchat created the Lens Studio which is a project that allows outside developers to design and create AR lenses and filters to be used on the social media platform (Constine, 2017b).

The success of filters and lenses on Snapchat has only intensified the competition amongst other social media platforms to integrate AR technologies into their sites. Introduced in December 2017, Facebook is looking to compete with Snapchat by introducing new AR effects into their camera effects for their original social media site (Kirkpatrick, 2017), their photosharing giant, Instagram (O'Kane, 2017), and their Messenger app (Statt, 2017). In addition, Facebook has opened its own AR development platform, called AR Studio, to compete with Snapchat's Lens Studio and the AR development programs supported by Apple and Google (Statt, 2017, para. 4).

The use of AR to enhance our experience of visually-based social media has not been without its issues. There have been two major issues that have arisen as we adopt and adapt to AR on social media platforms. The first issue could be called—in the kindest of terms—a seeming ignorance toward cultural sensitivity. The filters and lenses create fantastic augmentations of you and your surroundings, but as the technology is still so very new, most people are not asking the question "is this offensive?" when they release new filters. For example, Snapchat got in trouble twice in 2016, first in April, then in August, when it released two racist filters. The first was a Bob Marley filter that essentially gave the user a Rastafarian cap and dreadlocks and darkened their complexion to what amounted to blackface (Meyer, 2016). In August of the same year, Snapchat released what they called their "anime filter," that "…covered over a user's eyes and forehead with closed-eye slants while enlarging their teeth and reddening their cheeks" (Meyer, 2016, para. 5). However, as Zhu (2016) pointed out, hallmark features for anime characters are "…angled faces, spiky and colorful hair, large eyes, and vivid facial expressions" (para. 3); what Snapchat created was the literal stereotype of "yellowface." Snapchat took both filters down within a week of their debut, but this does pose a potential problem as the use of AR grows and each major social media platform has developed projects to allow outside programmers to create third party AR filters and lenses for their sites.

The second issue to arise in the development of AR technologies on social media platforms is a fugue on an issue that has plagued social media since its inception: privacy. One of the most interesting features that Snapchat has developed is their Snap Map. By accessing the Snap Map in your Snapchat app, you can see exactly where all of your friends are sending snaps from. Another feature that calls into question your privacy is the geofilters which allow you to post where exactly you are. This poses a couple of issues: first, it tells everyone you are connected to on Snapchat whether or not you are home and/or exactly where you are, which can leave you vulnerable to having your house burglarized, or you yourself being robbed, stalked, or kidnapped (Issawi, 2017). In addition, police are warning parents about the potential dangers of the Snap Map as it can be used by potential predators to track the location of unsuspecting children (Elise, 2017). Second, there are the potential psychological issues that have come up, including inflaming trust issues between spouses and significant others (who use it to track every movement of their one and only), and creating potential FOMO (and realistically, depression) for the people who realize they are the only person not invited to a gathering (Issawi, 2017). The issue of privacy on social media platforms is not new. Each platform has had to deal with it's own privacy issues—from check-ins that allow everyone, including potential burglars—to know that you will be in Cabo for a week, to location markers that tell your group where you are posting from. Augmented reality technologies are putting a new lens over what has been an age-old problem in relation to social media.

Augmenting Reality? Social Media, Fake News, and the Twitterer in Chief

No matter what your personal opinion is about the 45th President of the United States, no one can deny the influence Donald J. Trump has had using Twitter to try to change the conversation. No other president before him has taken to social media as proactively (and controversially) as he has. Trump has gone on record at least once to say that he credits his success on social media with his election as the President of the United States (McCaskill, 2017). In addition, the President sees social media in general--and Twitter specifically-- as a way for him to fight back against mainstream media. On December 30th, 2017 he tweeted:

"I use Social Media, not because I like to, but because it is the only way to fight a VERY dishonest and unfair 'press,' now often referred to as Fake News Media. Phony and non-existent 'sources' are being used more often than ever. Many stories & reports a pure fiction!" (Trump, as cited in Greenwood, 2017, para. 2)

Beyond showing the dangers of increasing the lengths of tweets to 280 characters, Trump's reliance upon social media to connect with the public-at-large also demonstrates a powerful new way these platforms are being used.

Figure 24.1

Percentage of Adults Who Use Specific Social Media Platforms

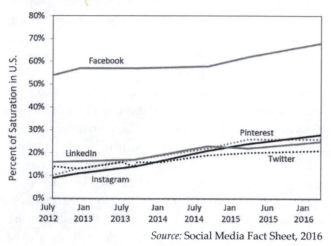

Source: Social Media Fact Sheet, 2016

One of the most surprising trends to encompass social media is how it has become a conduit for how people receive news and information. Since the inception of social media, there have been discussions around the social and cultural implications of individuals moving part of their lives and interactions to online platforms like Facebook, Twitter, and Instagram. From the beginning, social media was imagined as a place for individuals to connect with people who they either know in the real world or with people who have similar interests but are prohibited from meeting in the real world by geographical restrictions or other contributing factors. Now social media is being used in a way that is surprising to many: as a platform for "legitimate" news and information gathering.

According to the Pew Research Center, 67% of American adults reported that they get at least some of their news from social media sites (Shearer & Gottfried, 2017). On Twitter, the Commander in Chief's preferred platform, between 2016 and 2017, there was a 15% increase in users who got at least some of their

news from the social media site; part of this has been attributed to the President's proclivity to control the news cycle via his tweets (Shearer & Gottfried, 2017).

Facebook dominates as the social media platform of choice for news gathering by its users. Overall, nearly 45% of the entire population of the United States uses Facebook as a news source at least some of the time (Grieco, 2017). YouTube is gaining in popularity as a news site, a far second behind Facebook with 18% of the American population using it for news (Shearer & Gottfried, 2017). Twitter is an interesting study of news gathering: only 15% of adults are on Twitter, but fully 74% of those on the social media platform report using it for news gathering. (Shearer & Gottfried, 2017).

Figure 24.2

Percentage of Adults Who Use at Least One Social Media Site, by Age

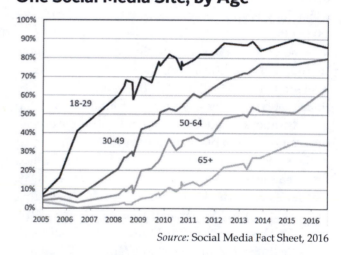

Source: Social Media Fact Sheet, 2016

The issue with using social media as a viable platform for news and information gathering comes from its inherent structure around user-generated content. This structure allows for endless possibilities and creativity, but does not always lend itself to truthfulness. As such, the concept of fake news and the augmenting of truth and reality has taken center stage on social media. Fringe social media platforms like 4chan or the subreddit /r/the_donald are used as megaphones to disseminate disinformation from websites like Breitbart News or InfoWars and vice versa—where conspiracies will start on social media, and then be reported as true by news sites.

One of the most insidious (and ridiculous) examples of the perpetuation of "fake news" by social media and fringe news sites was the #pizzagate conspiracy in 2016 (Aisch, Huang, & Kang, 2016). Users of the social media site 4chan concocted a theory that because some of the emails stolen from the Democratic National Committee included the term "cheese pizza," it was actually a code "c.p.," meaning child pornography. From there, the conspiracy spread to include Hillary Clinton, John Podesta, Clinton's campaign chairman, his brother Tony Podesta, and the Comet Ping Pong pizza restaurant in Washington, DC. Soon sites like Breitbart and InfoWars were reporting the story as true, relaying tales of child sex abuse, kill, and torture rooms, all being hushed by a massive government cover-up. The fringe news sites and social media ping-ponged the story back and forth, each time it picking up more steam and even wilder details until finally on December 4th, 2016, a man entered Comet Ping Pong with an assault rifle and a handgun. He was there to "liberate" the child slaves that were being trafficked by the Clintons and the Podestas through a torture room in the basement. He surrendered after he found no evidence of any wrongdoing (Aisch et al., 2016). Still today, the Pizzagate conspiracy pops up on /r/the_donald subreddit and 4chan. When everyone is screaming "fake news!" when they do not agree with it, it is hard to actually tell what is real and what is fake.

Another issue that is related to the concerted effort to spread disinformation via social media is the potential for it to be used maliciously against us by foreign entities. In December 2016, the United States Intelligence Community (IC) concluded that the Russian government was directly involved in creating electoral interference during the 2016 presidential election to try to unfairly sway the election to Donald Trump (Wells & Andrews, 2017). Since the release of that report by the IC, it has been determined that millions of users on platforms like Facebook, Twitter, and Google were exposed to Russian-driven propaganda during the election cycle. Twitter reported that "its users saw automated, election-related tweets from accounts tied to Russia approximately 288 million times between Sept. 1 and Nov. 15, 2016" (Wells & Andrews, 2017, para. 7). Many of these tweets on Twitter or posts on Facebook were generated by bots—computer programs that act and interact like actual human beings.

How social media has intersected news gathering by the general public has turned out to be one of the most surprising developments for the technology in the recent years. The reliance upon social media as a viable news source brings into focus the fact that the technology is maturing and it can no longer be written off as merely a series of social platforms. Both social media companies and their users are trying to better understand their own roles and responsibilities about the information that is shared and spread on these sites. It will only be a matter of time before we gain an understanding of how social media affect society, culture, and politics.

Current Status

Social media usage around the world continues to increase at a steady rate. Of the 7.476 billion people on the planet, 3.773 billion are active Internet users, and 2.789 billion, or 73.9%, of that population are active on at least one social media platform (Kemp, 2017). Signaling the shift of social media to our mobile devices, 2.549 billion social media users are accessing their profiles via mobile devices (Kemp, 2017). Between January 2016 and January 2017 there was 30% increase of social media users worldwide. China had the biggest increase of social media users, with 133.87 million users added in 2016 (Kemp, 2017). Cuba saw the largest percentage increase of new social media users, with an increase of 368% from 2015 to 2016 (Kemp, 2017). Worldwide, the most popular social media platforms are Facebook with 2.047 billion unique users and YouTube with 1.5 billion unique users. (Williams, 2017).

As of late 2016, approximately 69% of adults access at least one or more social media site in the United States (Social Media Fact Sheet, 2018). The demographic breakdowns of this group demonstrate that while growth has slowed, there has been an increase of social media usage across the board. The one area where there has actually been a decrease in social media usage is amongst people aged 18-29. According to the Pew Research Center, there was a 4% decrease between July 2015 to November 2016 in the use of social media platform by this group.

According to the Pew Research Center, one of the fastest growing groups adopting social media is the Hispanic population in the United States. As of late 2016, 74% of Hispanic adults had one or more social media profile—up from 65% the year before (Social Media Fact Sheet, 2018). The slowest growing demographic group based on ethnicity was the Caucasian population of the U.S., at 69% in 2016, up from 65% in 2015 (Social Media Fact Sheet, 2018). Women are more likely to be on social media than men, with 72% of women in the United States reporting to have one or more social media profiles, compared to 66% of men. (Social Media Fact Sheet, 2018).

As it is in the rest of the world, Facebook is still the most popular social media platform, with 68% of Americans reporting to have a Facebook profile—this works out to nearly 78% of the entire American online population (Greenwood, Perrin, & Duggan, 2016). Within the population of American social media users on Facebook, 76% report as accessing the site daily, and another 15% access it weekly (Social Media Fact Sheet, 2018). The next most popular social media sites behind Facebook are Instagram (28% of the population), Pinterest (26%), LinkedIn (25%), and Twitter (21%) (Social Media Fact Sheet, 2018.

Factors to Watch

The evolution of social media has been, on one level, relatively predictable. When the modern version of the industry was new in the late 1990s–early 2000s, the way people interacted on them was static—it was primarily text-based with the option for some pictures. As social media platforms have evolved, so too has how we interact across each platform. As new Internet and mobile technologies advanced, so too did social media platforms. The adoptions and integrations of new communication technologies into the existing social media sites has followed a trajectory that was anticipated successfully by many futurists within the field. What has been much less predictable, however, has been how individuals have chosen to use social media. Much like the evolution of the Internet, there has been a relatively large question mark attached to the unique ways each person will figure out how to use it. Looking at both the predictable trajectory and the unknown possibilities, some of the factors to watch for the evolution of social media over the next few years include more integration of AR and to a lesser degree virtual reality (VR), a continued increase of live-streaming as the new frontier of media sharing, and finally, an increased awareness of the impact of social media on society and politics.

Expect AR to become commonplace across social media platforms. As Snapchat, Facebook, and Instagram--in addition to Apple and Google—continue to shift focus and resources to AR technologies, there will be an increased integration of these technologies into our everyday social media use. Beyond the fun filters that give us bunny ears, or make us look like a singing pile of doo, there will be an increase in AR filters and lenses that will give the illusion that you are taking selfies with your best friend who lives thousands of miles away or that you are standing next to a celebrity (Patel, 2017) a la Tupac Shakir's hologram appearing with Snoop Dogg and Dr. Dre during Coachella.

Expect live streaming to become a norm on social media. Sites like Periscope started the trend of live-streaming your life, but quickly Facebook, Instagram, Snapchat, and YouTube are integrating live streaming into their everyday operations. Facebook Live, Instagram Live, and YouTube Live each provide users with a unique experience to live stream their lives. Like AR, live-streaming will become more commonplace across platforms as there is an increased adoption of the technologies.

Finally, expect there to be an increased awareness—and wariness—about the spread of information over social media. The Twitter proclivities of the President of the United States of America and the "fake news" gaslighting that is now a constant across social media platforms is creating a skepticism about what (and who) is real, and what is not. In addition, the social media interaction and behavior exhibited by the President has the potential for fundamentally changing how the leaders of the world interact with their constituents and other World leaders. We have reached a point, specifically in the United States, where tweeting vague threats about nuclear war, using racial slurs against perceived adversaries, and officially tweeting out nonsense (covfefe!) have become normalized. Social media is at a crossroads: if we can take what we know now and apply it to constructively changing how we disseminate and consume information, social media sites can become powerful platforms for social, cultural, and political change. If we do nothing, then social media has the potential to plunge us even further into the murky waters of fake news. We all must remember, however, that social media is truly user generated content, which means that we can control how it controls us.

Getting a Job

Social media, through sheer staying power, is a force to be reckoned with and as more businesses start to take the whole genre of sites seriously, there has been the creation of numerous new jobs. According to the *Daily Muse* (2016), whether you want to be the chief media officer of a company or a developer or anything in between, there are a few tips you should keep in mind when hitting the job market.

- First, be proficient in all forms of social media. It is not enough to be a Twitter monster or a Facebook fiend, knowing all of the social media apps, new and old, will help you stand out as a proficient employee (Daily Muse, 2016).

- Know the industry. This is similar to knowing the different forms of social media, but it is more than knowing how to operate the apps. Know about the history of the industry and the company you are applying to (Daily Muse, 2016). This shows your potential employers that you have a passion for the industry as a whole and their business in particular.

- Be familiar with the need-to-know programs (Daily Muse, 2016). Be proficient in photo programs, HTML, video editing, etc. Businesses are more likely to find you an attractive candidate if they can see that you are ready to hit the ground running as a contributing member of their team.

Businesses, large and small, are beginning to realize that in order to stay relevant, they are going to need to have a social media presence, which means that no matter what industry you are interested in, if you love social media, there will be a job for you.

Bibliography

About us. (2018). *LinkedIn.* Retrieved from http://press.linkedin.com/about

Aisch, G., Huang, J., & Kang, C. (2016, December 10). Dissecting the #pizzagate conspiracy theories. *The New York Times.* Retrieved from https://www.nytimes.com/interactive/2016/12/10/business/media/pizzagate.html

Arbel, T. (2016, February 11). Myspace still exists? Yes, and now Time Inc. owns it. *Seattle Times.* Retrieved from https://www.seattletimes.com/business/magazine-publisher-time-inc-buys-whats-left-of-myspace/

Augmented reality (2016). *Reality Technologies.* Retrieved from http://www.realitytechnologies.com/augmented-reality

Banks, M.A. (2007, January 1). The Internet, ARPANet, and consumer online. *All Business.* Retrieved from http://www.all-business.com/media-telecommunications/Internet-www/10555321-1.html.

Beaumont, C. (2008, November 25). The team behind Twitter: Jack Dorsey, Biz Stone and Evan Williams. *Telegraph.* Retrieved from http://www.telegraph.co.uk/technology/3520024/The-team-behind-Twitter-Jack-Dorsey-Biz-Stone-and-Evan-Williams.html.

Blumenthal, E. (2017, December 6). Ready for Vine 2.0? Co-founder teases new app. *USA Today.* Retrieved from https://www.usatoday.com/story/tech/talkingtech/2017/12/06/ready-vine-2-0-co-founder-teases-new-app/928300001/

boyd, d.m. & Ellison, N.B. (2008). Social networking sites: Definition, history, and scholarship. *Journal of Computer-Mediated Communication, 13,* 210-230.

BusinessWeek. (2005, July 29). MySpace: WhoseSpace?. *BusinessWeek.* Retrieved from http://www.businessweek.com/technology/content/jul2005/tc20050729_0719_tc057.htm.

Caslon Analytics. (2006). Caslon Analytics social networking services. *Caslon Analytics.* Retrieved from http://www.caslon.com.au/socialspacesprofile2.htm.

Chaykowski, K. (2016, August 1). Instagram, the $50 billion grand slam driving Facebook's future: The Forbes cover story. *Forbes.* Retrieved from https://www.forbes.com/sites/kathleenchaykowski/2016/08/01/instagram-the-50-billion-grand-slam-driving-facebooks-future-the-forbes-cover-story/#669957484a97

Colao, J. J. (2012, November 27). Snapchat: The biggest no-revenue mobile app since Instagram. *Forbes.* Retrieved from http://www.forbes.com/sites/jjcolao/2012/11/27/snapchat-the-biggest-no-revenue-mobile-app-since-instagram/.

Constine, J. (2017a, May 10). Snapchat hits a disappointing 166M daily users, growing only slightly faster. *TechCrunch*. Retrieved from https://techcrunch.com/2017/05/10/snapchat-user-count/

Constine, J. (2017b, December 14). Snapchat launches augmented reality developer platform Lens Studio. *TechCrunch*. Retrieved from https://techcrunch.com/2017/12/14/snapchat-developer-platform/

Costello, S. (2017, October 25). How to use augmented reality on iPhone. *Lifewire*. Retrieved from https://www.lifewire.com/augmented-reality-on-iphone-4138290

Daily Muse Editor. (2016). Your 5-minute guide to getting a job in social media. *The Muse*. Retrieved from https://www.themuse.com/advice/your-5minute-guide-to-getting-a-job-in-social-media

Darrow, B. (2016, June 13). Microsoft buying LinkedIn for $26.2 billion. *Fortune*. Retrieved from http://fortune.com/2016/06/13/microsoft-buying-linkedin/

Darrow, B. (2017, April 24). LinkedIn claims half a billion users. *Fortune*. Retrieved from http://fortune.com/2017/04/24/linkedin-users/

Dave, P. (2013, June 20). Video app Vine's popularity is spreading, six seconds at a time. *The Los Angeles Times*. Retrieved from http://articles.latimes.com/2013/jun/20/business/la-fi-vine-20130620.

Elise, A. (2017, June 27). Snapchat's 'Snap Map' may be dangerous for young users, police warn. *KCRA*. Retrieved from http://www.kcra.com/article/snapchats-snap-map-may-be-dangerous-for-young-users-police-warn/1022946

Etherington, D. (2017, September 25). Instagram now has 800 million monthly and 500 million daily active users. *TechCrunch*. Retrieved from https://techcrunch.com/2017/09/25/instagram-now-has-800-million-monthly-and-500-million-daily-active-users/

Fiegerman, S. (2014, December 3). Aliens in the valley: The complete history of Reddit, the front page of the internet. *Mashable*. Retrieved from http://mashable.com/2014/12/03/history-of-reddit/

Fisher, T. (2017, December 2). What is augmented reality?. *Lifewire*. Retrieved from https://www.lifewire.com/augmented-reality-ar-definition-4155104

Flegenheimer, M. (2017, May 31). What's a 'covfefe?' Trump tweet unites a bewildered nation. *The New York Times*. Retrieved from https://www.nytimes.com/2017/05/31/us/politics/covfefe-trump-twitter.html

Foxx, C. (2016, October 2017). Twitter axes Vine video service. *BBC News*. Retrieved from http://www.bbc.com/news/technology-37788052

Gannes, L. (2012, October 29). Fast-growing photo-messaging app Snapchat launches on Android. *All Things D*. Retrieved from http://allthingsd.com/20121029/fast-growing-photo-messaging-app-snapchat-launches-on-android/.

Greenwood, M. (2017, December 30). Trump: I use social media to 'fight' back against media. *The Hill*. Retrieved from http://thehill.com/homenews/administration/366891-trump-i-use-social-media-to-fight-back-against-media

Greenwood, S., Perrin, A., & Duggan, M. (2016, November 11). Social media update 2016. Pew Research Center. Retrieved from http://www.pewinternet.org/2016/11/11/social-media-update-2016/

Grieco, E. (2017, November 2). More Americans are turning to multiple social media sites for news. *Pew Research Center*. Retrieved from http://www.pewresearch.org/fact-tank/2017/11/02/more-americans-are-turning-to-multiple-social-media-sites-for-news/

Ingraham, N. (2017, April 18). Snapchat adds augmented reality emoji to your videos. *Endagaget*. Retrieved from https://www.engadget.com/2017/04/18/snapchat-3d-augmented-reality-emoji/

Introducing hashtags on Instagram. (2011 January 26). *Instagram*. Retrieved from http://blog.instagram.com/post/8755963247/introducing-hashtags-on-instagram

Issawi, D. (2017, June 23). Snapchat's newest update is potentially dangerous. *Dallas News*. Retrieved from https://www.dallasnews.com/opinion/commentary/2017/06/23/snapchats-newest-update-potentially-dangerous

Kemp, S. (2017, January 24). Digital in 2017: Global review. *We Are Social*. Retrieved from https://wearesocial.com/special-reports/digital-in-2017-global-overview

Kirkpatrick, F. (2017, April 18). Introducing camera effects platform. *Facebook for Developers*. Retrieved from https://developers.facebook.com/blog/post/2017/04/18/Introducing-Camera-Effects-Platform/

LaMonica, P. R. (2017, March 3). Snapchat is worth more than Delta, Target, and CBS. *CNN Money*. http://money.cnn.com/2017/03/03/investing/snapchat-market-value/index.html

Langer, A. (2013, June 15). Six things you didn't know about the Vine app. *Yahoo! Finance*. Retrieved from http://finance.yahoo.com/news/six-things-didnt-know-vine-192222105.html.

Levy, A. (2015). What can Vine be for Twitter Inc?. *The Motley Fool*. Retrieved from https://www.fool.com/investing/general/2015/11/08/what-can-vine-be-for-twitter-inc.aspx

McArthur, A. (2017, November 7). The real history of Twitter, in brief. *Lifewire*. Retrieved from https://www.lifewire.com/history-of-twitter-3288854

McCaskill, N. D. (2017, October 20). Trump credits social media for his election. *Politico*. Retrieved from https://www.politico.com/story/2017/10/20/trump-social-media-election-244009

Meyer, R. (2016, August 13). The repeated racism of Snapchat. *The Atlantic*. Retrieved from https://www.theatlantic.com/technology/archive/2016/08/snapchat-makes-another-racist-misstep/495701/

Milian, M. (2009, July 22). Friendster founder on social networking: I invented this stuff (updated). *The Los Angeles Times*. Retrieved from http://latimesblogs.latimes.com/technology/2009/07/friendster-jonathan-abrams.html.

Newton, C. (2017, May 9). Snapchat adds new creative tools as its rivalry with Instagram intensifies. *The Verge*. Retrieved from www.theverge.com/2017/5/9/15592738/snapchat-limitless-snaps-looping-videos-magic-eraser-emoji-drawing

New subreddits by date. (n.d). In *Reddit Metrics*. Retrieved from http://redditmetrics.com/history

Newsroom. (2018). *Facebook*. Retrieved from https://newsroom.fb.com/company-info/

Nickson, C. (2009, January 21). The history of social networking. *Digital Trends*. Retrieved from http://www.digital-trends.com/features/the-history-of-social-networking/.

O'Kane, S. (2017, May 16). Instagram adds augmented reality face filters. *The Verge*. Retrieved from www.theverge.com/2017/5/16/15643062/instagram-face-filters-snapchat-facebook-features

Orland, K. (2014, March 25). Facebook purchases VR headset maker Oculus for $2 billion [updated]. *Ars Technica*. Retrieved from https://arstechnica.com/gaming/2014/03/facebook-purchases-vr-headset-maker-oculus-for-2-billion/

Patel, D. (2017, September 27). 10 social-media trends to prepare for in 2018. *Entrepreneur*. Retrieved from https://www.entrepreneur.com/article/300813

Peckham, J. (2017, May 9). How to use Snapchat filters and lenses. *Techradar*. Retrieved from http://www.techra-dar.com/how-to/how-to-use-snapchat-filters-and-lenses

Perez, S. (2017, November 7). Twitter officially expands its character count to 280 starting today. *TechCrunch*. Retrieved from https://techcrunch.com/2017/11/07/twitter-officially-expands-its-character-count-to-280-starting-today/

Protalinski, E. (2011, November 30). Facebook increases status update character limit to 63,206. *ZDNet*. Retrieved from http://www.zdnet.com/blog/facebook/facebook-increases-status-update-character-limit-to-63206/5754.

Schauer, P. (2015, June 28). 5 biggest differences between social media and social networking. *Social Media Today*. Retrieved from http://www.socialmediatoday.com/social-business/peteschauer/2015-06-28/5-biggest-differences-between-social-media-and-social

Shearer, E., & Gottfried, J. (2017, September 7). News use across social media platforms 2017. *Pew Research Center*. Retrieved from http://www.journalism.org/2017/09/07/news-use-across-social-media-platforms-2017/

Social Media Fact Sheet. (2018, February 5). Pew Research Center. Retrieved from http://www.pewinternet.org/fact-sheet/social-media/

Statt, N. (2017, December 12). Facebook introduces new augmented reality effects for Messenger. *The Verge*. Retrieved from https://www.theverge.com/2017/12/12/16767664/facebook-messenger-world-effects-augmented-reality-developers

Stross, R. (2012, January 7). Sifting the professional from the personal. *The New York Times*. Retrieved from http://www.nytimes.com/2012/01/08/business/branchout-and-beknown-vie-for-linkedins-reach.html?_r=2.

Upbin, B. (2012, April 9). Facebook buys Instagram for $1 billion. Smart arbitrage. *Forbes*. Retrieved from http://www.forbes.com/sites/bruceupbin/2012/04/09/facebook-buys-instagram-for-1-billion-wheres-the-revenue/.

Vine. (2018). *Crunch Base*. Retrieved from http://www.crunchbase.com/company/vine.

Weber, H. (2015, October 29). LinkedIn now has 400M users, but only 25% of them use it monthly. *Venture Beat*. Retrieved from http://venturebeat.com/2015/10/29/linkedin-now-has-400m-users-but-only-25-of-them-use-it-monthly/

Wells, G., & Andrews, N. (2017, October 31). Five things about Russian influence on social media platforms. *The Wall Street Journal*. Retrieved from https://www.wsj.com/articles/five-things-about-russian-influence-on-social-media-platforms-1509484005

Williams, B. (2017, August 7). There are now over 3 billion social media users in the world—about 40 percent of the global population. *Mashable*. Retrieved from https://mashable.com/2017/08/07/3-billion-global-social-media-us-ers/#K.dv049zEaqu

Zhu, K. (2016, August 3). I'm deleting Snapchat, and you should too. *Medium*. Retrieved from https://medium.com/@katie/im-deleting-snapchat-and-you-should-too-98569b2609e4

25

Big Data

Tony R. DeMars, Ph.D.[*]

Overview

New buzzwords related to modern digital media include Big Data and the interrelated term the "Internet of Things" (IoT). Big Data refers to the volume of data created, distributed and stored by way of Internet connections. It is important because business organizations can use Big Data analytics to understand customers in order to create new products, services and experiences (How Big Data Can Help, n.d.). Big Data can be gathered in a variety of ways, including consumer surveys, browsing behavior, social media commenting, and credit card transactions (Delgado, 2017).

In the Internet of Things (IoT), tens of billions of Internet-connected things are expected by 2020 (Energize Your Business, 2015). For Big Data, a 2017 IBM report says we create 2.5 quintillion bytes of data every day, meaning 90% of the data in the world today was created in the past two years (10 Key Marketing Trends, 2017).

'Non-data companies' are using Big Data to better understand their customers and to maintain a competitive edge over their competition (What is Big Data, 2015). Analysts believe Big Data will change the way we live, work and think (Mayer-Schönberger & Cukier, 2013). In relation, consumers need to be better informed about how their data are being used and related privacy issues (Manyika et al, 2011).

Introduction

Having achieved ubiquity in many parts of industry and academia, Big Data now deserves greater attention in the field of communication technology, although early research on Big Data was more in the information technology (IT) field than in communication. The term Big Data only started to gain broad attention in recent years, but it is a long-evolving term when viewed from a wide perspective.

DeMauro, Greco, and Grimaldi (2015) acknowledge there is not an easy definition for Big Data and provide a survey of existing definitions. They note that the "fuel" of Big Data is information, adding, "One of the fundamental reasons for the Big Data phenomenon to exist is the current extent to which information can be generated and made available" (p. 98). Data analytics company SAS says

> Big Data is a term that describes the large volume of data—both structured and unstructured—that inundates a business on a day-to-day basis. But it's not the amount of data that's important. It's what organizations do with the data that matters. Big Data can be analyzed for insights that lead to better decisions and strategic business moves (Big Data:What it is, n.d.).

[*] Professor and Division Director, Mass Media & Journalism, Texas A&M University-Commerce; Commerce, Texas

Gandomi and Haider (2015), while focusing in a research project on the analytic methods used for Big Data, say a major challenge to defining the term is that the industry was ahead of academicians in recognizing Big Data's development and in becoming engaged in evaluating its emergence, recognition, and potential for the industry.

Laney (2001) is credited by some as providing an early definition that has continued to be applied to what constitutes Big Data and that is now commonly called "the three Vs," as described in more detail below. Laney's explanation connects the technological development to economic issues:

> The effect of the e-commerce surge, a rise in merger/acquisition activity, increased collaboration, and the drive for harnessing information as a competitive catalyst is driving enterprises to higher levels of consciousness about how data is managed at its most basic level (Laney, 2001, p. 1).

What does this really mean? As defined in one instance, Big Data is "large pools of data that can be captured, communicated, aggregated, stored and analyzed" (Manyika et al, 2011, p. iv). Dewey (2014, para. 1) defines Big Data as "in-field shorthand that refers to the sheer mass of data produced daily by and within global computer networks at a pace that far exceeds the capacity of current databases and software programs to organize and process." Yee (2017) notes how remarkable it is "this new form of predictive analysis has taken hold on the way people learn and companies do business" (p. 1).

When you use Facebook, Instagram or Snapchat, you contribute to datasets of Big Data. When you take a picture with your smart phone, your phone—through all its sensors to collect and store data—adds to the pool of stored data, and your online actions with this digital image add to the pool of stored data. When you add a video to YouTube, when you do a search on Google, when you shop on Amazon, when you watch Netflix, when your automotive telematics communicate your location through its wireless, connected system, and when you comment on someone's blog, you contribute to the ever-growing pool of Big Data. Humans and their Internet-related communication have created an extraordinary volume of data in recent years, far outdistancing the volume of data created in the entire history of the human previously (Marr, 2016; 10 Key Marketing Trends, 2017; Yee, 2017).

So where is this Big Data stored? There are two options: a company may own its own hardware—a data center—and do its own data storage and pay its own people to maintain the system, and by doing so maintains control of their data and equipment, or a company can outsource its data storage to a third-party provider—where the server is 'in the cloud,' connected to the Internet and with data stored and maintained by the third party (Angeles, 2013). In reality, most companies have multiple data centers in different locations in order to assure data availability and for security purposes. The decision of which option to use is based on business needs, costs and data security. It is also said that Big Data became important when the cost of storing it become less costly than deleting it (Hofacker, Malthouse & Sultan, 2016).

Big Data means we must become familiar with the next terms. (See Table 25.1)

While the term "growing exponentially" can start to sound cliché, in this case it is what has happened over the past two decades with data through the Internet. Even though we have just started becoming accustomed to terabyte data storage devices, the amount of data that will travel through the Internet by sometime within the year 2016 was expected to exceed one zettabyte for the first time (Kneale, 2016), so we must start becoming comfortable with the next terms for the size of sets of data.

Cisco (2017) says annual global Internet Protocol traffic will reach 3.3 zettabytes per year by 2021 (278 exabytes each month), adding that in 2016, the annual run rate for global IP traffic was 1.2 zettabytes per year.

Figure 25.1

Bytes and Bigger Bytes

Byte (8 bits)

1 byte	A single character
10 bytes	A single word
100 bytes	A telegram, tweet, or short SMS

Kilobyte (1,000 bytes)

1 kilobyte	A few paragraphs
2 kilobytes	Typewritten page
10 kilobytes	Encyclopedia page or static Web page
100 kilobytes	Low-resolution photograph
300 kilobytes	Average-resolution photograph (jpg) or 20 seconds of 8-bit mono audio
500 kilobytes	30-second audio/radio commercial announcement

Megabyte (1,000 kilobytes or 1 million bytes)

1 megabyte	A small novel, 1 minute of 8-bit mono audio or 1-minute stereo MP3
2 megabytes	High-resolution photograph, 7-megapixel, 2,832 x 2,128
5 megabytes	Complete works of Shakespeare, 30 seconds of TV-quality video, or 5-minute podcast
10 megabytes	One minute of uncompressed, CD-quality sound or 25 seconds of smartphone mp4 video
100 megabytes	One meter of shelved books or a two-volume encyclopedia
216 megabytes	Standard digital video, 720 x 480, 5:1 compression, 1 minute

Gigabyte (1,000 megabytes or 1 trillion bytes)

1 gigabyte	Pick-up truck filled with paper, 10 meters of shelved books, or a CD-quality symphony
5 gigabytes	One standard DVD (digital video disc or digital versatile disc)
15 gigabytes	High-definition DVD (holds 4 hours of HD video)
50 gigabytes	Floor of books
300 gigabytes	A full broadcast day, standard DV, 720 x 480, 5:1 compression

Terabyte (1,000 gigabytes)

1 terabyte	50,000 trees made into paper and printed
2 terabytes	Entire contents of an academic research library
20 terabytes	Printed collection of the U.S. Library of Congress

Petabyte (1,000 Terabytes)

3 petabytes	Total holdings, all media, U.S. Library of Congress
20 petabytes	Total production of hard-disk drives in 1995
200 petabytes	All printed material or production of digital magnetic tape in 1995

Exabyte (1,000 petabytes or 1 billion gigabytes)

5 exabytes	All words ever spoken by human beings up to the year 2000
12 exabytes	Total volume of information generated worldwide 1999
500 exabytes	Total digital content in 2009

Zettabyte (1,000 exabytes)

1.2 zettabytes	Estimated total amount of data produced in 2010

Yottabyte (1,000 zettabytes)

1 Yottabyte	Let your imagination run wild!	

Source: Adapted from Wilkinson, Grant, & Fisher, 2012

The bottom line is that the data stored and mined by companies have value. A report by Economist Intelligence Unit says "Big Data analysis, or the mining of extremely large data sets to identify trends and patterns, is fast becoming standard business practice" (Marr, 2016, p. 1). Consider these examples:

- New business growth opportunities are likely because of Big Data, and new business models to aggregate and analyze industry data for companies are emerging (McGuire et al, 2012).

- "Non-data companies" are using Big Data to better understand their customers and to maintain a competitive edge over their competition (What is Big Data, 2015).

- One major component of data growth is the Internet of Things (IoT), with tens of billions of Internet-connected things expected by 2020 (Energize Your Business, 2015).

- Analysts believe Big Data will change the way we live, work, and think (Mayer-Schönberger & Cukier, 2013).

- Consumers will need to be better informed as to how data about them are being used and be aware of related privacy issues (Manyika et al, 2011).

Background

The history of Big Data is as nebulous as its definition. Lohr (2013) acknowledged 2012 as being the first point of wide scale recognition of the importance of Big Data. In his *New York Times* column, he further attempted to trace a history of Big Data and decided it was John Mashey, chief scientist at Silicon Graphics in the 1990s, who deserved recognition as first understanding and articulating the phenomenon. Mashey was also recommended as the originator of the term by data analyst Douglas Laney from the Gartner Analysts company, who, as noted above and below, has also contributed to the early evolving definition and recognition of the importance of Big Data. Diebold (2012), also discussed by Lohr, attempted to some level to take credit for the term Big Data in the econometrics field, while recognizing others in other fields. Diebold says "Big Data the term is now firmly entrenched, Big Data the phenomenon continues unabated, and Big Data the

discipline is emerging," adding Big Data is "arguably the key scientific theme of our times" (pp. 2-3).

Although he did not specifically use the term Big Data, a pivotal point in the development of Big Data as a unique category of digital communication is the description in the early 2000s by Doug Laney of collected and organized data based on what are now commonly called "the three Vs" of Big Data, volume, velocity, and variety. The data analysis company SAS, begun as a project to analyze agricultural research at North Carolina State University, describes Laney's terms in a publicly available white paper released by SAS as a marketing tool (Big Data: What It Is, n.d.). They, plus others (What is Big Data, 2015) summarize Laney's now-mainstream definition known as the three Vs of Big Data. This model comes from the IT industry's effort to define what Big Data is and what it is not:

- **Volume:** The amount of data is immense. Each day 2.3 trillion gigabytes of new data is being created.

- **Velocity**: The speed of data (always in flux) and processing (analysis of streaming data to produce near or real time results).

- **Variety:** The different types of data, structured, (and) unstructured.

Veracity was suggested as a "fourth V of Big Data" by IBM, and since then others have suggested Big Data also is measured by Value and Viability (Powell, 2014).

Diebold (2012) investigated the origins of the term Big Data in industry, academics, computer science and statistics. His research led him to claim the first use of the term Big Data in a research paper, completed in the year 2000 (Diebold, 2000). Lohr (2013) disagrees with this assertion, and Diebold conceded the origins of the term to others in his 2012 article.

Press (2013) provides a comprehensive review of technological developments from the 1940s to current times that can be seen as the continued evolution of digital technology that ultimately emerged as Big Data around the turn of the century. This review includes Fremont Rider's 1944 book *The Scholar and the Future of the Research Library*, the 1960s Derek Price publication *Science Since Babylon*, and Arthur Miller's 1971 *The Assault on Privacy*, and then, into more current times, the 1997 Michael Lesk publication *How Much Information Is There in the World?*, the developments noted above regarding Mashey, Laney and Diebold, and the 2005 Tim O'Reilly article *What is Web 2.0?* In that article O'Reilly (2005) asserts that "data is the next Intel inside," but also discusses database management as a core competency of Web 2.0 companies and asks the question: "Who owns the data?" In using mapping technology development of that time as an example, O'Reilly states "The race is on to own certain classes of core data" (p. 3).

In 2007, a major progression of Big Data was the white paper released by John F. Gantz, David Reinsel and other researchers at IDC that was the first study to estimate and forecast the amount of digital data created and replicated each year. Gantz and Reinsel (2012) later released a video analysis that further explored the future of Big Data. Swanson and Gilder (2008) used the term "exaflood" to project toward what was expected in online data expansion, listing such items as movie downloads, P2P file sharing, video calling, cloud computing, Internet video, gaming, virtual worlds, non-Internet IPTV, business IP traffic, phone, email, photos, and music as some of the contributors to increased volumes of Internet data.

Bryant, Katz, and Lazowska (2008) compared Big Data's ability to transform the way companies, the health industry, government defense and intelligence, and scientific research operate to the way search engines changed how people access information. Cukier (2010) recognized "an unimaginably vast amount of digital information which is getting ever vaster more rapidly" (para. 3). In this article, Cukier notes "Scientists and computer engineers have coined a new term for the phenomenon: "Big Data" (para. 6).

By 2011, Big Data had gained significant recognition, as researchers from the McKinsey Global Institute published an analytical paper that confirmed Big Data's importance and offered what they called seven key insights. Looking at several major sectors of society, they projected that, with the correct use of their Big Data, a retailer could improve its profit margin by 60%, U.S. healthcare could create more than $300 billion in value each year, and, from services enabled by personal-location data, users could capture 600 billion in consumer surplus (Manyika et al, 2011).

A final publication in the emergence of Big Data noted by Press (2013) is the 2012 article *Critical Questions for big data* (boyd & Crawford, 2012). These scholars describe why they think Big Data is a flawed term. Comparing, for example, the data set of any one particular topic on Twitter to all the data contained within one census, where the census data is significantly larger, they thus suggest "Big Data is less about data that is big than it is about a capacity to search, aggregate, and cross-reference large data sets" (p. 663).

It is also important to understand that many Big Data databases are actually composed of numerous, separate databases that are linked together by some key variable—some unique identifier of a person, company, or other record. For example, to merge all of person's credit information, the key variable in the U.S. is usually a social security number, while an airline uses a frequent flier number to connect all of the information about each of its customers.

Recent Developments

Unlike many other communication technologies, Big Data is not one particular product category, set of tools or technological development. Rather, it is the accumulated data that is in essence a byproduct of the emergence of digital communication, networking and storage over the past few decades. Some recent industry developments and uses of Big Data can help us gain a better understanding of Big Data as it applies to communication technology. Big Data is being used by manufacturing, retail, industry, government, healthcare, banking and education (Big Data, n.d.).

Big Data is Changing Healthcare.

McDonald (2017) says healthcare, perhaps more than any other industry, is on the verge of a major transformation due to Big Data analytics. Analysts expect predictive analytics to help target interventions to patients who could most benefit (Carecloud, n.d.). Specific patient and environmental information could help improve decision-making about patient needs and services. Cooperation between government and healthcare systems could drive better access to and use of healthcare data. Interoperability of medical records could improve connections among multiple practitioners treating a single patient and improve flow of healthcare information to patients. Healthcare organizations can use analytics to mine data and improve health care and, as a result enhance financial performance and administrative decision-making (Big Data & Healthcare, 2015).

MapR Data Technologies' *Guide to Big Data in Healthcare* (McDonald, 2017) lists five current Big Data trends in healthcare:

- *Value-Based, Patient Centric Care,* using technology to provide optimal care,

- *The Healthcare Internet of Things,* or industrial Internet, where an increasing number of health monitoring systems will communicate between the patient and physician,

- *Reduction in Fraud, Waste and Abuse* through the volume of data storage and machine learning algorithms to review the data,

- *Predictive Analytics to Improve Outcomes,* allowing data analysis models of Electronic Health Records to provide early diagnoses and reduce mortality rates

- *Real Time Monitoring of Patients,* with wearable sensors and devices providing real-time and more efficient patient monitoring.

Big Data is Changing Business Practices

Arsenault (2017) says media companies adopt Big Data applications to further their market strategies and amass capital and/or audiences. Impact on business is of course the most significant issue related to the importance of Big Data. Sarma (2017) suggests that "most organizations now understand that if they capture all the data that streams into their businesses, they can deploy Big Data analytics to get significant leverage in understanding their customers, forecasting business trends, reducing operational costs, and realizing more profits" (p. 1). Sarma further says the advanced analytics provided by Big Data can provide lean management operations, improve target segmentation, make better business decisions, improve employee hiring and retention, and increase revenue and reduce cost.

Examples of business use of Big Data are readily available (6 Big Data Examples, n.d.).

- Wellmark Blue Cross/Blue Shield found that 6% of its members accounted for 50% of its inbound call volume. The company leveraged text from call logs and integrated that with transaction data to better control incoming calls.

- NCR uses Big Data to predict when certain devices in the field will fail.

- Dell Computers enhances the view of its customers and prospects in real time by connecting traditional transaction records with things like social media user names and email addresses.

Morgan (2015) notes as additional business examples:

- Bristol-Myers Squibb reduced the time it takes to run clinical trial simulations by 98% by extending its internally hosted grid environment into the Amazon Web Services Cloud

- Xerox used Big Data to reduce the attrition rate in its call centers by 20%

- Kroger personalized its direct mailer based on the shopping history of the individual customer to get a coupon return rate of 70%, compared to the average direct mail return rate of 3.7%

- Avis Budget implemented an integrated strategy to increase market share, yielding hundreds of millions of dollars in additional revenue, working through its IT partner CSC (now DXC Technology).

- CSC applied a model to predict lifetime value to Avis Budget's customer database.

- Avis Budget is using Big Data to forecast regional demand for fleet placements and pricing.

One of the best media-related Big Data examples comes from Netflix (Chowdhury, 2017). Did you know that, as you watch Netflix, Netflix is also watching you? With a user base of around 100 million, Netflix gathers an extraordinary amount of data. Certain data points ("events" in the world of Big Data Analytics) help predict success with specific content but also as a means of recommending content to specific users. Netflix analytics evaluate such events as searches, when a user watches a show, behaviors while watching, the device used to watch, and ratings of shows watched.

Arsenault (2017) says Big Data is shaping media networks globally in two primary ways: (1) media content is being digitized, creating new means of distribution, and (2) media organizations are attempting to compete in this new media world by using Big Data analytics for decision-making—tailoring content to specific audiences. Wheeler (2017) suggests television programming has undergone a renaissance because of Big Data. Wheeler reports that Americans still watch over five hours of television per day, and that over 50% of homes have Over the Top (OTT) subscription services such as Netflix and Hulu and says data is the key to television's profitability: "Advertisers have access to more audience information than ever, while networks and content providers use data, in addition to instinct, to guide programming decisions" (p. 1).

Big Data is Changing Defense by Government

Rossino (2015) says the Defense Advanced Research Projects Agency is investing significantly in Big Data, resulting in the development of distributed computing as a component of networked weapons systems. Research and development funding requests from all branches of the U.S. military continue to spend more each year on Big Data analysis, allowing development of better defenses in many areas, including cyber threats and insider threats. The U.S. Department of Defense spent more than $7 billion in 2017 on artificial intelligence, Big Data and cloud computing, a 32% increase over five years earlier. Experts suggest "the U.S. military can either lead the coming revolution, or fall victim to it" (Corrin, 2017, p. 1).

Big Data is Changing Education

Reyes (2015), says the shift from classroom to blended and online learning systems like Blackboard and Moodle inevitably leads to a place for Big Data analytics in education, although there are challenges in changing from traditional to learner-centered analytics. There are technological issues to address and resolve and there are ethical concerns, but research shows that Big Data will transform teaching and learning. Some, for example, envision a classroom where cameras constantly capture data about each student, while a Fitbit-like device tracks each student's reactions and development (Herold, 2016).

Current Status

IBM and Oracle are two of the largest companies involved in Big Data business models. IBM reported that ten billion mobile devices are expected to be in use by 2020 (Quick Facts, n.d.). Further, 294 billion emails are sent every day, there are over one billion Google searches every day, and trillions of sensors monitor, track, and communicate with each other, generating real-time data for the Internet of Things (IoT; See Chapter 12). Oracle suggests that by 2020, all industries will roll out IoT initiatives, including $4.6 trillion spent in the public sector, $1.9 trillion in logistics, and $326 billion in retail. Further, manufacturing is expected to invest $140 billion, and telematics, $22 billion. The Oracle report says 50% of IoT activity is centered in manufacturing, transportation, smart cities and consumer products, and that IoT will account for 4.4 trillion gigabytes of the data in the digital universe by the year 2020 (Energize Your Business, 2015). Other companies are involved in this growing business, with partnerships and consolidations expected. AT&T and IBM, as one example, have developed a strategic partnership in recent years (Butler, 2015).

Kneale (2016) provides several communication and media-related examples of the application of Big Data. The more than 2.6 billion smartphones worldwide are considered a marketer's dream—as they track almost everything you do—watch TV, or interact with apps, websites, and Facebook friends. Further, analyzing the data helps distinguish between what consumers say they do and what they really do. Instead of the old system of television programming based as much on intuition as research, programmers can now rely on the latest in machine learning and artificial intelligence. On the business decision-making side, correct use of Big Data analytics helps TV programmers know which programs will be able to generate profits from advertising revenue after considering their cost, compared to those that will cause a loss, no matter what.

A major player in Big Data analytics in recent years has been Apache Hadoop. Kerner (2017) says "The open source Apache Hadoop project provides the core framework on which dozens of other Big Data efforts rely" (p. 1). Hortonworks describes Hadoop as an "open source framework for distributed storage and processing of large sets of data on commodity hardware," enabling businesses to analyze large amounts of structured and unstructured data. (What is Apache, n.d., para. 1). Marr (2016) expects Hadoop to have a 58% annual growth rate in coming years. Hadoop is even a component of IBM and other companies' Big Data analytics services. The introduction of Apache Hadoop 3.0.0 in late 2017 was expected to make the application even more useful (Kerner, 2017). In particular, YARN, or Yet Another Resource Negotiator, creates significant improvements in scalability in data access as well as increased ability to incorporate machine learning and artificial intelligence. However, Heller (2017), prior to its late 2017, suggested Hadoop was declining in importance, as new trends and tools in Business Intelligence (BI) and Business Analytics (BA) emerge.

Tableau is another company offering Big Data analytics for clients, although not among the top 10 companies by revenue as reported in 2016 (Top 8 trends, 2016). Tableau reports a survey of more than 2000 Hadoop customers found only 3% expecting to do less with Hadoop in the next 12 months, 76% already using Hadoop expecting to use it more in the following three months, and about half of those not using Hadoop expecting to deploy it in the next 12 months. Asked about the importance of data as a strategic resource, 60% of professionals quizzed said data is generating revenue for their organizations and 83% believe data makes existing services and products more profitable (Marr, 2016).

Google and Amazon are two other major Big Data companies. Do you think Big Data is not really a part of your life? Google proves it is. Google is transparent when it comes to disclosure, even if you may not be concerned about how they use your online behavior to target you with advertising (Ads Help, n.d.). In their online support pages, Google reminds you that, as you browse the Internet, read email, shop online and do similar online activities, you will notice advertisements for items and content based on previous searches. Google uses tools like DoubleClick and Ad

Sense, but essentially all their apps, from Gmail to Chrome to YouTube, track your behaviors and your interests in some way (O'Reilly, 2012). Every time you do a Google search while logged in to your Gmail account, Google keeps track of your searches, and their tools connect advertisers with your interests. Google gives users a means of controlling how data about them is used. In Google Dashboard, you can view almost everything Google knows about you and manage your privacy settings. George Orwell was concerned Big Brother would be watching you; more accurately today, through Google, Big Data is watching you!

In 2015, Google added Google Cloud Dataflow and Google Cloud Pub/Sub to its existing BigQuery SQL-query based tools (Babcock, 2015). These services are used to analyze large data sets and data streams and put Google into full competition with Amazon Web Services (AWS). The AWS suite includes Amazon DynamoDB, Data Pipeline, Amazon Kinesis, and Hadoop-like service Elastic MapReduce. AWS is one of the platforms most widely used by other companies for Big Data analytics and decision-making. On its AWS promotional web page, Amazon says "Amazon Web Services provides a broad and fully integrated portfolio of cloud computing services to help...build, secure, and deploy...Big Data applications...(with) no hardware to procure, and no infrastructure to maintain and scale" (What is, n.d.)

The fact is, however, as of 2018, there are dozens upon dozens of Big Data companies, most of which the average consumer has no knowledge about. Columbus (2017), using GlassDoor.com data, lists a large number of these based on evaluating the best Big Data companies to work for today. Among the top ten companies on the list, Google is the only commonly known company. The first five on his list do only business analytics. The others on the list vary: providing business analytics, Big Data platforms, data management and integration, or data protection and security analytics. Companies named on this list include ThoughtSpot, Wavefront, and Kyvos Insights at the top, along with such companies as Dell Technologies, Splunk, Microsoft, and Amazon Web Services.

Factors to Watch

Heller (2017) suggests some of the most important trends in Big Data as of 2018 are:

- Self-service Business Intelligence (BI), where companies like Tableau and Domo allow managers to obtain up-to-date business information on demand

- Mobile dashboards, allowing managers to see BI updates while away from their desk

- R language, available as a free open source application, that is already incorporated into such commercial products as Microsoft Azure Machine Learning Studio

- Deep neural networks, said to come from some of the most powerful algorithms for evaluating deep layers of data, and including such deep learning tools as The Microsoft Cognitive Toolkit (CNTK 2.0) and MXNet (pronounced MixNet)

- TensorFlow from Google, an open source machine learning and neural network library.

Otherwise, the battle continues between the so-called NoSQL companies against relational database systems by companies like Oracle and IBM. SQL (structured query language) databases are table-based while NoSQL are document-based, where queries are based on a structure of documents. Smart devices, e-commerce and social media create data demands on businesses that relational database systems cannot support at the necessary speed and cost, resulting in the emergence of a variety of NoSQL databases (Clark, 2014). Once the NoSQL trend started, there was an assumption that NoSQL would take over, yet the battle between the two continues. NoSQL continues to gain traction, as relational databases are judged to be incompatible with changing business requirements and limited in providing affordable scalability (Bhatia, 2017). At the same time, Apache Spark is recognized as the leading platform for large-scale SQL and "has become one of the key Big Data distributed processing frameworks in the world…(and)…provides native bindings

for the Java, Scala, Python, and R programming languages, and supports SQL, streaming data, machine learning, and graph processing" (Pointer, 2017). Apache Spark is used by major technology giants like Apple and IBM and by governments, telecommunications companies, and banks. Apache Spark is expected to continue to be a significant platform in Big Data analytics.

Big Data also means we need to know as consumers how data are being used to monitor us within a society; in our businesses we need to know how to use data that our customers create and that represents who our customers are while still respecting their privacy and maintaining their trust. In healthcare, for example consumers will have to accept giving up some privacy in exchange for the benefits of data pooling (McGuire et al, 2012). Companies and governments will likewise have to find the balance between privacy and data security. The recent growth of digital assistants, including Google Assistant, Alexa, Siri and even Facebook's M create further concerns, as each uses a level of artificial intelligence to 'learn' more about you as you interact with the technology (Maisto, 2016), and of course generate even more data.

One example: some companies are developing plans to use product sensor data to identify precise consumer targets for post-sale services or for cross-selling. Would you object to companies monitoring how you use their products? Just as in healthcare, businesses will need to work with policy makers and consumers to find the correct balance for how data about individuals are used and the level of transparency provided to the consumer (Manyika et al, 2011). Marr (2016, p. 1) says "…customer trust is absolutely essential…people are becoming increasingly willing to hand over personal data in return for products and services that make their lives easier (but) that goodwill can evaporate in an instant if customers feel their data is being used improperly, or not effectively protected."

Getting a Job

One of the biggest current challenges for Big Data analytics is the limited number of people in the U.S. with the necessary skills to do the job. The people shortage expected in the coming years is for those ready to handle the statistics, machine learning, problem framing and data interpretation, and analytical and managerial skills required to generate Big Data analytics with the quality of results needed for correct business decision-making. (McGuire et al, 2012). The top five industries hiring people with Big Data skills include (1) Professional, Scientific and Technical Services, (2) the Information Technology industry, (3) Manufacturing, (4) Finance, and (5) Retail. The average annual salary is over $100,000 a year (Big Data Job, n.d.).

Bibliography

6 Big Data Examples From Big Global Brands (n.d.). Teradata. Retrieved from http://blogs.teradata.com/data-points/big-data-examples-from-global-brands/

10 Key Marketing Trends for 2017 (2017). Retrieved from https://public.dhe.ibm.com/common/ssi/ecm/wr/en/wrl12345usen/watson-customer-engagement-watson-marketing-wr-other-papers-and-reports-wrl12345usen-20170719.pdf

Ads Help (n.d.). Retrieved from https://support.google.com/ads/answer/1634057?hl=en

Angeles, S. (2013, August 26). Cloud vs. data center: What's the difference? Business News Daily. Retrieved from http://www.businessnewsdaily.com/4982-cloud-vs-data-center.html

Arsenault, A. H. (2017). The datafication of media: Big data and the media industries, International Journal of Media & Cultural Politics, 13: 1+2, 7–24, doi: 10.1386/macp.13.1-2.7_1

Babcock, C. (2015, August 13). Google adds big data services to cloud platform. Information Week. Retrieved from http://www.informationweek.com/big-data/big-data-analytics/google-adds-big-data-services-to-cloud-platform/d/d-id/1321743

Bhatia, R. (2017). NoSQL vs SQL—Which database type is better for big data applications. Analytics India Magazine, Retrieved from https://analyticsindiamag.com/nosql-vs-sql-database-type-better-big-data-applications/

Big Data: What it is and why it matters (n.d.). Retrieved from http://www.sas.com/en_us/insights/big-data/what-is-big-data.html#mdd-about-sas

Big Data & Healthcare Analytics Forum (2015). Retrieved from http://www.himss.org/Events/EventDetail.aspx?ItemNumber=42336

Big Data Job Opportunities In 2017 And The Coming Years (n.d.). Retrieved from https://acadgild.com/blog/big-data-job-opportunities-in-2017-and-the-coming-years/

boyd, d. & Crawford, K. (2012). Critical questions for big data. Information, Communication & Society,15(5), 662-679. doi: 10.1080/1369118X.2012.678878

Bryant, R. E., Katz, R. H. & Lazowska, E. D. (2008, December 22). Big-Data computing: Creating revolutionary breakthroughs in commerce, science and society. Retrieved from http://cra.org/ccc/wp-content/uploads/sites/2/2015/05/Big_Data.pdf

Butler, B. (2015 December 18). IBM will manage AT&T's hosted and cloud services as part of partnership. Network World. Retrieved from http://www.networkworld.com/article/3016926/cloud-computing/ibm-will-manage-atandt-s-hosted-and-cloud-services-as-part-of-partnership.html

Carecloud (n.d.). Healthcare will deliver on the promise of 'big data' in 2015. Retrieved from http://www.carecloud.com/healthcare-will-deliver-promise-big-data-2015/

Chowdhury, T. (2017, June 28). How Netflix uses big data analytics to ensure success. Retrieved from http://upxacademy.com/netflix-data-analytics/

Cisco (2017, June 7). The zettabyte era: Trends and analysis. Retrieved from https://www.cisco.com/c/en/us/solutions/collateral/service-provider/visual-networking-index-vni/vni-hyperconnectivity-wp.html

Clark, L. (2014, March). Using NoSQL databases to gain competitive advantage. ComputerWeekly.com. Retrieved from http://www.computerweekly.com/feature/NoSQL-databases-ride-horses-for-courses-to-edge-competitive-advantage

Columbus, L. (2017, May 20). The best big data companies And CEOs to work for in 2017 based on Glassdoor. Retrieved from https://www.forbes.com/sites/louiscolumbus/2017/05/20/the-best-big-data-companies-and-ceos-to-work-for-in-2017-based-on-glassdoor/#74d76294326e

Corrin, A. (2017, December 6). DoD spent $7.4 billion on big data, AI and the cloud last year. Is that enough? Retrieved from https://www.c4isrnet.com/it-networks/2017/12/06/dods-leaning-in-on-artificial-intelligence-will-it-be-enough/

Cukier (2010, February 25). Data, data everywhere. The Economist. Retrieved from http://www.econmist.com/node/15557443

Delgado, R. (2017, June 16). Improved decision making comes from connecting big data sources. Retrieved from http://data-informed.com/improved-decision-making-comes-from-connecting-big-data-sources/

DeMauro, A., Greco, M., & Grimaldi, M. (2015). What is big data? A consensual definition and a review of key research topics. AIP Conference Proceedings. 2015, Vol. 1644 Issue 1, 97-104.

Dewey, J. P. (2014, January). Big Data. Salem Press Encyclopedia.

Diebold, F.X. (2000). Big Data dynamic factor models for macroeconomic measurement and forecasting. Retrieved from http://citeseerx.ist.psu.edu/viewdoc/download;jsessionid=3DB75BD2F8565682589A257CE3F4150C?doi=10.1.1.12.1638&rep=rep1&type=pdf

Diebold, F.X. (2012). A personal perspective on the origin(s) and development of 'big data': The phenomenon, the term, and the discipline, second version. Retrieved from http://papers.ssrn.com/sol3/papers.cfm?abstract_id=2202843

Energize Your Business with IOT-Enabled Applications (2015). Retrieved from http://www.oracle.com/us/dm/oracle-iot-cloud-service-2625351.pdf

Gandomi, A., & Haider, M. (2015). Beyond the hype: Big data concepts, methods, and analytics. International Journal Of Information Management, 35(2), 137-144. doi:10.1016/j.ijinfomgt.2014.10.007

Gantz, J. & Reinsel, D. (2012, December). The digital universe in 2020: Big Data, bigger digital shadows, and biggest growth in the Far East. Retrieved from http://www.emc.com/leadership/digital-universe/2012iview/analyst-perspective-john-gantz-david-reinsel.htm

Heller (2017, August 7). 10 hot data analytics trends — and 5 going cold. Retrieved from https://www.cio.com/article/3213189/analytics/10-hot-data-analytics-trends-and-5-going-cold.html

Herold, B. (2016, January 11). The future of big data and analytics in K-12 education. Retrieved from https://www.edweek.org/ew/articles/2016/01/13/the-future-of-big-data-and-analytics.html

Hofacker, C. F., Malthouse, E.C. & Sultan, F. (2016). Big Data and consumer behavior: imminent opportunities, *Journal of Consumer Marketing*, 33(2), 89-97.

How Big Data Can Help You Do Wonders in Your Business (n.d.) Retrieved from https://www.simplilearn.com/how-big-data-can-help-do-wonders-in-business-rar398-article

Kerner, S. M. (2017, December 22). Apache Hadoop 3.0.0 boosts big data app ecosystem. Retrieved from http://www.enterpriseappstoday.com/data-management/apache-hadoop-3.0.0-boosts-big-data-app-ecosystem.html

Kneale, D. (2016, January 21). Big data is everywhere—now what to do with it? New tools unlock the secrets of consumer desire. Broadcasting and Cable. Retrieved from http://www.broadcastingcable.com/news/rights-insights/big-data-dream/147166

Laney, D. (2001, February 6). 3D data management: Controlling data volume, velocity, and variety. Retrieved from http://blogs.gartner.com/doug-laney/files/2012/01/ad949-3D-Data-Management-Controlling-Data-Volume-Velocity-and-Variety.pdf

Lohr, S. (2013, February 4). Searching for origins of the term 'big data.' New York Times, New York edition, B4.

Maisto, M. (2016, March 2). Siri, Cortana are listening: How 5 digital assistants use your data. Retrieved from https://www.informationweek.com/big-data/siri-cortana-are-listening-how-5-digital-assistants-use-your-data/d/d-id/1324507?

Manyika, J. & Chui, M., Brown, B., Bughin, J., Dobbs, R., Roxburgh, C., & Byers, A.H. (2011). Big Data: The next frontier for innovation, competition, and productivity, Washington, D.C.: McKinsey Global Institute. Retrieved from http://www.mckinsey.com/business-functions/business-technology/our-insights/big-data-the-next-frontier-for-innovation

Marr, B. (2016, January 13). Big data facts: How many companies are really making money from their data? Forbes.com. Retrieved from: http://www.forbes.com/sites/bernardmarr/2016/01/13/big-data-60-of-companies-are-making-money-from-it-are-you/#773b9f194387 (or 2015?)

Mayer-Schönberger, V. & Cukier, K. (2013). Big Data: A revolution that will transform how we live, work, and think. Boston: Houghton Mifflin Harcourt.

McDonald C. (2017, February 13). 5 big data trends in healthcare for 2017. Retrieved from: https://mapr.com/blog/5-big-data-trends-healthcare-2017/

McGuire, T., Manyika, J., Chui, M. (2012, July/August). Why big data is the new competitive advantage. Ivy Business Journal. Retrieved from http://iveybusinessjournal.com/publication/why-big-data-is-the-new-competitive-advantage/

Morgan, L. (2015, May 27). Big Data: 6 real-life business cases. Retrieved from https://www.informationweek.com/software/enterprise-applications/big-data-6-real-life-business-cases/d/d-id/1320590?

O'Reilly, T. (2005, September 30). What is web 2.0? Retrieved from http://www.oreilly.com/pub/a/web2/archive/what-is-web-20.html

O'Reilly, D. (2012, January 30). How to prevent Google from tracking you. CNET. Retrieved from http://www.cnet.com/how-to/how-to-prevent-google-from-tracking-you/

Pointer, I. (2017, November 13). What is Apache Spark? The big data analytics platform explained. Retrieved from https://www.infoworld.com/article/3236869/ analytics/what-is-apache-spark-the-big-data-analytics-platform-explained.html

Powell, T. (2014, August 25). The fourth V of big data. Retrieved from http://researchaccess.com/2014/08/the-fourth-v-of-big-data/

Press, G. (2013, May 9). A very short history of Big Data. Retrieved from http://www.forbes.com/sites/gilpress/2013/05/09/a-very-short-history-of-big-data/#4a0f418b55da

Quick Facts and Stats on Big Data (n.d.) IBM big data and analytics hub. Retrieved from http://www.ibmbigdata-hub.com/gallery/quick-facts-and-stats-big-data

Reyes, J. (2015). The skinny on big data in education: Learning analytics simplified. Techtrends: Linking Research & Practice To Improve Learning, 59(2), 75-80.

Rossino, A. (2015, September 9). DoD's big bets on big data. Retrieved from http://www.c4isrnet.com/story/military-tech/blog/business-viewpoint/2015/08/25/dods-big-bets-big-data/32321425/

Sarma, R. (2017, March 22). 5 business impacts of advanced analytics and visualization. Retrieved from https://resources.zaloni.com/blog/5-business-impacts-of-advanced-analytics-and-visualization

Swanson, B. & Gilder, G. (2008, January 29). Estimating the exaflood: The impact of video and rich media on the Internet— A 'zettabyte' by 2015? Retrieved from http://www.discovery.org/a/4428

Top 8 trends for 2016: Big Data (2016). Retrieved from http://www.tableau.com/sites/default/files/media/top8bigdata-trends2016_final_2.pdf?ref=lp&signin=33c5bb5269eb06f9f31274c1a6610036&1[os]=mac%20os

What is Apache Hadoop? (n.d.). Hortonworks. Retrieved from http://hortonworks.com/hadoop/

What is Big Data (n.d.). Retrieved from https://aws.amazon.com/big-data/what-is-big-data/

What is Big Data? And why is it important to me? (2015, March 31). Retrieved from http://www.gennet.com/big-data/big-data-important/

Wheeler, C. (2017, June 1). The future of TV is today: How TV is using Big Data. Retrieved from https://www.clarityinsights.com/blog/television-big-data-analytics

Wilkinson, J.S., Grant, A. E., & Fisher, D. (2012). Principles of convergent journalism (2nd ed.) New York: Oxford University Press.

Yee, S. (2017, February 8). The Impact Of Big Data: How It Is Changing Everything, Including Corporate Training. Retrieved from https://elearningindustry.com/ impact-of-big-data-changing-corporate-training

Conclusions

26

Other New Technologies

Jennifer H. Meadows, Ph.D.[*]

Introduction

Every two years, the editors of this book struggle to decide which chapters to include. There are always more options than there is space because new communication technologies develop almost daily. This chapter outlines some of these technologies deemed important, but not ready for a full chapter. The 17th edition of this book may cover these technologies in full chapters just as digital signage was once featured in an "Other New Technologies" chapter. Among the choices, three technologies stand out: artificial intelligence, blockchain and cryptocurrency, and 5G.

Artificial Intelligence

"Alexa, what's the weather today?" That's a question commonly asked in homes across the nation. Amazon's Echo personal assistant is so helpful because it is driven by artificial Intelligence or AI. AI is defined as "The development of computers capable of tasks that typically require human intelligence (Simonite, 2018). Getting computers to think like people is another way to think of AI. We don't function via programming, we learn how to do things. So, a key to artificial intelligence is machine learning—the ability to train computers by example rather than programming.

The concept of AI has been around since the mid-1960's. Computer scientists developed computers able to play games, and slowly computers were developed that could complete tasks that previously were thought to be limited to humans. AI is having a moment in 2018, as the technology is progressing in innovative, useful ways, and it is getting into the hands of consumers.

One key to this growth is the development of deep learning. Deep learning is an advanced form of machine learning where data is filtered through self-adjusting networks of math loosely inspired by neurons in the brain (Simonite, 2018). These networks are called artificial neural networks.

Advances in AI are literally making our devices smarter and more responsive across many areas including health care, national security, education, and marketing. For example, in South Korea, scientists have developed an AI algorithm to review digital mammography images and detect breast cancer (New AI, 2018).

---

[*] Professor and Chair, Department of Media Arts, Design, and Technology, California State University, Chico (Chico, California)

Image detection, voice recognition, and text generation are all applications enhanced by developments in deep learning. Google Clips is a smart camera driven by AI that scans the environment and uses machine learning to capture "great pictures." The $249 camera learns to recognize familiar faces including pets and will capture images of them—automatically (Google Clips, 2018). There is no need for a person behind the camera. How does it work? The camera captures images in 7 second "clips." It has learned to identify humans, dogs and cats, and once captured, clips can be noted as suggested clips—ones of familiar people or pets. Often reviewers note that aside of a few good candid shots, there doesn't seem to be a whole lot of use for the camera—maybe not yet (Heater, 2018).

Translation is another useful application of AI. Imagine traveling and having seamless, real-time translation of different languages. This type of AI is already in place with services such as Google Translate but AI is enhancing translation with better, more nuanced translations (Lewis-Kraus, 2016). Microsoft announced in March 2018 that it had perfected English to Chinese translation and vice versa though deep neural networks that mimic human behavior. The company uses deliberation networks that "mimic the repetitive process people use to revise writing" (Fingas, 2018).

Cryptocurrency and Blockchain

Cryptocurrency has become a resource for business and commerce, with Bitcoin becoming the most widely-accepted cryptocurrency in 2017. Defining cryptocurrency can be confusing and there is a lot of misleading information out there about the term. In 2018, cryptocurrency was officially added to the Meriam- Webster dictionary"

"Any form of currency that only exists digitally, that usually has no central issuing or regulating authority but instead uses a decentralized system to record transactions and manage the issuances of new units, and that relies on cryptography to prevent counterfeiting and fraudulent transactions" (Merriam-Webster, 2018).

Let's break that down. Cryptocurrency is only digital and does not have a physical presence. Traditional currencies like the U.S. dollar are issued and regulated, usually by governments. In the case of the dollar, the U.S. Federal Reserve Bank issues and regulates the currency. A cryptocurrency such as Bitcoin doesn't have that central regulator, rather it is regulated through a decentralized system of users. Cryptocurrencies work outside the control of any government, and users don't have to reveal their identities.

The lack of governmental control offers cryptocurrencies some true advantages over traditional currencies. No banks are involved so nobody automatically takes a cut of transactions. Because International currency exchange and payments are not subject to the same regulations, users can buy and sell things anonymously (What is Bitcoin, n.d.)

A decentralized network keeps track of all transactions in a digital ledger using a technology called blockchain. Blockchain also allows transactions to occur without having to go through a traditional payment network such as American Express or MasterCard. Iansiti and Lahani (2017) define blockchain as "an open, distributed ledger that can record transactions between two parties efficiently and in a verifiable and permanent way." These authors have identified five basic principles of blockchain technology.

- Distributed Database—everyone has access and no party controls the data

- Peer-to-Peer Transmission

- Transparency with Pseudonymity—each user or node has a unique identified but users can remain anonymous

- Irreversibility of Records—Once transactions are entered they cannot be altered because they linked to other transactions.

- Computational Logic (Iansiti & Lahani, 2017).

Together, these factors create a type of database that can offer security and permanence, both of which are critical for cryptocurrency. But it should be noted that blockchain technology can be applied to any database that requires high security and transaction capability with numerous entities having simultaneous access.

How is the worth determined for a cryptocurrency? The currency is worth what people decide it is worth, and this can lead to massive fluctuations and manipulation. For example, Bitcoin's value went from $1,000 to $20,000 per Bitcoin in 2017, then back down to under $8,000 per coin. Bitcoin's value continues to be volatile (Kleinman, 2018).

Introduced in 2008 by a mysterious person calling themselves Satoshi Nakmoto, Bitcoin is now the most recognizable cryptocurrency in the world. To purchase Bitcoin people can use an exchange that sells Bitcoins for traditional currency using a credit card or bank account or they can use an anonymous service that allow people connect and use cash (Popper, 2017). Peer-to-peer transactions are also possible using mobile apps. Users then store their Bitcoins in a digital wallet on their computer or in the cloud.

New Bitcoins are made through a computer process called "mining." that requires computers to solve complex mathematical problems. Bitcoin creation involves tremendous computational power and energy to point where, as of mid-2018, the cost of mining was often more than the worth of the coins yielded.

Because they are not regulated, Bitcoin value is quite volatile, such that many have compared involvement in Bitcoin exchanges to gambling. As of mid-2018, Japan, China, and Australia have begun to consider different regulatory frameworks for cryptocurrencies, which is interesting because the point of these currencies was to get away from governmental regulation of currency.

In addition, the Wild West nature of the Bitcoin market has brought out some shady characters and organizations making false claims about the value of Bitcoin investments and understating the risks. As of mid-March 2018, both Facebook and Google have banned cryptocurrency advertising (D'Onfro, 2018)

Of course, Bitcoin isn't the only cryptocurrency out there. Others include Ethereum Stellar, Dash, Tether, NEO, and EOS.

5G

Several of the chapters in this book have mentioned 5G networks. Those of us with a cellphone will be familiar with the terms 3G and 4G. These terms describe the type of high speed wireless network used by different providers and are based upon standards such as LTE-Advanced (4G) that allow users to enjoy services such as gaming and video streaming.

Nowadays the term 5G is being thrown around a lot without having a good definition— much like when 4 G was first introduced. The International Telecommunications Union (ITU) is the United Nations' agency for information and communication technologies and allocates global spectrum and develops technical standards (ITU, n.d.). In 2017, the ITU issued minimum specifications for 5G called IMT-2020. IMT-2020 specifies that 5G must have a downlink peak data rate of 20 Gbps and an upload rate of 10 Gbps for base stations, while consumers should expect minimum download speeds of 100Mbps and upload speeds of 50Mbps (Triggs, 2017).

The 3G Partnership Project (3GPP) unites cellular network standards organizations. The group signed off on the first 5G cellular standard 5G NR (Non-Standalone 5G New Radio) in December 2017 (First 5G, 2017). The standard is due for completion in June 2018.

5G deployment should move quickly with 1.2 billion people expected to have access by 2025 (Auchard & Nellis, 2018). The stakes are high as 5G isn't just important to cellphone users, but also to the Internet of Things (IoT). In March 2018, the Trump administration even blocked a takeover of U.S. owned Qualcomm by China owned Broadcom over concerns about 5G technology control. National security concerns were cited as the reason for the executive order and at the root is who has the lead in mobile technology (Cheng, 2018).

Conclusions

This chapter provides just a brief overview of some other new technologies. Who know, maybe the next edition of this book will dedicate entire chapters to the technologies discussed here. What's important to remember, though, is that technology is always changing, and you can keep up with these changes using tools your learned in this book.

Bibliography

Auchard, E. & Nellis, S. (2018). What is 5G and who are the major players. Reuters. Retrieved from https://www.reuters.com/article/us-qualcomm-m-a-broadcom-5g/what-is-5g-and-who-are-the-major-players-idUSKCN1GR1IN

Cheng, R. (2018). Why Trump blocked Qualcomm-Broadcom: It's all about 5G. Cnet. Retrieved from https://www.cnet.com/news/why-trump-blocked-qualcomm-broadcom-its-all-about-5g/

D'Onfro, J. (2018). Google will ban all cryptocurrency-related advertising. CNBC. Retrieved from https://www.cnbc.com/2018/03/13/google-bans-crypto-ads.html

Fingas, J. (2018). Microsoft says its AI can translate Chinese as well as a human. Engadget. Retrieved from https://www.engadget.com/2018/03/14/microsoft-ai-can-translate-chinese-as-well-as-a-human/

First 5G. (2017) 3GPP. Retrieved from http://www.3gpp.org/news-events/3gpp-news/1929-nsa_nr_5g

Google Clips (2018) Google. Retrieved from https://store.google.com/us/product/google_clips?hl=en-US

Heater, B. (2018). Google clips review. Techcrunch. Retrieved from https://techcrunch.com/2018/02/27/google-clips-review/

Iansiti, M & Lakhani, K (2017). The truth about blockchain. *Harvard Business Review*. Retrieved from https://hbr.org/2017/01/the-truth-about-blockchain

ITU (n.d.) About International Telecommunication Union. Retrieved from https://www.itu.int/en/about/Pages/default.aspx

Kleinman, J. (2018). Why Bitcoin's Price is so Volatile. Lifehacker. Retrieved from https://lifehacker.com/why-bitcoin-s-price-is-so-volatile-1822143846

Lewis-Kraus, G. (2016). The Great A.I. Awakening. *The New York Times*. Retrieved from https://www.nytimes.com/2016/12/14/magazine/the-great-ai-awakening.html

Merriam-Webster (2018). Definition of cryptocurrency. Retrieved from https://www.merriam-webster.com/dictionary/cryptocurrency

New AI algorithm for breast cancer screening. (2018). HeathManagement.org. Retrieved from https://healthmanagement.org/c/imaging/news/new-ai-algorithm-for-breast-cancer-screening

Popper, N. (2017). What is Bitcoin? All about the mysterious digital currency. *The New York Times*. Retrieved from https://www.nytimes.com/2017/05/15/business/all-about-bitcoin-the-mysterious-digital-currency.html

Simonite, T. (2018). The WIRED Guide to Artificial Intelligence. *Wired*. Retrieved from https://www.wired.com/story/guide-artificial-intelligence/

Triggs, R. (2017). 5G specifications sets 100Mbps. Android Authority. Retrieved from https://www.androidauthority.com/5g-spec-defined-20gbps-download-752063/

What is Bitcoin? (n.d.) CNN Tech. Retrieved from http://money.cnn.com/infographic/technology/what-is-bitcoin/

27

Your Future & Communication Technologies

August E. Grant, Ph.D.[*]

This book has introduced you to a range of ideas on how to study communication technologies, given you the history of communication technologies, and detailed the latest developments in about two dozen technologies. Along the way, the authors have told stories about successes and failures, legal battles and regulatory limitations, and changes in lifestyle for the end user. Authors have also offered tips on how you can get a job working with that technology.

So what can you do with this information? If you're entrepreneurial, you can use it to figure out how to get rich. If you're academically-inclined, you can use it to inform research and analysis of the next generation of communication technology. If you're planning a career in the media industries, you can use it to help choose the organizations where you might work, or to find new opportunities for your employer or for yourself.

More importantly, whether you are in any of those groups or not, you are going to be surrounded by new media for the rest of your life. The cycle of innovation, introduction, and maturity of media almost always includes a cycle of decline as well. As new communication technologies are introduced and older ones disappear, your media use habits will change. What you've learned from this book should help you make decisions on when to adopt a new technology or drop an old one. Of course, those decisions depend upon your personal goals—which might be to be an innovator, to make the most efficient use of your personal resources (time and money), or to have the most relaxing lifestyle.

This chapter explores a few ways you can apply this information to improve your ability to use and understand these technologies—or simply to profit from them.

---

[*] J. Rion McKissick Professor of Journalism, School of Journalism and Mass Communications, University of South Carolina (Columbia, South Carolina).

Researching Communication Technologies

The speed of change in the communication industries makes the study of the technologies discussed in this text kind of a "spectator sport," where anyone willing to pay the price of admission (simple Internet access) can watch the "game," predicting winners and losers and identifying the "stars" who help lead their teams to success. There are numerous career opportunities in researching these technologies, ranging from technology journalist and professor to stock market analyst and government regulator.

As detailed in Chapter 3, there are many ways to conduct this type of research. The most basic is to provide descriptive statistics regarding users of a technology, along with statistics on how and when the technology is used.

But greater insight can usually be obtained by applying one or more theories to help understand WHY the technology is being used by those doing so. Diffusion theory helps you understand how an innovation is adopted over time among members of a social system. Effects theories help you understand how and why a technology impacts the lives of individuals, the organizations involved in the production and distribution of the technology, and the impact on social systems.

In order to do this type of research successfully, you need three sets of knowledge:

- History of communication technologies

- Understanding of major theories of technology adoption, use, and effects

- Ability to apply appropriate quantitative and qualitative research methods

With these three, you will be prepared for any of the careers mentioned above, and you will have the chance to enjoy observing the competition as you would any other game that has so much at stake with some of the most interesting players in the world!

Making Money from Communication Technologies

You have the potential to get rich from the next generation of technologies. Just conduct an analysis of a few emerging technologies using the tools in the "Fundamentals" section of the book; choose the one that has the best potential to meet an unmet demand (one that people will pay for); then create a business plan that demonstrates how your revenues will exceed your expenses from creating, producing, or distributing the technology.

Regardless of the industry you want to work in, there are five simple steps to creating a business plan:

1) **Idea:** Every new business starts with an idea for a new product or service, or a new way to provide or enhance an existing business model. Many people assume that having a good idea is the most important part of a business plan, but nothing could be further from the truth. The idea is just the first step in the process. Sometimes the most successful ideas are the most innovative, but other times the greatest success goes to the team that makes a minor improvement on someone else's idea that enables a simpler or less expensive option. Applying Roger's five attributes of an innovation (compatibility, complexity, relative advantage, trialability, and observability, discussed in Chapter 3) can help you understand how innovative your idea is.

2) **Team:** Bringing together the right group of people to execute the idea is the next step in the process. The most important consideration in putting together a team is creating a list of competencies needed in your organization, and then making sure the management team includes one expert in each area. Make sure that you include experts in marketing and sales as well as production and engineering. The biggest misstep in creating a team is bringing together a group of people who have the same background and similar experiences. At this stage you're looking for diversity in knowledge and experience. Once you've identified the key members of the team, you can create the organizational structure, identifying the departments and divisions within

each department (if applicable) and providing descriptions for each department and responsibilities for each department head. It is also critical to estimate the number of employees in each position.

3) **Competitive Analysis:** The next step is describing your new company's primary competition, detailing the current strengths and weaknesses of competing companies and their product or service offerings. Consider the resources and management team, and agility of the company as factors that will influence how they will compete with you. In the process, this analysis needs to identify the challenges and opportunities they present for your company.

4) **Finance:** The most challenging part of most business plans is the creation of a spreadsheet that details how and where revenues will be generated, including projections for revenues during the first few years of operation. These revenues are balanced in the spreadsheet with estimates of the expected expenses (e.g., personnel, technology, facilities, and marketing). This part of your business plan also has to detail how the initial "start-up" costs will be financed. Some entrepreneurs choose to start with personal loans and assets, growing a business slowly and using initial revenues to finance growth—a process known as "bootstrapping." Others choose to attract financing from venture capitalists and angel investors who receive a substantial ownership stake in the company in return for their investment. The bootstrapping model allows an entrepreneur to keep control—and a greater share of profits—but the cost is much slower growth and more limited financial resources. Once the initial plan is created, study every expense and every revenue source to refine the plan for profitability.

5) **Marketing Plan:** The final step is creating a marketing plan for the new company that discusses the target market and how you will let those in the target know about your product or service. The marketing plan must identify the marketing and promotional strategies that will be utilized to attract consumers/audiences/users to your product or service, as well as the costs of the plan

and any related sales costs. Don't forget to discuss how both traditional and new media will be employed to enhance the visibility and use of your product, then follow through with a plan for how the product or service will be distributed and priced.

Conceptually, this five-step process is deceptively easy. The difficult part is putting in the hours needed to plan for every contingency, solve problems as they crop up (or before they do), make the contacts you need in order to bring in all of the pieces to make your plan work, and then distribute the product or service to the end users. If the lessons in this book are any indication, two factors will be more important than all the others: the interpersonal relationships that lead to organizational connections—and a lot of luck!

Here are a few guidelines distilled from 30-plus years of working, studying, and consulting in the communication technology industries that might help you become an entrepreneur:

- **Ideas are not as important as execution**. If you have a good idea, chances are others will have the same idea. The ones who succeed are the ones who have the tools, vision, and willingness to work long and hard to put the ideas into action.

- **Protect your ideas**. The time and effort needed to get a patent, copyright, or even a simple non-disclosure agreement will pay off handsomely if your ideas succeed.

- **There is no substitute for hard work**. Entrepreneurs don't work 40-hour weeks, and they always have a tool nearby to record ideas and keep track of contacts.

- **There is no substitute for time away from work**. Taking one day a week away from the job gives you perspective, letting you step back and see the big picture. Plus, some of the best ideas come from bringing in completely unrelated content, so make sure you are always scanning the world around you for developments in the arts, technology, business, regulation, and culture.

- **Who you know is more important than what you know**. You can't succeed as a solo act in the communication technology field. You have to a) find and partner with or hire people who are better than you in the skill sets you don't have, and b) make contacts with people in organizations that can help your business succeed.

- **Keep learning**. Study your field, but also study the world. The technologies that you will be working with have the potential to provide you access to more information than any entrepreneur in the past has had. Use the tools to continue growing.

- **Create a set of realistic goals**. Don't limit yourself to just one goal, but don't have too many. As you achieve your goals, take time to celebrate your success.

- **Give back**. You can't be a success without relying upon the efforts of those who came before you and those who helped you out along the way. The best way to pay it back is to pay it forward.

This list was created to help entrepreneurs, but it may be equally relevant to any type of career. Just as the communication technologies explored in this book have applications that permeate industries and institutions throughout society, the tools and techniques explored in this book can be equally useful regardless of where you are or where you are going.

Keep in mind along the way that working with and studying communication technologies is not simply a means to an end, but rather a process that can provide fulfillment, insight, monetary rewards, and pure joy. Those people who have the fortune to work in the communication technology industries rarely have a chance to become bored, as there is always a new player in the game, a new device to try, and a great chance to discuss and predict the future.

In the process, be mindful of all of the factors in the communication technology ecosystem—it's not the hardware that makes the biggest difference, but all of the other elements of the communication technology ecosystem that together define this ecosystem.

The one constant you will encounter is change, and that change will be found most often at the organizational level and in the application of system-level factors that impact your business. As discussed in the first chapter of this book, new, basic technologies that have the potential to change the field come along infrequently, perhaps once every ten or twenty years. But new opportunities to apply and refine these technologies come along every day. The key is training yourself to spot those opportunities and committing yourself to the effort needed to make your ideas a success.

Index